Springer
Tokyo
Berlin
Heidelberg
New York
Hong Kong
London
Milan
Paris

Y. Saijo, A.F.W. van der Steen (Eds.)

Vascular Ultrasound

With 156 Figures, Including 28 in Color

Yoshifumi Saijo, M.D., Ph.D.
Department of Medical Engineering and Cardiology
Institute of Development, Aging and Cancer, Tohoku University
4-1 Seiryo-machi, Aoba-ku, Sendai 980-8575, Japan

Antonius Franciscus Wilhelmus van der Steen, Ph.D.
Thorax Center, Erasmus Medical Center Rotterdam
PO Box 1738, 3000 DR Rotterdam, The Netherlands

ISBN 4-431-70328-4 Springer-Verlag Tokyo Berlin Heidelberg New York

Library of Congress Cataloging-in-Publication Data

Vascular Ultrasound / [edited by] Y. Saijo, A.F.W. van der Steen.
p. ; cm.
Includes index.
ISBN 4-431-70328-4 (alk. paper)
1. Blood-vessels--Ultrasonic imaging. I. Saijo, Y. (Yoshifumi), 1962- II. Steen, A. F. W. van der (Antonius Franciscus Wilhelmus), 1964-
[DNLM: 1. Cardiovascular Diseases--ultrasonography. 2. Ultrasonography, Interventional--methods. WG 141 V331 2003]
RC691.6.U47V375 2003
616.1'307543--dc22

2003057290

Printed on acid-free paper

Typesetting: Camera-ready by the editors and authors

SPIN: 10855423

To those who inspired us

Motonao Tanaka

Klaas Bom

And to our families

Terumi
Naoya
Fumika

Minjon
Inez
Our yet-to-be-born child

Foreword

I have been asked to write a short foreword to this book. As a matter of fact the two editors have nicely and candidly introduced themselves and their book in their preface, to which I have little to add. It is true that the format and the content of this monograph differ in many aspects from those of a conventional "textbook of IVUS." Not only is this monograph a glimpse into the future, but many of the chapters have been written by bioengineers, deeply involved in the clinical field. The motto of this monograph could be "High tech of the future, but down to Earth." I believe that this quite unique monograph will be an eye-opener for many interventional cardiologists who should pay attention to some of the new developments and IVUS tools, which are going to play a pivotal role in the assessment of new physio-pathology of the coronary and peripheral vasculature and circulation. I can only commend the two editors and their coauthors for their remarkable achievement. Their work will be appreciated by the readers of this book, whatever their discipline.

Patrick W. Serruys, M.D., Ph.D., FACC, FESC
Professor in Interventional Cardiology
Thorax Center
Erasmus Medical Center Rotterdam
The Netherlands

Preface

About Ton:
It was October 1993 in Innsbruck, Austria, that Dr. van der Steen and I first met each other at the 8th Congress of the European Federation of Societies for Ultrasound in Medicine and Biology (EFSUMB). At that time, we presented two different papers on acoustic microscopy in the same session. After the session with some constructive discussions and comments, he might have said to me, "I am going to the department of cardiology, Erasmus University, Rotterdam." As my understanding of English conversation was poor (it is not at all excellent even now), I misunderstood him to be a cardiologist with great knowledge of ultrasound engineering. This really great "cardiologist" made me study very hard about ultrasound engineering and signal processing besides my clinical activities in cardiology, especially coronary interventional therapy and echocardiography, ever since that meeting.

Yoshifumi Saijo

About Yoshi:
When I started to work on acoustic microscopy of biological tissue, I found out that there was a group in Sendai, Japan, that was doing seminal work in this field. My admiration grew for the technical development that they had achieved. By the time I first met somebody from the Sendai group, I had done extensive studies on the effects of tissue preparation. During a session at the EFSUMB in 1993 in which Dr. Saijo presented, my criticism of his handling of tissue was more aggressive than I had intended. To my surprise I had a very pleasant discussion with him after the session. Since that date I have been running into him at all the technical ultrasound conferences. When I asked him years later why he attended all those meetings and how he could cope with simultaneously running a cardiology practice and keeping up with engineering, his reply was simply: "You can do it, so why shouldn't I?" It was only then that I realized that he had thought that I was a cardiologist as well.

Ton van der Steen

Around the beginning of this century the editors started realizing that a cardiologist working in basic ultrasonics and an engineer in applied physics, working in a clinical cardiology setting, may be able to bridge the gap between fundamental ultrasonics and clinical application. It was the beginning of the idea of publishing this international and interdisciplinary monograph entitled *Vascular Ultrasound.*

Over the last decades, percutaneous coronary intervention therapy has developed dramatically, and the number of cases has nearly doubled year by year. Besides percutaneous balloon angioplasty, "new devices" for interventional therapy such as the cutting balloon, directional coronary atherectomy, and stents, being bare, radioactive, or drug coated, have become clinically available. Also, the scale of diagnostic devices has been extended continuously. Coronary angiography using X-ray was the main diagnostic method in the age of classical coronary intervention. In the late 1980s, coronary angioscopy was clinically applied first as a new imaging device, although it was not successful around that period because it only visualized the surface morphology of the arterial wall. The clinical significance of intravascular ultrasound (IVUS) was first reported in 1989. It has advantages over angiography and angioscopy because IVUS can visualize and assess the pathology beneath the surface of coronary arteries. Today, IVUS has achieved its position as a useful clinical tool to measure both the inner and outer diameter of coronary arteries and the expansion of the stent to the arterial wall.

In the field of vascular biology, the concept that inflammation plays an important role in the progression of atherosclerosis has become very popular. Development of molecular biology together with classical pathology has proven this concept in vitro. Acute myocardial infarction was believed to be the end stage of severe narrowing of the coronary arteries. In 1985 Dr. M.J. Davies and Dr. E. Falk reported independently the epoch-making discovery that plaque rupture was the main cause of acute coronary syndrome, and that the plaque that ruptures usually is not the most narrowing or culprit lesion. Not only the autopsy specimen, but also clinical findings have shown that plaque rupture leads to acute myocardial infarction. The great progress of vascular biology in the 1990s finally made clinical cardiologists believe that inflammation and its relation to plaque rupture are the main cause of acute coronary syndrome, initiating the hunt for the "vulnerable plaque".

This book contains five sections. In the first section a pathologist, a cardiologist, and a vascular surgeon disclose their wish list with regards to vascular ultrasound. In the second section the current technical potential of IVUS is illustrated. The third section describes clinical applications of IVUS, including diagnosis, therapy guidance, and vulnerable plaque detection. In the final sections, carotid ultrasound scanning is addressed, followed by low- and high-frequency acoustic microscopy of vascular tissue.

The editors intend this monograph to introduce novel techniques to medical doctors and researchers in the field of medical ultrasound, thus narrowing the gap between fundamental and clinically applied vascular ultrasound.

Yoshifumi Saijo, M.D., Ph.D.
Cardiologist
Department of Medical Engineering and Cardiology
Institute of Development, Aging and Cancer
Tohoku University
Sendai, Japan

Antonius Franciscus Wilhelmus van der Steen, Ph.D.
Professor in Biomedical Engineering in Cardiology
Thorax Center
Erasmus Medical Center Rotterdam
The Netherlands

Contents

Part 3: Clinical Applications of IVUS

Part 4: Carotid Scanning

Part 5: Acoustic Microscopy of Vascular Tissue

Part 1

Introduction to Vascular Ultrasound

What do cardiologists want from vascular ultrasound?

Hideaki Kaneda, Yasuhiro Honda, Paul G. Yock, Peter J. Fitzgerald

Stanford University School of Medicine, 300 Pasteur Drive, Room H3554, Stanford, CA94305, USA

Summary: Although the value of coronary angiography remains unquestioned, investigations have revealed many deficiencies inherent in this approach. Intravascular ultrasound has several important advantages over contrast angiography and may provide new diagnostic and therapeutic insights for coronary disease. The tomographic orientation of ultrasound enables visualization of the entire circumference of the vessel wall. Intravascular ultrasound (IVUS) can interrogate tissues beneath the luminal border. IVUS studies have demonstrated a new dimension to the development of arteriosclerosis: arterial remodeling. In ambiguous lesions, IVUS permits lesion quantification effecting on clinical decision-making. IVUS has proven useful in selecting the device most suitable for a specific patient or lesion, leading to better clinical outcome. Information from IVUS imaging may be used to predict short - and long-term outcomes, particularly following stenting. IVUS has proven useful in clarifying the mechanisms of benefit and refining the techniques in brachytherapy. IVUS studies have shown the mechanisms of lumen enlargement with balloon angioplasty, atherectomy, and stent implantation. IVUS studies have also shown the mechanisms of restenosis. Therapeutic ultrasound (thrombolysis with adjunctive ultrasound and ultrasound-assisted local gene therapy) is promising.

Key words: ultrasonics, imaging, coronary disease, angioplasty, stents

Accurate intravascular imaging has become a reality over the past 10 years. Imaging in the early 90's provided an insight into some of the macroscopic endovascular structures, but it wasn't until mid to late 90's that this technology advanced to provide optimal intravascular imaging. Clarification of plaque subtypes, vessel size, dissection, thrombus, and various other subtitles "inside" the peripheral and coronary vessel. As frequency increased and signal processing techniques improved, coupled with fast computational speed, the ability to reconstruct and provide the interventionists with a true look " inside the artery" allowed several diagnostic and interventional endpoints studies to be conducted.

These studies over the last 3 to 5 years have provided the interventionists with an armament of rationales for using this imaging format as a diagnostic tool, guidance for device choice and endpoint assessment of various therapeutic interventions.

In particular, the community at large has learned that remolding has not only impacted how we treat these segmental obstructions with mechanical tools but has provided insight into the vascular response of these target segments. Remolding has been associated with the biological active plaque that often responds poorly to pure mechanical interventions. With the advent of drug eluting platform, the combination of the mechanical and biological process has been particular well accepted by these remodeled coronary segments.

IVUS has allowed not only the magnitude but distribution of intimal hyperplasia to be accessed accurately- thus providing critical evaluation of different drug eluting platforms. Being able to track these changes in a given patient with intravascular ultrasound, particularly from a three-dimensional stand point, will help us understand the optimal drug eluting platform to treat coronary segments and expanded lesion subsets optimally. In addition safety profile and long-term durability within these vessel segments are needed and IVUS will be critical to their evaluation. As we adopt this technology to "many patients" up front we need to make sure that these drugs brought directly to the endovascular surface constitute a safe profile long-term.

In summary interventionists need additional information from inside the artery together with the road map of the angiogram to allow optimal treatment for focal coronary artery disease. With the emergence of noninvasive technology (MRA, MSCT) to diagnose non-obstructive CAD, intravascular ultrasound may have an increasingly important role in clarifying focal disease as well as classifying the diffuse nature of atherosclerosis in general for both the coronary and peripheral vascular system.

Diagnostic Applications

Arterial Remodeling

The term arterial remodeling refers to alterations in vascular dimensions during the development of atherosclerosis. The phenomenon of remodeling was initially described by the pathologist Glagov [1]. As noted, the lumen area increases with atherosclerotic lesion development up to a stenosis of <40%. However, as lesion stenoses advance above 40%, the lumen area diminishes markedly, because remodeling is less obvious. This phenomenon represents a compensatory mechanism to preserve lumen area. IVUS studies have illustrated a correlation between external elastic membrane (EEM) and plaque area and have confirmed positive remodeling (Figure 1) in early atherosclerotic lesions in vivo [2, 3]. In addition, IVUS studies have demonstrated the presence of negative remodeling (arterial wall shrinkage, Figure 2) which also may contribute to severe luminal stenosis at diseased segments [4-6]. It is important to realize that when remodeling is defined by comparing lesion to reference EEM areas in most IVUS studies of de novo atherosclerosis, there is an inherent assumption that the ref erence EEM area represents the original vessel size and atherosclerotic lesions are usually focal. However, the variability of remodeling pattern throughout atherosclerotic segments has been reported [7].

While multicenter randomized lipid-lowering trials, using both angiographic and clinical assessment, showed a negligible improvement of luminal caliber, (typically an absolute difference of only 1% to 3%), these same studies yielded a staggering 25% to 75% reduction in acute events, including myocardial infarction [8-10]. Thus, IVUS may shed new light on the differences between angiographic and clinical end points. Preliminary data suggest that as plaques stabilize and become smaller, reverse modeling occurs; that is, the EEM area decreases but the lumen size remains unchanged [11]. Similar findings have been reported in familial hypercholesterolemia patients treated with low density lipoprotein apheresis [12]. With the ability to interrogate tissues beneath the luminal border, IVUS studies have demonstrated a new dimension to the development of arteriosclerosis: arterial remodeling.

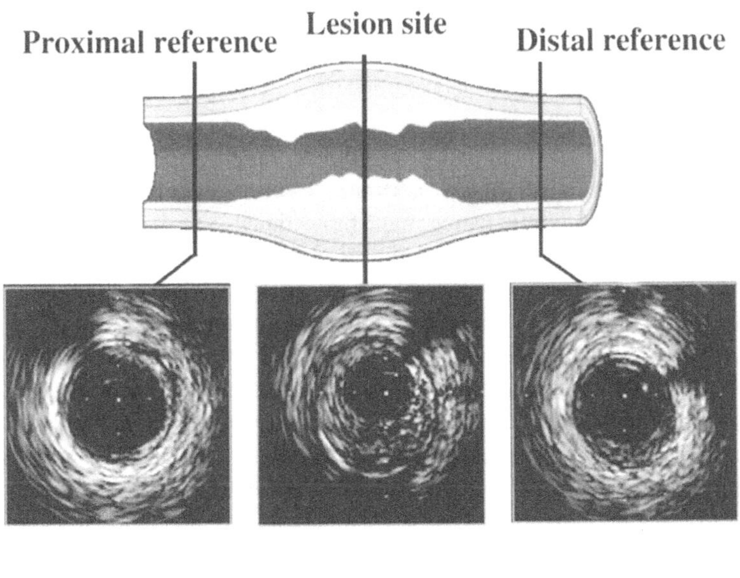

Figure 1: Positive remodeling. Vessel (external elastic membrane) area at lesion site is larger than that at reference site

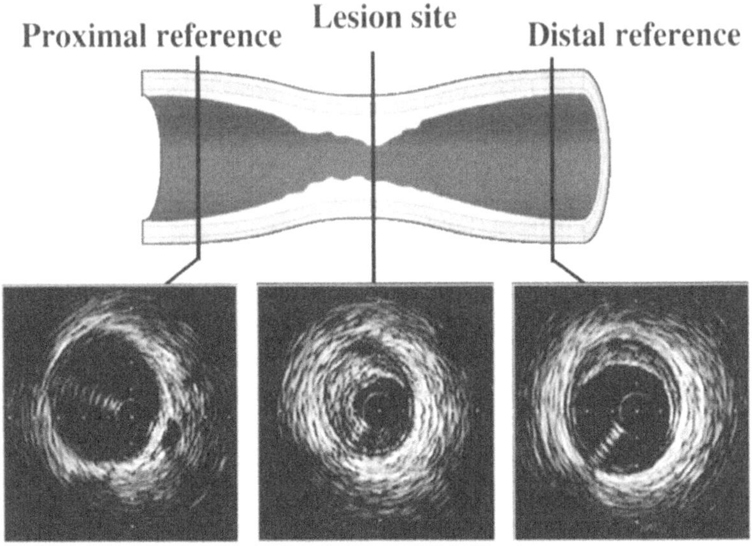

Figure 2: Negative remodeling. Vessel (external elastic membrane) area at lesion site is smaller than that at reference site

Ambiguous lesions

Despite improvements in x-ray equipment, angiographers still commonly encounter lesions that elude accurate characterization, despite thorough examination using multiple radiographic projections.

In ostial and bifurcation lesions, the stenosis may be obscured by overlapping contrast-filled structures. Angiographic assessment of left main coronary disease represents a particularly challenging clinical problem (Figure 3-1,2) [13]. The 3 major anatomic factors affecting angiographic left main evaluation are: (1) Aortic cusp opacification or "streaming" of contrast may obscure the ostium; (2) short vessel length may leave no normal segment for comparison, and (3) the distal left main artery may be concealed by bifurcation or trifurcation. Angiography underestimates stenosis severity, particularly in vessels with 50-75% narrowing [14, 15], which is precisely the range in which left main decisions are most crucial. Ultrasound can help overcome these confounding factors[16]. As reported by Abizaid et al. [17], IVUS measurement of the left main coronary artery lesion site minimum lumen diameter was the most important quantitative predictor of cardiac events.

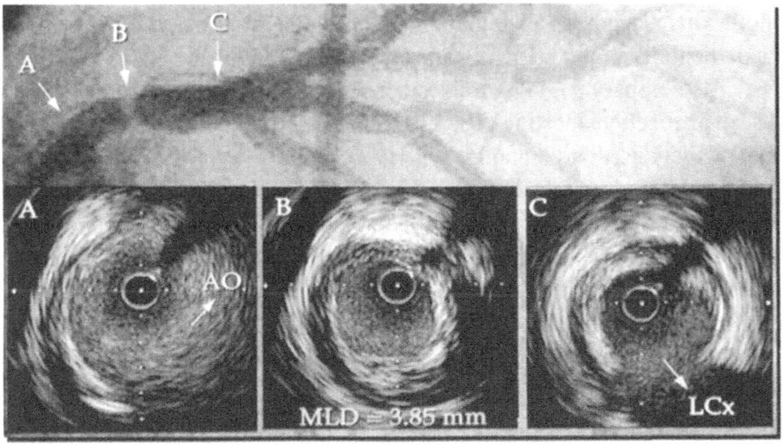

3-1

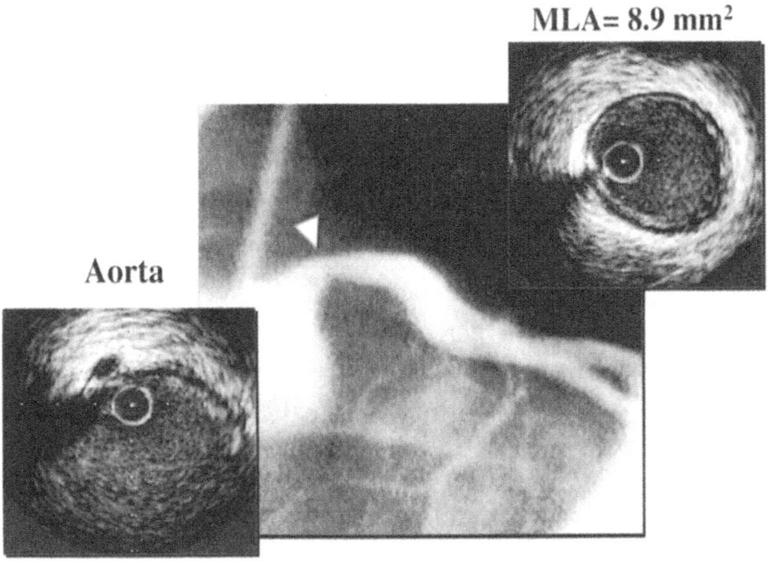

3-2

Figure 3: Left Main Assessment. 3-1A: Ostium of left main coronary artery. AO denotes aorta. 3-1B: Although coronary angiogram reveals a possible severe narrowing at the left main coronary artery, ultrasound images show a mild stenosis with a 3.85mm minimum lumen diameter (MLD). 3-1C: Ostium of left circumflex artery. 3-2: Although coronary angiogram reveals a possible severe narrowing at the left main ostium, ultrasound images show a mild stenosis with an 8.9mm^2 minimum lumen area (MLA).

Intermediate stenoses (angiographic severity ranging from 50% to 75%, Figure 4) are particularly problematic in patients whose symptomatic status is difficult to assess. The coronary angiogram does not reliably predict the results of the stress ^{201}Tl scan or functional testing [18, 19]. Neither a 50% nor 70% diameter stenosis, the most widely used angiographic criteria for stenosis severity, reliably predicted the presence of a reversible defect on the stress ^{201}Tl scan [20]. On the other hand, IVUS measurements have shown a good correlation with functional testing. Comparing IVUS measurements and stress myocardial perfusion single-photon emission computed tomography, lesion lumen area and 3 IVUS-derived stenosis indices showed sensitivities and specificities ranging between 80% and 90% [21]. In a study by Abizaid et al. [22], a good correlation between the lesion lumen area by IVUS and the coronary flow reserve by Doppler technique was found, with further correlation with noninvasive evidence of ischemia [20] and clinical events after catheter intervention [23]. Anatomic parameters obtained by IVUS showed a significant correlation to fractional flow reserve values [24], correlating with noninvasive assessments of ischemia, such as exercise-induced ECG changes or positron emission tomography [25, 26]. Finally, long-term follow-up after IVUS-guided deferred interventions in patients with de novo intermediate native artery lesions showed a low event rate [27]. Thus, in ambiguous lesions, IVUS permits lesion quantification affecting clinical decision-making.

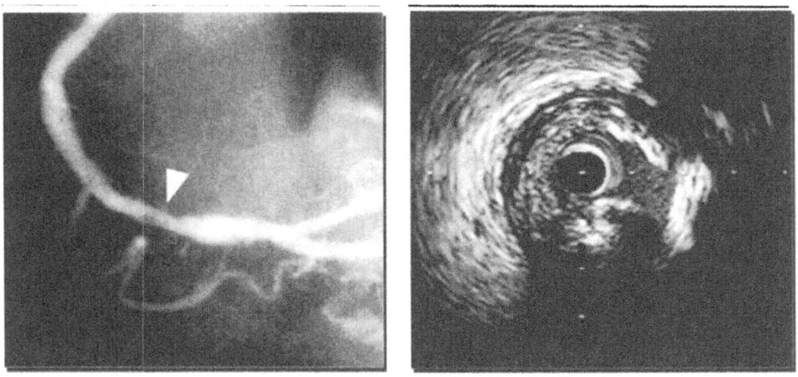

MLD 1.8 mm, MLA 3.2 mm^2

Figure 4: Intermediate Lesion. Although coronary angiogram reveals an intermediate lesion at the right coronary artery, ultrasound image shows a calcified plaque accumulation with a 1.8mm minimum lumen diameter (MLD) and a 3.2mm^2 minimum lumen area (MLA).

Association with Clinical Presentation

Indeed, several IVUS studies have suggested that certain plaque types are significantly associated with unstable clinical presentations. Compared to patients with stable angina, patients with unstable angina had more echolucent lesions but fewer calcified and mixed plaques [28-30]. IVUS permitted differentiation between pulsatile, low echogenic intraluminal material, suggesting thrombus and mural highly-echogenic atherosclerotic plaque in patients with acute myocardial infarction (AMI) [31]. Large vessels with lipid pool–like images from pre-interventional IVUS are at high risk for no reflow phenomenon after primary intervention for AMI [32].

In a study using angioscopy and IVUS, complex atherosclerotic lesions were found predominantly in compensatory enlarged arterial segments, whereas smooth lesions were found predominantly in shrunken arterial segments [33]. Positive remodeling and larger plaque areas were associated with unstable clinical presentation, whereas negative remodeling was more common in patients with stable clinical presentation [34]. A pathology study has shown that coronary artery plaques with positive remodeling have higher lipid content and macrophage count, which are both markers of plaque vulnerability [35].

With regard to plaque instability, pathophysiologic processes such as inflammation may exert adverse effects throughout the coronary vasculature, therefore resulting in multiple unstable lesions. This was shown in an IVUS study analyzing the 3 coronary arteries of patients with acute coronary syndrome, revealing that at least 1 plaque rupture was found somewhere other than on the culprit lesion in most patients [36]. Thus, these studies provide additional insight into the association between vascular and plaque morphologies and their subsequent clinical presentation.

Transplant Vasculopathy

In the first year following transplantation, coronary disease represents the major cause of death, and is often clinically silent because the heart is denervated. From a clinical management standpoint, conventional functional testing does not reveal ischemia until the disease is advanced, and angiography, performed annually largely for surveillance, may not effectively detect the disease owing to its diffuse nature. Accordingly, IVUS has emerged as the optimal method for early detection [37-39].

Transplant vasculopathy is characterized by diffuse angiographic luminal narrowing, which is frequently not amenable to revascularization. Recently, IVUS has shown that in addition to progressive intimal thickening, inward or inadequate outward remodeling is common in transplanted hearts, and the importance of its contribution to lumen loss increases with time from transplantation [40]. Despite diffuse endotheliopathy, some remodeling in response to hemodynamic stimuli appears to persist [41]. Although sympathectomy prevents inward remodeling after angioplasty [42], the effect of transplant denervation on remodeling is unknown. Although drugs such as diltiazem and HMG-CoA-reductase inhibitors reduce angiographic progression of transplant vasculopathy, IVUS studies have attributed this to reduction in intimal growth rather than prevention of inward remodeling [43, 44]. Another IVUS study demonstrated that in vivo IVUS tissue characterization can differentiate vasculopathic adventitial tissue changes in an animal model of allograft rejection [45]. Thus, these studies provide additional insight into the pathogenesis of transplant vasculopathy.

Interventional Applications

Preinterventional Imaging

Some operators use IVUS to select the device most suitable for a specific patient or lesion. Device selection for intervention depends on several factors, but is mostly determined by target lesion morphology / characterisitics (vessel size, plaque distribution, depth and extent of calcification, and the presence of thrombi or dissections), which are better characterized by IVUS than by angiography. In two large prospective series, IVUS changed the management strategy (primarily decisions to perform or defer interventions or choice of interventional device) in 20% of examinations performed immediately before coronary intervention [46, 47].

Several studies have demonstrated the benefits of intravascular imaging before the implantation of coronary stents. In heavily calcified lesions, stenting results in a relatively small and asymmetrical acute lumen gain [48]. In ostial lesions, ultrasound can identify whether the lesion involves the "true" ostium or if it spares the first 1 or 2 mm. Ultrasound imaging may prove very helpful in the treatment of instent restenosis, often demonstrating inadequate stent expansion at the initial procedure [49], and proper balloon sizing may lead to an improved lumen area. In the setting of intracoronary radiation, IVUS is useful for vessel sizing to determine proper dose, catheter size, and treatment length [50].

Most recently, in the setting of drug-eluting stents, characterizing plaque type may be more important because potential influences of plaque type (especially calcium) on the pharmacokinetics of drug-eluting stents remain yet to be fully investigated. Compared to virtually eliminating neointimal hyperplasia within the stent, efficacy of drug-eluting stents at the peri-stent segments, particularly at the proximal edge, is less striking. Incomplete coverage of diseased segments may be an explanation. IVUS assessment of lesion length, typically underestimated by angiography, dictates the exact length of stent necessary to completely scaffold a lesion. IVUS has proven useful in selecting the most suitable device for a specific patient or lesion, leading to better clinical outcome.

Mechanisms of Lumen Enlargement

Because of its ability to directly visualize coronary arteries in vivo, intravascular ultrasound has been used to study the mechanisms of lumen enlargement with balloon angioplasty [51, 52] directional coronary atherectomy [53, 54], rotational atherectomy [55, 56] and stent implantation [57, 58].

With balloon angioplasty, lumen enlargement is the result of combined plaque dissection, arterial expansion (presumably, limited by media-adventitial elasticity), and plaque redistribution [52]. As such, plaque compression, as previously thought, does not appear to be the primary mechanism of balloon angioplasty. Both vessel stretch and dissection are uncommon after directional coronary atherectomy, with plaque removal being the major mechanism for improved lumen area after this procedure [53]. High speed rotational atherectomy causes lumen enlargement by selective ablation of hard, especially calcific, atherosclerotic plaque with little tissue disruption and rare arterial expansion [55]. Although first-generation stent designs, especially the Palmaz-Schatz stent, precluded consistent imaging of peristent arterial structures, second-generation stent designs, especially the Multi-Link stent, permit routine measurements of EEM [57]. The mechanisms of lumen enlargement after stenting involve (1) significant longitudinal redistribution of plaque from the lesion into the reference segments, (2) vessel expansion, and (3) either plaque embolization or compression [57].

When balloon angioplasty is used to treat instent restenosis, lumen enlargement is achieved through a combination of additional stent expansion and tissue extrusion outside the stent segment [59]. On the other hand, when cutting balloon angioplasty is used to treat instent restenosis, lumen enlargement is almost entirely the result of intimal hyperplasia extrusion through stent struts and, to a lesser extent, axial redistribution of intimal hyperplasia into adjacent reference segments [60, 61]. Thus, information from IVUS imaging provided insight into lumen enlargement mechanisms seen with balloon angioplasty, atherectomy, and stent implantation.

Predictor of Complications and Restenosis

In the early 1990s, studies were initiated to determine whether intravascular imaging could predict clinical and angiographic outcome after intervention. The goal of these investigations was to determine whether morphological features, such as the presence or extent of dissection, or morphometric features (e.g., lumen size or plaque burden) were related to outcome.

Complications

In contemporary practice, dissection leading to acute closure is extremely unusual, since virtually all significant dissections are treated by coronary stents, clearly the

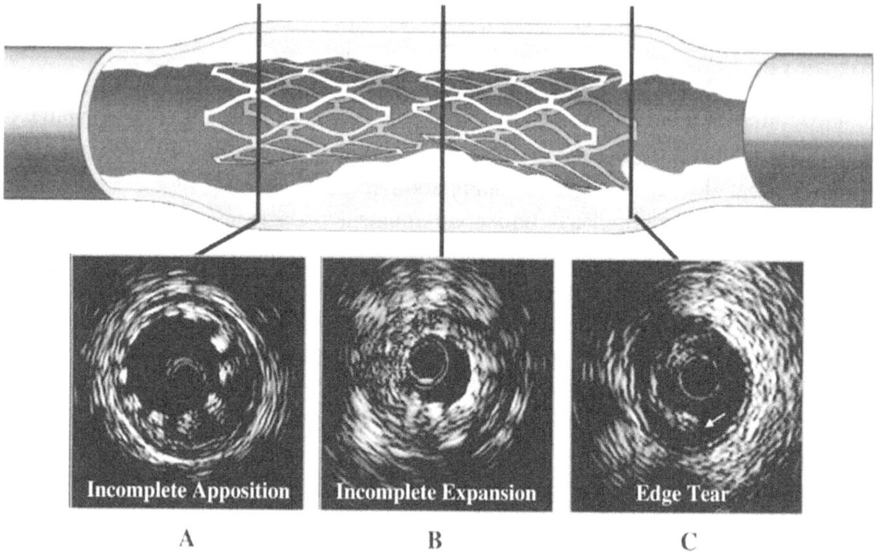

Figure 5: IVUS-detected problems with stent deployment. A: Incomplete apposition. B: Incomplete expansion. C: Edge dissection (indicated by white arrow). (Adapted from Honda, et al. Topol text book of interventional cardiology 4th edition. 893-917)

most important tool for treating dissection, averting acute closure, and reducing the risk of subsequent ischemic complications [62]. Dissection is still an important cause of ischemic complications, but usually occurs as a manifestation of edge dissections following stenting (Figure 5C), which may predispose the patient to stent thrombosis [63]. When stents are used to treat dissections, ultrasound often reveals the involvement of a longer vessel segment than was appreciated angiographically. Routine IVUS identifies angiographically-unapparent edge dissections in approximately 11% of the patients [64]. Most non-obstructive residual dissections after percutaneous coronary intervention are "favorable" and not seen as needing stent implantation [65, 66]. Although resolution appears to be the natural history of IVUS-detected dissections after intravascular brachytherapy in most cases, significant impairment of vessel wall healing was noted with this treatment modality [67, 68].

Intravascular ultrasound imaging has played a pivotal role in understanding and optimizing the benefits of stent therapy. Several studies suggest that angiography often fails to identify inadequate stent deployment (incomplete apposition and incomplete expansion, Figure 5A, B), readily evident by IVUS [49, 69-71]. Currently, no strict consensus exists regarding optimal ultrasound procedural endpoints for stent implantation. However, most operators strictly advocate complete apposition of stent struts to the vessel wall, because of the thrombotic risk from protrusion of stent struts into the blood field [72]. This interventional endpoint carries over into the current drug-eluting stent era, where incomplete apposition and asymmetric stent expansion may be of particular importance. Potentially, both incomplete apposition and asymmetric stent expansion may result in heterogenous drug delivery, which could lead to non-uniform efficacy or focal overdose toxicity, depending on the pharmacologic property of the agent used [73].

IVUS studies have shown some unusual findings after brachytherapy (the first catheter-based technology to modulate the biological response): unhealed dissection [67, 68], late stent incomplete apposition [74, 75], and the "black hole" phenomenon [76, 77]. Although these findings occur with increased frequency in irradiated patients, none are unique to brachytherapy [78], and their exact clinical consequence is unknown. Overall, these findings represent valuable lessons to be remembered in the new era of drug-eluting stents [79].

Restenosis

Final instent cross-sectional area (CSA) by IVUS is a powerful predictor of restenosis [80, 81]. When IVUS measurements were entered into the multivariate model, post-interventional quantitative coronary angiographic parameters (minimal lumen diameter or percent diameter stenosis) did not predict any of the follow-up angiographic end points. IVUS predictors of restenosis at the stent margins include smaller reference vessel and lumen size, larger plaque burden at the reference segments, and smaller final in-stent lumen area at the margin [82].

The amount of instent neointimal proliferation has been shown to correlate with the pre-interventional [81, 83, 84] or post-interventional plaque burden [85, 86], although some studies have not shown that residual plaque burden after stenting was predictive of angiographic restenosis [87, 88]. Mintz et al. also reported a relationship between the amount of residual plaque after non-stent intervention and restenosis [89]. For cases using ß-radiation therapy and balloon angioplasty, residual plaque burden, delivered dose, and tissue composition played a fundamental role in the volumetric outcome at 6-month follow-up [90].

Pre-intervention arterial remodeling is associated with restenosis after nonstent coronary intervention [91, 92] and stent implantation [93, 94]. Positive lesion-site remodeling is associated with a higher long-term target lesion revascularization rate after a nonstent interventional procedure [91]. Plaque CSA in the minimal lumen CSA at pre-intervention and intimal hyperplasia CSA at follow-up were significantly larger in the adequate remodeling group than in the inadequate remodeling group [93]. Target vessel revascularization rate (TVR) at 9-month follow-up was significantly higher in the positive remodeling group.

IVUS studies demonstrate less restenosis after IVUS-guided stenting than angiography-guided stenting [95, 96]. Arteries which appear small by angiography are often larger by IVUS, allowing the use of larger stents and balloons. IVUS can also identify poor stent expansion, edge dissection and plaque prolapse, and can ensure a large final lumen CSA, which is the best predictor of restenosis. In the CRUISE (Can Routine Ultrasound Influence Stent Expansion) study, operators changed therapy after viewing the IVUS images in 48% of patients; restenosis at 1 year was <10% if the final IVUS CSA exceeded $9mm^2$ [95]. Colombo et al. showed intravascular ultrasound-guided spot stenting for the treatment of long coronary lesions was associated with good acute outcome [97]. A strategy of IVUS-guided angioplasty with provisional stenting reduced the stent rate by half, with similar 6-month angiographic IVUS and clinical outcome compared with stent implantation [98]. Thus, information from IVUS imaging is useful to predict short- and long-term outcomes and its use leads to better clinical outcome.

Mechanisms of Restenosis

Nonstent Intervention

Recently, negative remodeling has been implicated in restenosis after nonstent intervention [89, 99]. Over 70% of lumen loss was attributable to the decrease in EEM area, whereas the neointimal area accounted for only 23% of the loss [89]. Segments with positive remodeling at 6-month follow-up after balloon angioplasty showed an increase in plaque volume, whereas those with vessel shrinkage had a plaque reduction [100]. Similar findings have been previously reported [101]. In the multivariate model, residual plaque burden had an influence on negative remodeling, whereas dissection had an effect on total vessel size increase [100].

Stent

Unlike the restenotic response to angioplasty or atherectomy, which is a mixture of arterial remodeling and neointimal growth, stent restenosis is primarily due to neointimal proliferation. While self-expanding coronary stents [102-106] continue to enlarge during the follow-up period, most studies using current balloon-expandable stents [104, 107-115] remain unchanged. As expected, late lumen loss directly correlated with the degree of instent neointimal growth with no significant change of stent area. Lumen loss at the stent edge without brachytherapy is also due primarily to intimal proliferation [114].

In vivo IVUS measurement of intimal hyperplasia (IH) volumes correlated strongly with ex vivo histomorphometry [116]. An IVUS study has shown that 29% of the Palmaz-Schatz stent volume is filled with intimal hyperplasia at 6 months [114]. Costa et al. [112] studied 2 new generation tubular stents; IH volume occupied 24% of Multilink stents and 26% of NIR stents. An ultrasound analysis demonstrated that degree of instent intimal hyperplasia is independent of final stent lumen size, perhaps explaining higher restenosis rates in smaller vessels and poorly expanded stents [110].

Neointima formation is often observed at specific locations in the stented segment [117], similar to predisposing locations of de novo atherosclerotic lesions or allograft vasculopathy (at the outer wall of the flow divider [118-120], on the inner wall of curved coronary segments [121], or on the myocardial side of the vessel wall [122]). A number of risk factors, such as thrombus formation [123] and endothelial dysfunction [124], are related to restenosis, but their relationship to a specific intimal hyperplasia distribution is unknown. Localizing factors that have been studied include plaque burden and wall stress.

As mentioned above, the amount of instent neointimal proliferation has been shown to correlate with the plaque burden [81, 83-86]. Atherosclerotic plaque may be a source for cells and growth factors involved in the restenotic process. It is possible that greater plaque area and thickness necessitate more aggressive balloon dilations to achieve adequate stent expansion. These aggressive dilations may result in more media disruption and macrophage infiltration. Studies have shown instent neointimal hyperplasia was related to aggressiveness of the stent implantation technique measured by balloon/artery ratio [111] or change of EEM before intervention to stent implantation [125].

Neointimal hyperplasia distribution after Wallstent implantation has been found to be related to shear stress distribution, such that when compared with average values, low shear stress regions were accompanied by maximal intimal thickness and high shear stress regions by minimal intimal thickness [126]. Studies have explored the ability of shear stress to modulate atherogenesis via its effects on endothelial-mediated alterations in coagulation, leukocyte and monocyte migration, smooth muscle growth, lipoprotein uptake and metabolism, and endothelial cell survival [127].

While some studies [108, 128-130] have suggested the existence of positive, peri-stent arterial remodeling, others have not [109, 128]. Although a study [108] of Palmaz-Schatz stent implantation has shown greater tissue growth within the stent correlated weakly with tissue growth surrounding the stent, only less than half of the image slices could be evaluated because of poor EEM visualization through the Palmaz-Schatz stent. Positive remodeling of the vessel exterior to a coronary stent occurs to a variable degree after ACS Multi-Link stent implantation [130]. There is a distinct trade-off between positive remodeling and in-stent hyperplasia: in segments with less peri-stent remodeling, intrastent neointimal proliferation is greater and accompanied by more significant late lumen loss.

Instent restenosis

When compared to immediately after balloon angioplasty and/or ablation for instent restenosis, lumen and EEM CSA decreased 42 minutes later [131]. This early lumen loss due to tissue reintrusion was greater in longer lesions and those with greater instent tissue burden, thus contributing to recurrence. On the other hand, in a study of cutting balloon angioplasty for instent restenosis, there was no significant change in lumen or EEM CSA at 24 hr post-intervention compared to the acute phase results [61].

Radiation

The mechanisms of brachytherapy in balloon angioplasty [90, 132-134], stenting [128], and treatment of instent restenosis [135-141] are dose related: (1) decreased neointimal hyperplasia within stent, (2) positive remodeling in nonstented segments, and (3) either no effect or positive remodeling along irradiated edges and uninjured reference segments. At the edges, where the dose is less, and especially in the presence of geographic mi ss [142, 143], there is a combination of increased neointimal hyperplasia and either absence of positive remodeling or frank negative remodeling that contribute to lumen loss.

Drug-eluting stents

Several new drug-eluting stents (paclitaxel [144, 145], QP2: a taxane analogue [146, 147], sirolimus [148-150]), or everolimus [151]) have shown a minimal amount of neointimal proliferation (1% to 14% cross-sectional narrowing) in the stented segment, comparable to the data reported for radioactive stents (7% to 17%)[128]. Late lumen loss at the reference sites adjacent to the QuaDS-QP2 stent was acceptable and predominantly due to plaque proliferation [152]. Thus, information from IVUS imaging provides additional insight into mechanisms of restenosis following several kinds of percutaneous coronary intervention.

Future Directions

Therapeutic Ultrasound

Clinical experience with catheter-based therapeutic ultrasound has shown that acoustic energy has the ability to induce effective thrombolysis without injuring the surrounding vessel [153-155]. The potential therapeutic use of external high-intensity focused ultrasound (HIFU) was first explored by Lynn et al. in 1942.They demonstrated that HIFU induced localized tissue damage at a focal point within the body with no effect on the surface or on the overlying or surrounding tissue. Recently, therapeutic HIFU has been successfully used in neurosurgery, ophthalmology, urology, and oncology [156-158]. In an earlier feasibility study, Rosenschein et al. showed that the use of focused shock waves as an acoustic energy source provided effective thrombolysis with no damage to the artery [159]. External high-intensity focused ultrasound thrombolysis using optimal pulsed-wave parameters for periods of 300 seconds appears to be a safe and effective method to induce thrombolysis [160].

Studies dating back to the mid-1970s have demonstrated that ultrasound can reduce mammalian cell viability [161]. In the past decade, investigators have reported that high-energy cavitational/mechanical ultrasound is capable of inhibiting both in vitro smooth muscle cell (SMC) migration and adhesion [162, 163]. More recently, a study by Lawrie and colleagues [164] showed that noncavitational ultrasound energy can directly reduce SMC proliferation in culture, suggesting the potential of ultrasound therapy in vivo to limit neointimal growth after stenting. In the swine peripheral model, intravascular sonotherapy seemed to decelerate cellular proliferation and decrease instent hyperplasia [165]. Therefore, intravascular sonotherapy may be an effective form of nonionizing energy to reduce instent restenosis. On the basis of these favorable preclinical data, several human trials have been initiated in the United States and Europe. However, despite the favorable preclinical results, the results of European Sonotherapy Prevention of Arterial Hyperplasia (EuroSPAH) and Sound Wave Inhibition of Neointimal Growth (SWING) recently reported failed to demonstrate superiority of intravascular sonotherapy over the cont rol group in terms of clinical outcome [166].

Adjunctive ultrasound exposure was associated with enhanced transgene expression in vascular smooth muscle cells (VSMCs) and endothelial cells (ECs) and reduced VSMC but not EC proliferation in vitro, which suggests that ultrasound-assisted local gene therapy has potential as an antirestenotic therapy [164]. Endovascular ultrasound enhanced vascular gene delivery and increased the efficiency of nonviral platforms to levels previously attained only by adenoviral strategies [167]. An ultrasound transfection method with echo contrast microbubble enhanced transfection efficiency of naked plasmid DNA into blood vessels without any apparent toxicity [168].

The ability to deliver high frequency mechanical energy to the tip of an endovascular catheter has demonstrated initial feasibility in the recanalization of vascular occlusions [169]. An additional application for externally applied ultrasound is wound closure. Therapeutic ultrasound represent future applications and seem promising for the treatment of patients with coronary artery disease.

Each new therapeutic technology goes through an IVUS stage, where the performance of the device is evaluated and refined by direct visualization of the results. Many operators will continue to use IVUS as a clinical tool, because the operator needs more detailed information in order to proceed in more complicated situation with new technology.

References

1. Glagov S, Weisenberg, E, Zarins, CK, et al. (1987) Compensatory enlargement of human atherosclerotic coronary arteries. N Engl J Med 316:1371-5.
2. Hermiller JB, Tenaglia, AN, Kisslo, KB, et al. (1993) In vivo validation of compensatory enlargement of atherosclerotic coronary arteries. Am J Cardiol 71:665-8.

3. Losordo DW, Rosenfield, K, Kaufman, J, et al. (1994) Focal compensatory enlargement of human arteries in response to progressive atherosclerosis. In vivo documentation using intravascular ultrasound. Circulation 89:2570-7.
4. Pasterkamp G, Wensing, PJ, Post, MJ, et al. (1995) Paradoxical arterial wall shrinkage may contribute to luminal narrowing of human atherosclerotic femoral arteries. Circulation 91:1444-9.
5. Nishioka T, Luo, H, Eigler, NL, et al. (1996) Contribution of inadequate compensatory enlargement to development of human coronary artery stenosis: an in vivo intravascular ultrasound study. J Am Coll Cardiol 27:1571-6.
6. Mintz GS, Kent, KM, Pichard, AD, et al. (1997) Contribution of inadequate arterial remodeling to the development of focal coronary artery stenoses. An intravascular ultrasound study. Circulation 95:1791-8.
7. Iyisoy A, Schoenhagen, P, Balghith, M, et al. (2002) Remodeling pattern within diseased coronary segments as evidenced by intravascular ultrasound. Am J Cardiol 90:636-8.
8. Brown BG, Zhao, XQ, Sacco, DE, et al. (1993) Arteriographic view of treatment to achieve regression of coronary atherosclerosis and to prevent plaque disruption and clinical cardiovascular events. Br Heart J 69:S48-53.
9. Brown BG, Zhao, XQ, Sacco, DE, et al. (1993) Lipid lowering and plaque regression. New insights into prevention of plaque disruption and clinical events in coronary disease. Circulation 87:1781-91.
10. (1994) Randomised trial of cholesterol lowering in 4444 patients with coronary heart disease: the Scandinavian Simvastatin Survival Study (4S). Lancet 344:1383-9.
11. Nissen SE (2000) Rationale for a postintervention continuum of care: insights from intravascular ultrasound. Am J Cardiol 86:12H-17H.
12. Matsuzaki M, Hiramori, K, Imaizumi, T, et al. (2002) Intravascular ultrasound evaluation of coronary plaque regression by low density lipoprotein-apheresis in familial hypercholesterolemia: the Low Density Lipoprotein-Apheresis Coronary Morphology and Reserve Trial (LACMART). J Am Coll Cardiol 40:220-7.
13. Isner JM, Kishel, J, Kent, KM, et al. (1981) Accuracy of angiographic determination of left main coronary arterial narrowing. Angiographic--histologic correlative analysis in 28 patients. Circulation 63:1056-64.
14. Arnett EN, Isner, JM, Redwood, DR, et al. (1979) Coronary artery narrowing in coronary heart disease: comparison of cineangiographic and necropsy findings. Ann Intern Med 91:350-6.
15. Marcus ML, Skorton, DJ, Johnson, MR, et al. (1988) Visual estimates of percent diameter coronary stenosis: "a battered gold standard". J Am Coll Cardiol 11:882-5.
16. Hermiller JB, Buller, CE, Tenaglia, AN, et al. (1993) Unrecognized left main coronary artery disease in patients undergoing interventional procedures. Am J Cardiol 71:173-6.
17. Abizaid AS, Mintz, GS, Abizaid, A, et al. (1999) One-year follow-up after intravascular ultrasound assessment of moderate left main coronary artery disease in patients with ambiguous angiograms. J Am Coll Cardiol 34:707-15.
18. White CW, Wright, CB, Doty, DB, et al. (1984) Does visual interpretation of the coronary arteriogram predict the physiologic importance of a coronary stenosis? N Engl J Med 310:819-24.
19. Kern MJ, Donohue, TJ, Aguirre, FV, et al. (1993) Assessment of angiographically intermediate coronary artery stenosis using the Doppler flowire. Am J Cardiol 71:26D 33D.

20.Heller LI, Cates, C, Popma, J, et al. (1997) Intracoronary Doppler assessment of moderate coronary artery disease: comparison with 201Tl imaging and coronary angiography. FACTS Study Group. Circulation 96:484-90.
21.Nishioka T, Amanullah, AM, Luo, H, et al. (1999) Clinical validation of intravascular ultrasound imaging for assessment of coronary stenosis severity: comparison with stress myocardial perfusion imaging. J Am Coll Cardiol 33:1870-8.
22.Abizaid A, Mintz, GS, Pichard, AD, et al. (1998) Clinical, intravascular ultrasound, and quantitative angiographic determinants of the coronary flow reserve before and after percutaneous transluminal coronary angioplasty. Am J Cardiol 82:423-8.
23.Serruys PW, di Mario, C, Piek, J, et al. (1997) Prognostic value of intracoronary flow velocity and diameter stenosis in assessing the short- and long-term outcomes of coronary balloon angioplasty: the DEBATE Study (Doppler Endpoints Balloon Angioplasty Trial Europe). Circulation 96:3369-77.
24.Takagi A, Tsurumi, Y, Ishii, Y, et al. (1999) Clinical potential of intravascular ultrasound for physiological assessment of coronary stenosis: relationship between quantitative ultrasound tomography and pressure-derived fractional flow reserve. Circulation 100:250-5.
25.De Bruyne B, Baudhuin, T, Melin, JA, et al. (1994) Coronary flow reserve calculated from pressure measurements in humans. Validation with positron emission tomography. Circulation 89:1013-22.
26.De Bruyne B, Bartunek, J, Sys, SU, et al. (1995) Relation between myocardial fractional flow reserve calculated from coronary pressure measurements and exerciseinduced myocardial ischemia. Circulation 92:39-46.
27.Abizaid AS, Mintz, GS, Mehran, R, et al. (1999) Long-term follow-up after percutaneous transluminal coronary angioplasty was not performed based on intravascular ultrasound findings: importance of lumen dimensions. Circulation 100:256-61.
28.Hodgson JM, Reddy, KG, Suneja, R, et al. (1993) Intracoronary ultrasound imaging: correlation of plaque morphology with angiography, clinical syndrome and procedural results in patients undergoing coronary angioplasty. J Am Coll Cardiol 21:35-44.
29.Bocksch W, Schartl, M, Beckmann, S, et al. (1995) Intravascular ultrasound imaging in patients with acute myocardial infarction. Eur Heart J 16 Suppl J:46 -52.
30.Beckman JA, Ganz, J, Creager, MA, et al. (2001) Relationship of clinical presentation and calcification of culprit coronary artery stenoses. Arterioscler Thromb Vasc Biol 21:1618-22.
31.Bocksch WG, Schartl, M, Beckmann, SH, et al. (1994) Intravascular ultrasound imaging in patients with acute myocardial infarction: comparison with chronic stable angina pectoris. Coron Artery Dis 5:727-35.
32.Tanaka A, Kawarabayashi, T, Nishibori, Y, et al. (2002) No-reflow phenomenon and lesion morphology in patients with acute myocardial infarction. Circulation 105:2148-52.
33.Smits PC, Pasterkamp, G, de Jaegere, PP, et al. (1999) Angioscopic complex lesions are predominantly compensatory enlarged: an angioscopy and intracoronary ultrasound study. Cardiovasc Res 41:458-64.
34.Schoenhagen P, Ziada, KM, Kapadia, SR, et al. (2000) Extent and direction of arterial remodeling in stable versus unstable coronary syndromes: an intravascular ultrasound study. Circulation 101:598-603.

35.Varnava AM, Mills, PG, Davies, MJ (2002) Relationship between coronary artery remodeling and plaque vulnerability. Circulation 105:939-43.
36.Rioufol G, Finet, G, Ginon, I, et al. (2002) Multiple atherosclerotic plaque rupture in acute coronary syndrome: a three-vessel intravascular ultrasound study. Circulation 106:804-8.
37.Little WC, Constantinescu, M, Applegate, RJ, et al. (1988) Can coronary angiography predict the site of a subsequent myocardial infarction in patients with mild-to-moderate coronary artery disease? Circulation 78:1157-66.
38.Tuzcu EM, Hobbs, RE, Rincon, G, et al. (1995) Occult and frequent transmission of atherosclerotic coronary disease with cardiac transplantation. Insights from intravascular ultrasound. Circulation 91:1706-13.
39.Mehra MR, Ventura, HO, Stapleton, DD, et al. (1995) Presence of severe intimal thickening by intravascular ultrasonography predicts cardiac events in cardiac allograft vasculopathy. J Heart Lung Transplant 14:632-9.
40.Lim TT, Liang, DH, Botas, J, et al. (1997) Role of compensatory enlargement and shrinkage in transplant coronary artery disease. Serial intravascular ultrasound study. Circulation 95:855-9.
41.Allen-Auerbach M, Schoder, H, Johnson, J, et al. (1999) Relationship between coronary function by positron emission tomography and temporal changes in morphology by intravascular ultrasound (IVUS) in transplant recipients. J Heart Lung Transplant 18:211-9.
42.Lamawansa MD, Wysocki, SJ, House, AK, et al. (1997) Morphometric changes seen in balloon-injured porcine iliac arteries: the influence of sympathectomy on intimal hyperplasia and remodelling. Eur J Vasc Endovasc Surg 13:43-7.
43.Mehra MR, Ventura, HO, Smart, FW, et al. (1995) An intravascular ultrasound study of the influence of angiotensin- converting enzyme inhibitors and calcium entry blockers on the development of cardiac allograft vasculopathy. Am J Cardiol 75:853-4.
44.Wenke K, Meiser, B, Thiery, J, et al. (1997) Simvastatin reduces graft vessel disease and mortality after heart transplantation: a four-year randomized trial. Circulation 96:1398-402.
45.Jeremias A, Kolz, ML, Ikonen, TS, et al. (1999) Feasibility of in vivo intravascular ultrasound tissue characterization in the detection of early vascular transplant rejection. Circulation 100:2127-30.
46.Mintz GS, Pichard, AD, Kovach, JA, et al. (1994) Impact of preintervention intravascular ultrasound imaging on transcatheter treatment strategies in coronary artery disease. Am J Cardiol 73:423-30.
47.Lee DY, Eigler, N, Luo, H, et al. (1995) Effect of intracoronary ultrasound imaging on clinical decision making. Am Heart J 129:1084-93.
48.Hoffmann R, Mintz, GS, Popma, JJ, et al. (1998) Treatment of calcified coronary lesions with Palmaz-Schatz stents. An intravascular ultrasound study. Eur Heart J 19:1224-31.
49.Castagna MT, Mintz, GS, Leiboff, BO, et al. (2001) The contribution of "mechanical" problems to in-stent restenosis: An intravascular ultrasonographic analysis of 1090 consecutive in-stent restenosis lesions. Am Heart J 142:970-4.
50.Mintz GS, Weissman, NJ, Fitzgerald, PJ (2001) Intravascular ultrasound assessment of the mechanisms and results of brachytherapy. Circulation 104:1320-5.
51.Potkin BN, Keren, G, Mintz, GS, et al. (1992) Arterial responses to balloon coronary angioplasty: an intravascular ultrasound study. J Am Coll Cardiol 20:942-51.

52.Mintz GS, Pichard, AD, Kent, KM, et al. (1996) Axial plaque redistribution as a mechanism of percutaneous transluminal coronary angioplasty. Am J Cardiol 77:427-30.
53.Tenaglia AN, Buller, CE, Kisslo, KB, et al. (1992) Mechanisms of balloon angioplasty and directional coronary atherectomy as assessed by intracoronary ultrasound. J Am Coll Cardiol 20:685-91.
54.Matar FA, Mintz, GS, Farb, A, et al. (1994) The contribution of tissue removal to lumen improvement after directional coronary atherectomy. Am J Cardiol 74:647-50.
55.Kovach JA, Mintz, GS, Pichard, AD, et al. (1993) Sequential intravascular ultrasound characterization of the mechanisms of rotational atherectomy and adjunct balloon angioplasty. J Am Coll Cardiol 22:1024-32.
56.Dussaillant GR, Mintz, GS, Pichard, AD, et al. (1996)Effect of rotational atherectomy in noncalcified atherosclerotic plaque: a volumetric intravascular ultrasound study. J Am Coll Cardiol 28:856-60.
57.Ahmed JM, Mintz, GS, Weissman, NJ, et al. (2000) Mechanism of lumen enlargement during intracoronary stent implantation: an intravascular ultrasound study. Circulation 102:7-10.
58.Maehara A, Takagi, A, Okura, H, et al. (2000) Longitudinal plaque redistribution during stent expansion. Am J Cardiol 86:1069-72.
59.Mehran R, Mintz, GS, Popma, JJ, et al. (1996) Mechanisms and results of balloon angioplasty for the treatment of in- stent restenosis. Am J Cardiol 78:618-22.
60.Ahmed JM, Mintz, GS, Castagna, M, et al. (2001) Intravascular ultrasound assessment of the mechanism of lumen enlargement during cutting balloon angioplasty treatment of in-stent restenosis. Am J Cardiol 88:1032-4.
61.Montorsi P, Galli, S, Fabbiocchi, F, et al. (2002) Mechanism of cutting balloon angioplasty for in-stent restenosis: An intravascular ultrasound study. Catheter Cardiovasc Interv 56:166-73.
62.Bergelson BA, Fishman, RF, Tommaso, CL (1997) Abrupt vessel closure: changing importance, management, and consequences. Am Heart J 134:362-81.
63.Schuhlen H, Hadamitzky, M, Walter, H, et al. (1997) Major benefit from antiplatelet therapy for patients at high risk for adverse cardiac events after coronary Palmaz-Schatz stent placement: analysis of a prospective risk stratification protocol in the Intracoronary Stenting and Antithrombotic Regimen (ISAR) trial. Circulation 95:2015-21.
64.Sheris SJ, Canos, MR, Weissman, NJ (2000) Natural history of intravascular ultrasound detected edge dissections from coronary stent deployment. Am Heart J 139:59-63.
65.Albertal M, Van Langenhove, G, Regar, E, et al. (2001) Uncomplicated moderate coronary artery dissections after balloon angioplasty: good outcome without stenting. Heart 86:193-8.
66.Nishida T, Colombo, A, Briguori, C, et al. (2002) Outcome of nonobstructive residual dissections detected by intravascular ultrasound following percutaneous coronary intervention. Am J Cardiol 89:1257-62.
67.Meerkin D, Tardif, JC, Bertrand, OF, et al. (2000) The effects of intracoronary brachytherapy on the natural history of postangioplasty dissections. J Am Coll Cardiol 36:59-64.
68.Kay IP, Sabate, M, Van Langenhove, G, et al. (2000) Outcome from balloon induced coronary artery dissection after intracoronary beta radiation. Heart 83:332-7.
69.Goldberg SL, Colombo, A, Nakamura, S, et al. (1994) Benefit of intracoronary ultrasound in the deployment of Palmaz-Schatz stents. J Am Coll Cardiol 24:996-1003.

70.Nakamura S, Colombo, A, Gaglione, A, et al. (1994) Intracoronary ultrasound observations during stent implantation. Circulation 89:2026-34.
71.Kiemeneij F, Laarman, G, Slagboom, T (1995) Mode of deployment of coronary Palmaz-Schatz stents after implantation with the stent delivery system: an intravascular ultrasound study. Am Heart J 129:638-44.
72.Uren NG, Schwarzacher, SP, Metz, JA, et al. (2002) Predictors and outcomes of stent thrombosis: an intravascular ultrasound registry. Eur Heart J 23:124-32.
73.Lambert TL, Dev, V, Rechavia, E, et al. (1994) Localized arterial wall drug delivery from a polymer-coated removable metallic stent. Kinetics, distribution, and bioactivity of forskolin. Circulation 90:1003-11.
74.Kozuma K, Costa, MA, Sabate, M, et al. (1999) Late Stent Malapposition Occurring After Intracoronary Beta-Irradiation Detected by Intravascular Ultrasound. J Invasive Cardiol 11:651-655.
75.Mintz GS, Weissman, NJ, Pappas, C, et al. (2000) Positive remodeling, regression of in-stent neointimal hyperplasia, and late stent malapposition in the absence of brachytherapy. Circulation 102:E111.
76.Castagna MT, Mintz, GS, Weissman, N, et al. (2001) "Black hole": echolucent restenotic tissue after brachytherapy. Circulation 103:778.
77.Kay IP, Wardeh, AJ, Kozuma, K, et al. (2001) The pattern of restenosis and vascular remodelling after cold-end radioactive stent implantation. Eur Heart J 22:1311-7.
78.Shah VM, Mintz, GS, Apple, S, et al. (2002) Background incidence of late malapposition after bare-metal stent implantation. Circulation 106:1753-5.
79.Serruys PW, Degertekin, M, Tanabe, K, et al. (2002) Intravascular ultrasound findings in the multicenter, randomized, double-blind RAVEL (RAndomized study with the sirolimus-eluting VElocity balloon-expandable stent in the treatment of patients with de novo native coronary artery Lesions) trial. Circulation 106:798-803.
80.Kasaoka S, Tobis, JM, Akiyama, T, et al. (1998) Angiographic and intravascular ultrasound predictors of in-stent restenosis. J Am Coll Cardiol 32:1630-5.
81.Hoffmann R, Mintz, GS, Mehran, R, et al. (1998) Intravascular ultrasound predictors of angiographic restenosis in lesions treated with Palmaz-Schatz stents. J Am Coll Cardiol 31:43-9.
82.Hoffmann R, Mintz, GS, Kent, KM, et al. (1997) Serial intravascular ultrasound predictors of restenosis at the margins of Palmaz-Schatz stents. Am J Cardiol 79:951-3.
83.de Smet BJ, Kuntz, RE, van der Helm, YJ, et al. (1997) Relationship between plaque mass and neointimal hyperplasia after stent placement in Yucatan micropigs. Radiology 203:484-8.
84.Shiran A, Weissman, NJ, Leiboff, B, et al. (2000) Effect of preintervention plaque burden on subsequent intimal hyperplasia in stented coronary artery lesions. Am J Cardiol 86:1318-21.
85.Prati F, Di Mario, C, Moussa, I, et al. (1999) In-stent neointimal proliferation correlates with the amount of residual plaque burden outside the stent: an intravascular ultrasound study. Circulation 99:1011-4.
86.Hibi K, Suzuki, T, Honda, Y, et al. (2002) Quantitative and spatial relation of baseline atherosclerotic plaque burden and subsequent in-stent neointimal proliferation as determined by intravascular ultrasound. Am J Cardiol 90:1164-7.
87.Hong MK, Park, SW, Lee, CW, et al. (2002) Relation between residual plaque burden after stenting and six-month angiographic restenosis. Am J Cardiol 89:368-71.

88.Casserly IP, Aronow, HD, Schoenhagen, P, et al. (2002) Relationship between residual atheroma burden and neointimal growth in patients undergoing stenting: analysis of the atherectomy before MULTI- LINK improves lumen gain and clinical outcomes trial intravascular ultrasound substudy. J Am Coll Cardiol 40:1573-8.
89.Mintz GS, Popma, JJ, Pichard, AD, et al. (1996) Arterial remodeling after coronary angioplasty: a serial intravascular ultrasound study. Circulation 94:35 -43.
90.Sabate M, Marijnissen, JP, Carlier, SG, et al. (2000) Residual plaque burden, delivered dose, and tissue composition predict 6-month outcome after balloon angioplasty and beta-radiation therapy. Circulation 101:2472-7.
91.Dangas G, Mintz, GS, Mehran, R, et al. (1999) Preintervention arterial remodeling as an independent predictor of target-lesion revascularization after nonstent coronary intervention: an analysis of 777 lesions with intravascular ultrasound imaging. Circulation 99:3149-54.
92.Okura H, Hayase, M, Shimodozono, S, et al. (2001) Impact of pre-interventional arterial remodeling on subsequent vessel behavior after balloon angioplasty: a serial intravascular ultrasound study. J Am Coll Cardiol 38:2001-2005.
93.Endo A, Hirayama, H, Yoshida, O, et al. (2001) Arterial remodeling influences the development of intimal hyperplasia after stent implantation. J Am Coll Cardiol 37:70-5.
94.Okura H, Morino, Y, Oshima, A, et al. (2001) Preintervention arterial remodeling affects clinical outcome following stenting: an intravascular ultrasound study. J Am Coll Cardiol 37:1031-5.
95.Fitzgerald PJ, Oshima, A, Hayase, M, et al. (2000) Final results of the Can Routine Ultrasound Influence Stent Expansion (CRUISE) study. Circulation 102:523-30.
96.Oemrawsingh PV, Mintz, GS, Schalij, MJ, et al. (2003) Intravascular ultrasound guidance improves angiographic and clinical outcome of stent implantation for bng coronary artery stenoses: final results of a randomized comparison with angiographic guidance (TULIP Study). Circulation 107:62-7.
97.Colombo A, De Gregorio, J, Moussa, I, et al. (2001) Intravascular ultrasoundguided percutaneous transluminal coronary angioplasty with provisional spot stenting for treatment of long coronary lesions. J Am Coll Cardiol 38:1427-33.
98.Schiele F, Meneveau, N, Gilard, M, et al. (2003) Intravascular Ultrasound-Guided Balloon Angioplasty Compared With Stent: Immediate and 6-Month Results of the Multicenter, Randomized Balloon Equivalent to Stent Study (BEST). Circulation 107:545-51.
99.Kimura T, Kaburagi, S, Tamura, T, et al. (1997) Remodeling of human coronary arteries undergoing coronary angioplasty or atherectomy. Circulation 96:475-83.
100.Costa MA, Kozuma, K, Gaster, AL, et al. (2001) Three dimensional intravascular ultrasonic assessment of the local mechanism of restenosis after balloon angioplasty. Heart 85:73-9.
101.von Birgelen C, Mintz, GS, de Vrey, EA, et al. (1998) Atherosclerotic coronary lesions with inadequate compensatory enlargement have smaller plaque and vessel volumes: observations with three dimensional intravascular ultrasound in vivo. Heart 79:137-42.
102.Kay IP, Sabate, M, Van Langenhove, G, et al. (2000) The ESSEX (European Scimed Stent Experience) study. Catheter Cardiovasc Interv 50:419-25.
103.Kobayashi Y, Honda, Y, Christie, GL, et al. (2001) Longterm vessel response to a self-expanding coronary stent: a serial volumetric intravascular ultrasound analysis from the ASSURE Trial.A Stent vs. Stent Ultrasound Remodeling Evaluation. J Am Coll Cardiol 37:1329-34.

104.Nageh T, de Belder, AJ, Thomas, MR, et al. (2001) A randomised trial of endoluminal reconstruction comparing the NIR stent and the Wallstent in angioplasty of long segment coronary disease: results of the RENEWAL Study. Am Heart J 141:971-6.
105.Konig A, Schiele, TM, Rieber, J, et al. (2002) Stent designrelated coronary artery remodeling and patterns of neointima formation following selfexpanding and balloon-expandable stent implantation. Catheter Cardiovasc Interv 56:478 -86.
106.Yu ZX, Tamai, H, Kyo, E, et al. (2002) Comparison of the self-expanding Radius stent and the balloon- expandable Multilink stent for elective treatment of coronary stenoses: a serial analysis by intravascular ultrasound. Catheter Cardiovasc Interv 56:40 -5.
107.Painter JA, Mintz, GS, Wong, SC, et al. (1995) Serial intravascular ultrasound studies fail to show evidence of chronic Palmaz-Schatz stent recoil. Am J Cardiol 75:398-400.
108.Hoffmann R, Mintz, GS, Popma, JJ, et al. (1996) Chronic arterial responses to stent implantation: a serial intravascular ultrasound analysis of Palmaz-Schatz stents in native coronary arteries. J Am Coll Cardiol 28:1134-9.
109.Mudra H, Regar, E, Klauss, V, et al. (1997) Serial follow-up after optimized ultrasound-guided deployment of Palmaz- Schatz stents. In-stent neointimal proliferation without significant reference segment response. Circulation 95:363 -70.
110.Hoffmann R, Mintz, GS, Pichard, AD, et al. (1998) Intimal hyperplasia thickness at follow-up is independent of stent size: a serial intravascular ultrasound study. Am J Cardiol 82:1168-72.
111.Hoffmann R, Mintz, GS, Mehran, R, et al. (1999) Tissue proliferation within and surrounding Palmaz-Schatz stents is dependent on the aggressiveness of stent implantation technique. Am J Cardiol 83:1170-4.
112.Costa MA, Sabate, M, Kay, IP, et al. (2000) Three-dimensional intravascular ultrasonic volumetric quantification of stent recoil and neointimal formation of two new generation tubular stents. Am J Cardiol 85:135 -9.
113.Kuriyama N, Kobayashi, Y, Kuroda, N, et al. (2002) Effect of coronary stent overexpansion on lumen size and intimal hyperplasia at follow-up. Am J Cardiol 89:1297-9.
114.Weissman NJ, Wilensky, RL, Tanguay, JF, et al. (2001) Extent and distribution of in-stent intimal hyperplasia and edge effect in a non-radiation stent population. Am J Cardiol 88:248-52.
115.Nakamura M, Nishikawa, H, Mukai, S, et al. (2001) Impact of coronary artery remodeling on clinical presentation of coronary artery disease: an intravascular ultrasound study. J Am Coll Cardiol 37:63-9.
116.Mehran R, Mintz, GS, Hong, MK, et al. (1998) Validation of the in vivo intravascular ultrasound measurement of in- stent neointimal hyperplasia volumes. J Am Coll Cardiol 32:794-9.
117.Fontaine AB, Spigos, DG, Eaton, G, et al. (1994) Stentinduced intimal hyperplasia: are there fundamental differences between flexible and rigid stent designs? J Vasc Interv Radiol 5:739-44.
118.Kimura BJ, Russo, RJ, Bhargava, V, et al. (1996) Atheroma morphology and distribution in proximal left anterior descending coronary artery: in vivo observations. J Am Coll Cardiol 27:825-31.
119.Pethig K, Kofidis, T, Heublein, B, et al. (2001) Impactof vascular branching sites on focal progression of allograft vasculopathy in transplanted hearts. Atherosclerosis 158:155-60.

120.Balghith MA, Schoenhagen, P, Foody, JM, et al. (2003) Atherosclerotic plaque distribution in the left anterior descending coronary artery as assessed by intravascular ultrasound. Am J Cardiol 91:443-5.
121.Tsutsui H, Yamagishi, M, Uematsu, M, et al. (1998) Intravascular ultrasound evaluation of plaque distribution at curved coronary segments. Am J Cardiol 81:977-81.
122.Jeremias A, Huegel, H, Lee, DP, et al. (2000) Spatial orientation of atherosclerotic plaque in non-branching coronary artery segments. Atherosclerosis 152:209-15.
123.Rogers C, Parikh, S, Seifert, P, et al. (1996) Endogenous cell seeding. Remnant endothelium after stenting enhances vascular repair. Circulation 94:2909 -14.
124.van Beusekom HM, Whelan, DM, Hofma, SH, et al. (1998) Longterm endothelial dysfunction is more pronounced after stenting than after balloon angioplasty in porcine coronary arteries. J Am Coll Cardiol 32:1109-17.
125.Koyama J, Owa, M, Sakurai, S, et al. (2000) Relation between vascular morphologic changes during stent implantation and the magnitude of in-stent neointimal hyperplasia. Am J Cardiol 86:753-8.
126.Wentzel JJ, Krams, R, Schuurbiers, JC, et al. (2001) Relationship between neointimal thickness and shear stress after Wallstent implantation in human coronary arteries. Circulation 103:1740-5.
127.Traub O, Berk, BC (1998) Laminar shear stress: mechanisms by which endothelial cells transduce an atheroprotective force. Arterioscler Thromb Vasc Biol 18:677-85.
128.Kay IP, Sabate, M, Costa, MA, et al. (2000) Positive geometric vascular remodeling is seen after catheter-based radiation followed by conventional stent implantation but not after radioactive stent implantation. Circulation 102:1434-9.
129.Yoshitomi Y, Kojima, S, Yano, M, et al. (2002) Peristent tissue proliferation of multilink stents is dependent on preprocedural plaque area: a serial intravascular ultrasound analysis. Angiology 53:303-11.
130.Nakamura M, Yock, PG, Bonneau, HN, et al. (2001) Impact of peri-stent remodeling on restenosis : a volumetric intravascular ultrasound study. Circulation 103:2130 -2.
131.Shiran A, Mintz, GS, Waksman, R, et al. (1998) Early lumen loss after treatment of in-stent restenosis: an intravascular ultrasound study. Circulation 98:200 -3.
132.Meerkin D, Tardif, JC, Crocker, IR, et al. (1999) Effects of intracoronary betaradiation therapy after coronary angioplasty: an intravascular ultrasound study. Circulation 99:1660-5.
133.Sabate M, Serruys, PW, van der Giessen, WJ, et al. (1999) Geometric vascular remodeling after balloon angioplasty and beta- radiation therapy: A three-dimensional intravascular ultrasound study. Circulation 100:1182-8.
134.Kozuma K, Costa, MA, Sabate, M, et al. (2000) Three-dimensional intravascular ultrasound assessment of noninjured edges of beta-irradiated coronary segments. Circulation 102:1484-9.
135.Teirstein PS, Massullo, V, Jani, S, et al. (1997) Catheter-based radiotherapy to inhibit restenosis after coronary stenting. N Engl J Med 336:1697-703.
136.Waksman R, White, RL, Chan, RC, et al. (2000) Intracoronary gammaradiation therapy after angioplasty inhibits recurrence in patients with in-stent restenosis. Circulation 101:2165-71.
137.Ahmed JM, Mintz, GS, Waksman, R, et al. (2000) Safety of intracoronary gamma radiation on uninjured reference segments during the first 6 months after treatment of in stent restenosis: a serial intravascular ultrasound study. Circulation 101:2227 -30.

138.Ahmed JM, Mintz, GS, Waksman, R, et al. (2001) Serial intravascular ultrasound analysis of edge recurrence after intracoronary gamma radiation treatment of native artery in-stent restenosis lesions. Am J Cardiol 87:1145-9.
139.Ahmed JM, Mintz, GS, Waksman, R, et al. (2001) Serial Intravascular Ultrasound Analysis of the Impact of Lesion Length on the Efficacy of Intracoronary gamma-Irradiation for Preventing Recurrent In-Stent Restenosis. Circulation 103:188-191.
140.Mintz GS, Weissman, NJ, Teirstein, PS, et al. (2000) Effect of intracoronary gamma-radiation therapy on in-stent restenosis: An intravascular ultrasound analysis from the gamma-1 study. Circulation 102:2915-8.
141.Bhargava B, Mintz, GS, Mehran, R, et al. (2000) Serial volumetric intravascular ultrasound analysis of the efficacy of beta irradiation in preventing recurrent in-stent restenosis. Am J Cardiol 85:651-3, A10.
142.Sabate M, Costa, MA, Kozuma, K, et al. (2000) Geographic miss: a cause of treatment failure in radio-oncology applied to intracoronary radiation therapy. Circulation 101:2467-71.
143.Hong MK, Park, SW, Moon, DH, et al. (2002) Impact of geographic miss on adjacent coronary artery segments in diffuse in-stent restenosis with beta-radiation therapy: angiographic and intravascular ultrasound analysis. Am Heart J 143:327-33.
144.Grube E, Silber, S, Hauptmann, K, et al. (2003) TAXUS I Six- and Twelve-Month Results From a Randomized, Double-Blind Trial on a Slow-Release Paclitaxel-Eluting Stent for De Novo Coronary Lesions. Circulation 107:38-42.
145.Hong MK, Mintz, GS, Lee, CW, et al. (2003) Paclitaxel coating reduces in-stent intimal hyperplasia in human coronary arteries: a serial volumetric intravascular ultrasound analysis from the ASian Paclitaxel-Eluting Stent Clinical Trial (ASPECT). Circulation 107:517-20.
146.Honda Y, Grube, E, de La Fuente, LM, et al. (2001) Novel drugdelivery stent: intravascular ultrasound observations from the first human experience with the QP2 eluting polymer stent system. Circulation 104:380-3.
147.Kataoka T, Grube, E, Honda, Y, et al. (2002) 7-hexanoyltaxol-eluting stent for prevention of neointimal growth: an intravascular ultrasound analysis from the Study to COmpare REstenosis rate between QueST and QuaDSQP2 (SCORE). Circulation 106:1788-93.
148.Sousa JE, Costa, MA, Abizaid, A, et al. (2001) Lack of Neointimal Proliferation After Implantation of Sirolimus-Coated Stents in Human Coronary Arteries : A Quantitative Coronary Angiography and Three-Dimensional Intravascular Ultrasound Study. Circulation 103:192-195.
149.Morice MC, Serruys, PW, Sousa, JE, et al. (2002) A randomized comparison of a sirolimus-eluting stent with a standard stent for coronary revascularization. N Engl J Med 346:1773-80.
150.Ako J, Morino, Y, Sonoda, S, et al. (2002) Effect of Sirolimus Eluting Stents on Neointimal Hyperplasia: A Serial In-Stent IVUS Interim Analysis from the SIRIUS Trial (abstract). Circulation 106:II-520.
151.Grube E, Gerckens, U, Buellesfeld, L, et al. (2002) First Human Experience Using a New Everolimus Stent Coating: Early Findings of the FUTURE Trial (abstract). Am J Cardiol 90:71H.
152.Kataoka T, Grube, E, Hur, SH, et al. (2001) Intravascular Ultrasound Stent Edge Analysis Following Drug Eluting Stent Implantation: "Candy Without the Wrapper" (abstract). Circulation 104:II-770.

153.Rosenschein U, Roth, A, Rassin, T, et al. (1997) Analysis of coronary ultrasound thrombolysis endpoints in acute myocardial infarction (ACUTE trial). Results of the feasibility phase. Circulation 95:1411-6.
154.Rosenschein U, Gaul, G, Erbel, R, et al. (1999) Percutaneous transluminal therapy of occluded saphenous vein grafts: can the challenge be met with ultrasound thrombolysis? Circulation 99:26-9.
155.Ng KY, Liu, Y (2002) Therapeutic ultrasound: its application in drug delivery. Med Res Rev 22:204-23.
156.Fischer PD, Narayanan, K, Liang, MD (1992) The use of high-frequency ultrasound for the dissection of small-diameter blood vessels and nerves. Ann Plast Surg 28:326-30.
157.Valtot F, Kopel, J, Haut, J (1992) Therapeutic ultrasound for the treatment of glaucoma. Bull Soc Belge Ophtalmol 244:181-6.
158.Huber P, Debus, J, Jenne, J, et al. (1996) [Therapeutic ultrasound in tumor therapy. Principles, applications and new developments]. Radiologe 36:64-71.
159.Rosenschein U, Yakubov, SJ, Guberinich, D, et al. (1992) Shockwave thrombus ablation, a new method for noninvasive mechanical thrombolysis. Am J Cardiol 70:1358-61.
160.Rosenschein U, Furman, V, Kerner, E, et al. (2000) Ultrasound imaging-guided noninvasive ultrasound thrombolysis: preclinical results. Circulation 102:238-45.
161.Kaufman GE, Miller, MW, Griffiths, TD, et al. (1977) Lysis and viability of cultured mammalian cells exposed to 1 MHz ultrasound. Ultrasound Med Biol 3:21-5.
162.Alter A, Rozenszajn, LA, Miller, HI, et al. (1998) Ultrasound inhibits the adhesion and migration of smooth muscle cells in vitro. Ultrasound Med Biol 24:711-21.
163.Lejbkowicz F, Zwiran, M, Salzberg, S (1993) The response of normal and malignant cells to ultrasound in vitro. Ultrasound Med Biol 19:75-82.
164.Lawrie A, Brisken, AF, Francis, SE, et al. (1999) Ultrasound enhances reporter gene expression after transfection of vascular cells in vitro. Circulation 99:2617-20.
165.Fitzgerald PJ, Takagi, A, Moore, MP, et al. (2001) Intravascular sonotherapy decreases neointimal hyperplasia after stent implantation in swine. Circulation 103:1828-31.
166.Kuntz RE, Giugliano, GR, Caputo, R, et al. (2002) Intravascular Sonotherapy to Prevent In-Stent Restenosis: Results of the Combined Randomized SWING and EuroSPAH Trials (abstract). Circulation 106:II-482.
167.Amabile PG, Waugh, JM, Lewis, TN, et al. (2001) High-efficiency endovascular gene delivery via therapeutic ultrasound. J Am Coll Cardiol 37:1975-80.
168.Taniyama Y, Tachibana, K, Hiraoka, K, et al. (2002) Local delivery of plasmid DNA into rat carotid artery using ultrasound. Circulation 105:1233-9.
169.Cannon LA, John, J, LaLonde, J (2001) Therapeutic ultrasound for chronic total coronary artery occlusions. Echocardiography 18:219-23.

What pathologists want from vascular ultrasound

Gerard Pasterkamp[1], Erling Falk[2]

[1]Department of Cardiology, Utrecht University Medical Center, Heidelberglaan 100, 3584 CX Utrecht, the Netherlands, [2]Department of Cardiology, Aarhus University Hospital (Skejby) DK-8200 Aarhus, Denmark

Abstract. The human anatomy as shown by the pathologist and anatomist has been used as the gold standard for validation of imaging modalities like B mode ultrasound of the carotid artery and intravascular ultrasound imaging of the coronary artery. Sometimes the imaging technique functions as an "eye opener" for the pathologist when human arterial anatomy is placed in a 3 dimensional perspective. Transversal slicing of arteries has long strengthened the concept that atherosclerotic disease and restenosis after angioplasty are solely due to plaque mass and neointimaformation, respectively. Visualization in an axial direction and serial studies using intravascular ultrasound revealed that the vessel size , e.g. arterial remodeling, is another important determinant of arterial occlusive disease.

The last few years imaging modalities are being developed to visualize the plaque that will give rise to clinical events: the vulnerable, rupture prone plaque. Again, post mortem obtained specimen will help validating these imaging modalities. Subsequently, the imaging techniques may learn the pathologists the history of the lesions they have encountered in patients that died of coronary artery disease.

Key words: ultrasound, remodelling, atherosclerosis, plaque rupture, restenosis.

Introduction

The pathologist searches for causes and mechanisms of disease. For this, tissue is being macroscopically examined and investigated on a histochemical and cellular level. Imaging techniques like ultrasound have been developed to study pathological tissues *in vivo*. The title of this paper may therefore seem controversial. If pathologists have already obtained the pathological tissue, what do they need ultrasound for? Thus, the answer to the question raise d in the title of this paper would be: nothing. The pathologist may not need a visualization technique to diagnose, or improve the knowledge of disease. Nevertheless, the departments of pathology and anatomy are often involved in diagnostic research in which a visualization technique is validated (1-6). The human anatomy as shown by the pathologist and anatomist is the best gold standard for validation of imaging modalities although sometimes the technique functions as an "eye opener" when human arterial anatomy is placed in a 3 dimensional perspective (7) or when physiological parameters have been taken into account (8). Thus, the title of this paper could also have been reversed: what do investigators working with vascular ultrasound want from the pathologist?

Pathological studies are hampered by their cross-sectional study design that do not allow causal inferences. Vascular ultrasound may help to understand the pathogenesis of the process underlying the atherosclerotic disease. One major field in cardiovascular research that is in search for a tool to understand pathological observations is the acute coronary syndrome. It was the pathologist who were the first to describe the tissue substrate of myocardial infarction: a ruptured or eroded plaque rich in inflammatory cells with thrombus overlying the lipid rich plaque (9-12). Serial studies using vascular imaging techniques will be needed to show which parameters are causally related with rupture of the cap and subsequent clinical events. Unfortunately, the invasive nature makes the currently used intravascular ultrasound techniques unpractical for routinely serial imaging for this purpose.

This chapter will give an overview of the use of vascular ultrasound in which post mortem or *in vivo* obtained arterial tissue is being studied ex vivo.

Pathological validation of B-mode ultrasound carotid artery imaging

Atherosclerosis is considered a general process, leading to both diffuse subclinical arterial changes and localized lesions causing clinical syndromes. This view is strengthened by observations from population studies, where diffuse intimal thickening in the common carotid artery as visualized by B-mode ultrasound, constituting an important component of the measured intima-media thickness (IMT), appears to be a pretty good marker of the total atherosclerotic burden (13,14). Moreover, ultrasound assessment of the carotid IMT has been claimed to have value as an 'ultrasound biopsy' predicting clinical outcome (15). Measurements of the IMT of the carotid artery are now being used as a surrogate or intermediate end point to assess the severity of generalized atherosclerotic disease in both population studies and clinical trials. This non-invasive diagnostic tool has gained particular interest since several studies have revealed that the extent of extracranial carotid intimal thickening is correlated with coronary artery stenosis (16) and clinical manifestations of coronary (17,18) and femoral (19) artery disease.

Surprisingly, histopathologic validation studies of carotid B-mode ultrasound are scarce despite the widespread use of the technique. Moreover, controversial reports exist on the validation of the double line pattern in the carotid B-mode ultrasound appearance (20,21). Persson et al. (22) compared the IMT as observed in the ultrasound image using microscopic measurements as gold standard. A fair correlation was observed between ultrasound and microscopy measurements of the intima media thickness (r=0.84) with a slope of the regression line that approximated 1.0. However, the intercept (0.107 mm) indicated that ultrasound may slightly overestimate the thickness of the medial-intimal layer. In other validation studies carotid endarteriectomy specimens (ie, real plaques, not just diffuse intimal thickening) were examined (23,24). In a study by Schenk et al (24) 289 endarteriectomy specimen were studied and compared with data acquired with angiography and ultrasound. In 64% of cases a discrepancy of > 1.0 mm was observed between residual lumen measured in the ultrasound image and in the pathology specimen. Also in 64% of cases a discrepancy was observed between ultrasound and pathology determined lesion width. The authors subscribed part of the discrepancies to artifactual distortion of the pathologic specimen and loss of perfusion pressure during fixation. The authors concluded that the use of pathological endarteriectomy specimen as a gold standard has its limitations. Thus, this validation study was not successful due to methodological problems. The last statement of the authors (24) still holds: "Since there is much current interest in the capability of noninvasive diagnostic procedures to characterize atherosclerotic plaque components as well as plaque progression and regression, an awareness of problems associated with the evaluation of morphologic features will become increasingly important".

Validation studies of B-mode ultrasound confront the clinician with another issue that is often being raised: what does a visualization technique contribute to the wellbeing of the individual patient? In population studies, the intima media thickness on ultrasound may be predictive for the occurrence of adverse events. However, one should realize that the observed significant differences among groups of patients often go beyond the resolution of the imaging modality which limits the use for risk stratification in the individual person.

Histo-pathologic validation of intravascular ultrasound

Intravascular ultrasound (IVUS) has the ability to study arterial wall morphology in vivo. Muscular arteries, like the coronary and femoral arteries, are characterized by a three layered wall appearance in an IVUS image. The first histopathological validation studies reported that the inner echodense layer represents the intima, the middle hypoechoic layer the media and the outer echodense structure the adventitia (25-27) (Figures 1, 2). Absence of this three-layered appearance is reported when the intimal layer is too thin (28) or when the elastin or collagen content within the media is high (25,27,29) (Figure 1).

The initial assumption that the inner echodense layer solely represents the intimal layer was modified after a carefully performed post mortem analysis by

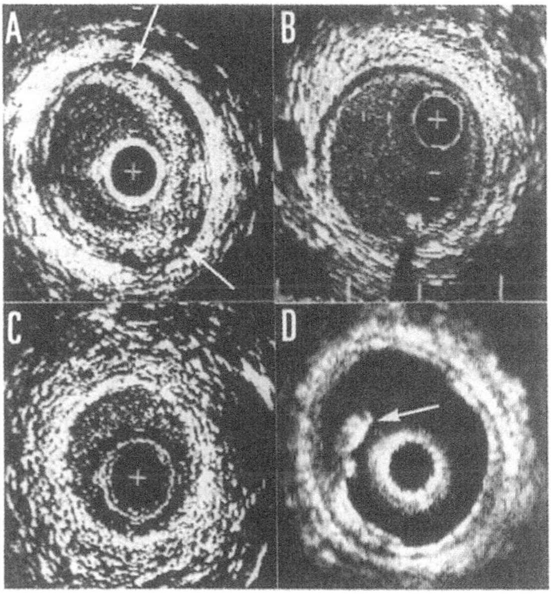

Figure 1. A. In vivo obtained intravascular ultrasound image of a left main coronary artery segment demonstrating a three layered wall appearance. The white arrows represent the interface between the echolucent media and the echodense intima. B. In vivo obtained intravascular ultrasound image of a femoral artery segment demonstrating a three layered wall appearance. C. In vivo obtained intravascular ultrasound image of a non muscular artery in which the 3 layered appearance is not evident. D. In vitro obtained intravascular ultrasound image of a renla artery. an intimal flap is visualized.

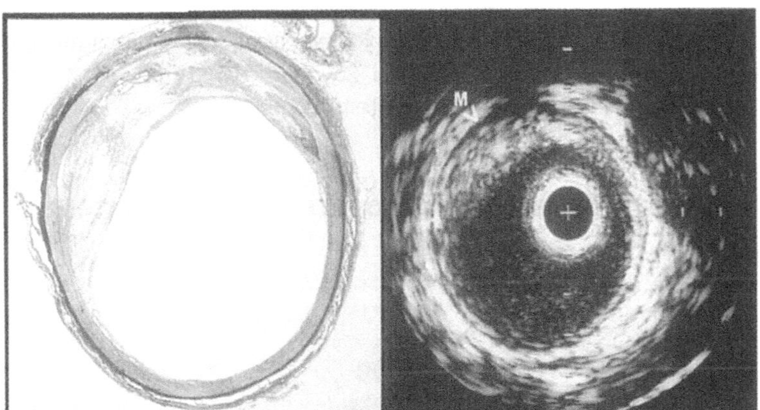

Figure 2. Histologic validation of ultrasound image obtained from the iliac artery of in atherosclerotic pig model (with courtesy of B.J.G.L. de Smet, Utrecht).

Porter et al. (29). In post mortem obtained coronary arteries they observed that the median difference between intimal and inner echogenic thickness was larger than the median difference between inner echogenic thickness and intimal+medial thickness. Thus, the inner echogenic layer of human coronary plaque reflects the intimal and at least part of the adjacent medial layer and, consequently, the echolucent layer by ultrasound represents only the outermost section of the media. Moreover, the most accurate transition between the inner echodense and echolucent layer was associated with the collagen content within the media: a higher collagen content was associated with more echodense ultrasound reflection. Fitzgerald et al. (28) showed that a three layered appearance represented the presence of an average intimal thickening of at least 178 μm at 30 MHz.

IVUS was found to be capable of visualizing the damage done by angioplasty procedures (29). It appeared that angiography underestimated the prevalence of intimal ruptures and dissections after balloon angioplasty (30). However, following balloon angioplasty even the high number of dissections and plaque ruptures as visualized with IVUS might be an underestimation. In a post mortem study, Potkin et al (31) found that a plaque tear was present in one or more histologic cross-sections in 25 of 26 patients (96%) with prior PTCA. Most in vivo IVUS studies do not approximate this number of tears found after balloon angioplasty which might be explained by the catheter compressing the dissection flap against the arterial wall. Computerized image processing facilitates the detection of plaque fractures and dissections after angioplasty (32).

The presence of plaque fractures and dissections also explains the poor correlation between IVUS and angiography luminal measurements after balloon angioplasty. The angiogram depicts a 2D silhouette of the lumen where circumferential dissections may be interpreted as part of the luminal area. This was also the rationale for multicenter studies that were designed to demonstrate the predictive value of IVUS for the development of restenosis. Surprisingly, the presence of dissections on IVUS was not predictive for the development of restenosis 6 months after balloon angioplasty (33).

Arterial remodeling in ultrasound and post-mortem studies

De novo atherosclerotic remodeling

Armstrong et al. (34) were the first to report that the arterial wall adapts to experimentally induced plaque formation in primates. Glagov et al. (35) observed a positive relation between plaque mass and vessel size in cross-sections obtained from 136 human left main coronary arteries. Lumen area was not found to reduce (it sometimes even increased) until, on average, 40% of the vessel cross-sectional area was occupied by plaque. At that point the artery would be maximally enlarged and further plaque increase would encroach upon the lumen. McPherson et al. (36) were among the first to use ultrasound to confirm the findings of Glagov et al. (35) *in vivo*. The inquiry into *de novo* vascular remodeling regained interest with the progressive availability of IVUS catheters. Several investigators have confirmed by IVUS (37-39) the observations made by Glagov et al. (35).

IVUS studies are mostly performed in patients suffering from clinically manifest atherosclerotic disease. In our group, we have assessed the extent and type of arterial remodeling in femoral artery segments with advanced as well as early atherosclerotic lesion development using IVUS and post mortem obtained specimen, respectively (40,41). In these studies, cross-sections were studied at regular intervals including the stenotic site and the cross-section which contained the least amount of plaque. The cross-section with the least amount plaque was considered being the reference, assuming that it had been least affected by remodeling. We found that divergent types of arterial remodeling may occur in response to increasing plaque mass: not only compensatory enlargement, but also failure of enlargement and even constrictive remodeling of the artery was observed (40-42). Even more surprising was the calculated negative relation between the type of remodeling and the percentage luminal stenosis (40): enlargement was mainly observed in arterial segments that showed mild or no luminal narrowing. On the other hand, constrictive remodeling was mainly observed in arterial segments that exhibited severe luminal narrowing. Compensatory enlargement will prevent and constrictive remodeling will accelerate luminal narrowing in the atherosclerotic femoral artery. Apparently, constrictive remodeling is an important mechanism in individual stenosis formation (40). Wong et al. (43) showed that an equivalent plaque burden in the coronary artery may lead to different clinical manifestations among patients: some arteries failed to enlarge and in other arteries an actual reduction of vessel area was observed at the most stenotic site. Nishioka et al. (44) compared the vessel area of the coronary artery at the lesion site with the vessel area at a proximal and distal reference site. They observed a smaller vessel area at the stenotic site compared with the proximal and distal reference site in 9/35 lesions.

In collaboration with Stanford University we recently studied 609 coronary artery lesions and adjacent reference sites, that appeared normal on the angiogram, using IVUS (45,46). For each lesion, differences in lumen area, plaque area and vessel area were calculated by comparison with the reference sites. Mean changes in plaque area and vessel area were 5.27+/- 3.20 mm^2 and -0.74+/-2.93 mm^2, respectively. Both, axial changes in plaque area and vessel area were normally distributed with a comparable width of the 95% confidence intervals ([-1.02 mm^2 ,+11.56 mm^2] and [-6.44 mm^2 ,+5.04 mm^2], respectively). Constrictive or expansive remodeling should therefore not be considered as separate entities in atherosclerotic disease but as tail ends of a normally distributed process. In only 37% of the lesions, the decrease of lumen area was fully explained by an increase of the plaque area and part of the increase in plaque area was compensated for by an increase of the vessel area. In 42% of the lesions an increase of plaque area explained 50-100% and a decrease in vessel area explained 0-50% of the narrowing of the lumen area. In 21% of the lesions vessel area decrease was the major determinant of lumen area decrease. Moreover, in 5% of the lesions plaque area was smaller at the lesion site compared with the reference sites, implying that lumen loss could be fully explained by a decrease in vessel area. Thus, the axial decrease in vessel area significantly contributes to luminal narrowing in the majority of patients with coronary artery disease remodeling.

De novo atherosclerotic remodeling and plaque vulnerability

An example that ultrasound and post mortem studies complement each other can be found in the recent observation that enlarged lesions hide plaques that are more prone to rupture and may lead to clinical events. Angiographic luminal diameter narrowing is not a predictive value for the onset of plaque rupture with subsequently superimposed thrombus formation (47,48). Intravascular ultrasound studies revealed that plaque mass is often hidden beneath the angiographically normal luminal surface. Almost by definition these lesions must have been compensatory enlarged since plaque mass is evident without subsequent luminal narrowing. We therefore hypothesized that histopathological markers for plaque vulnerability are more frequent in compensatory enlarged segments which might explain the higher incidence of plaque ruptures at minor angiographic stenoses. In a post mortem study of 50 femoral arteries we stained for immunohistological markers that are often related with plaque rupture. The presence of inflammatory cells, lack of smooth muscle cells and collagen and large percentage of atheroma in the plaque were found to be associated with a large plaque area and expansive remodeling. On the other hand, constrictive remodeling was associated with a stable plaque phenotype. Within the same arterial segments, differences in lumen area were not consistently associated with changes in plaque architecture (49).

These observations were extended in a post mortem study in coronary arteries (50). In that study we also demonstrated that compensatory enlarged lesions contained more matrix metalloproteinases (MMP) 2 and 9. Thus collagen degrading enzymes that play a role in plaque destabilization, were more prevalent in expansively remodeled lesions.

It is well appreciated that the pathological substrate of acute coronary syndromes is the atheromatous plaque with a thin fibrous cap and infiltration of large numbers of inflammatory cells (10). To close the triangle, we studied the relation between local arterial remodeling visualized with IVUS and the clinical syndrome. IVUS revealed that the culprit lesion in patients entering the hospital with either myocardial infarction or instable angina was often compensatory enlarged while stable angina patients mostly showed a constrictive remodeling on IVUS (51). This observation has been confirmed by other groups (52). More recently, Dangas et al. (53) demonstrated that preintervention expansive remodeling is an independent predictor of target lesion revascularization after non-stent intervention. Lesion based revascularisation was necessary in 31% and 20% of the compensatory and constrictive remodeled lesions, respectively. In addition, Wexberg et al. (53) demonstrated that enlargement was associated with major adverse cardiovascular events after intervention of patients suffering from stable angina. These results confirm that remodeling in *de novo* atherosclerosis is a double edged sword: on the one hand expansive remodeling prevents the lumen from being occluded by atherosclerotic plaque formation. On the other hand, however, expansively remodeled lesions may comprise a group of lesions that is highly vulnerable and prone to rupture.

Remodeling after balloon angioplasty: the erroneous interpretation of "stenosis"

To determine the percentage luminal narrowing in ischemic cardiovascular disease, the choice of the reference site is of crucial importance. In angiography, a longitudinal view of the artery is obtained and the reference site is always located proximal or distal from the obstructed arterial segment. Intravascular ultrasound offers a transverse view of the artery. Thereby, it is possible to measure the area

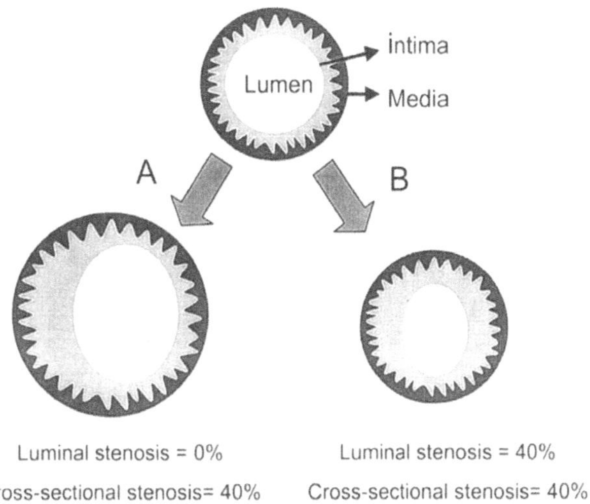

Figure 3. The influence of vascular remodeling on the calculation of the percentage luminal stenosis. Arterial enlargement (A) results in preservation of the lumen (luminal stenosis = 0%). However, the classical definition of luminal stenosis used in pathology ((lumen area / vessel area)*100) would result in the calculation of a significant percentage stenosis. The pathological cross-sectional definition of luminal stenosis is only correct when the vessel area dose not change during the atherosclerotic process (B).

circumscribed by the echolucent media and use this media bounded area as a reference as if it were a measure of the original lumen if no plaque was present. Also post mortem, the local area encompassed by the internal elastic lamina (IEL-area) is used as a reference in the transverse plane in histological measurements of the percentage luminal narrowing. However, the local IEL-area may be an improper reference because during the atherosclerotic process or reocclusion after balloon intervention the arterial wall may have been subjected to remodeling in the form of compensatory enlargement or constriction (35,40,55,56). Using the

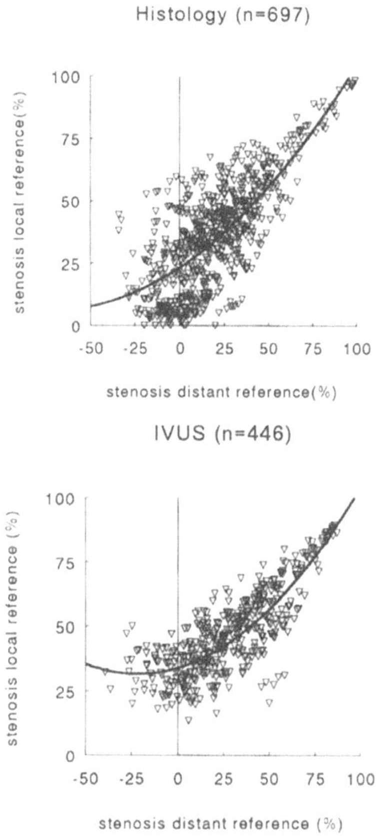

Figure 4. The relation between the calculated percentage stenosis when the local vessel area in the same cross-section is used as a reference (see Figure 3) and the calculated percentage stenosis when an angiographic like distant reference is used to calculated percentage stenosis. Note that the perçentages calculated stenosis may strongly differ depending on which definition for the calculation of percentage luminal stenosis is used. While the lumen area has not changed (distant reference = 0), the cross-sectional definition of percentage stenosis may reveal large percentages luminal stenosis. Differences can be explained by local changes in vessel area (arterial remodeling).

local IEL-area in the same cross-section as a reference, the effect of arterial wall remodeling on the calculation of the percentage luminal narrowing is disregarded (6) (Figures 3,4). The traditional pathologic definition of luminal stenosis (assuming that the IEL-area represents the original lumen) heavily overestimates true stenosis when compensatory enlargement is evident. It is not exceptional that over 50% area stenosis is reported because >50% of the IEL area is occupied by plaque, while the lumen is not narrowed. On the other hand when the artery shrinks, like during restenosis after balloon angioplasty, the intimal hyperplasia may erroneously be considered as the only determinant of restenosis. This "cross-sectional" view explains the discrepancy between results of animal intervention studies in which intimal hyperplasia was inhibited while the same intervention did not result in a decrease of restenosis rates in clinical studies. In the animal studies the local IEL area was mostly used as the reference and intimal hyperplasia was the endpoint, thereby not taking the vessel area changes into account. In the human studies the lumen after intervention was considered as the reference and lumen loss the endpoint which directly reflects both increase of intimal area and a decrease in vessel area.

Imaging of the vulnerable plaque

The fine structure and composition of the atherosclerotic lesion rather than the degree of stenosis are currently considered to be the important determinants for acute clinical events, together with the absence of collateral circulation (11,57,58), Modalities capable of characterizing the tissue of the atherosclerotic lesion may help to understand its natural history and detect lesions with high risk for acute events.

The typical advanced atherosclerotic lesion is characterized by a core of extracellular lipid with an overlaying fibrous collagen rich cap. The lipid core may contain layers of fibrous tissue. Atheromatous lesions with a sclerotic fibrous cap may transform into a complex type of lesion by rupture or erosion of the fibrous cap with subsequent formation of a thrombus. The lesion that is rupture prone is not clearly defined but several morphological and immunological determinants specific for the vulnerable plaque have been reported (10,12). In order to evaluate the applicability of imaging techniques that are potentially capable of predicting which lesions may rupture, it is necessary to understand what specific features of the vulnerable lesion each of these techniques reveal. Grossly, there are three major, interrelated determinants of a plaque's vulnerability to rupture: thickness of the fibrous cap, the size of the atheroma and local inflammation. We will discuss the capability of IVUS to assess these pathohistological determinants of plaque rupture. Studies have been reported in which IVUS was used to identify the vulnerable plaque (59). However, IVUS criteria for vulnerable plaques have scarcely been validated. In contrast, histological validation studies report low sensitivities for IVUS in detecting lipid rich lesions (60). Recently, de Korte et al.

(61) published results on the histopathological validation of IVUS elastography. The latter technique will be discussed in another chapter but also IVUS echodensity was compared for different plaque components. In femoral and coronary arteries they observed no differences in echogenecity between fibrous, fibro-fatty and fatty plaques compared with the normal vessel wall ($p=0.89$). Elastography, however, was capable in discriminating between different plaque components. Prati et al (62) used a 40 MHz ultrasound system and found that IVUS was capable of identifying 19/30 fatty plaques with a sensitivity of 65% and a specificity of 95%. Thus, with increasing IVUS frequencies the resolution and subsequent lipid-demarcating capabilities may improve.

A fibrous cap thickness of less than 150 μm has reported to be critical to withstand circumferential stresses (63). The resolution of most IVUS systems (100-200 μm) is too low to accurately discriminate between fragile and stable fibrous caps. Finally, the third determinant of the vulnerable plaque, inflammation, cannot be detected using IVUS. In summary, with the current resolutions the three histopathological determinants of the vulnerable plaque cannot be detected accurately with IVUS. IVUS derived techniques like elastography may prove to have more potential in this area.

Before a modality that aims to characterize the vulnerable plaque will have diagnostic or prognostic value, several hurdles need to be taken (64). First, validation of the imaged features like cap thickness, plaque type and inflammation on an histologic resolution level is obligatory. Interpretations of imaged features like a “lipid lake” or a “fibrous cap” must not only be based on the priori knowledge of the expected localization of that feature within the image but also on histology like validation. Secondly, the modality should be made safely applicable for in vivo studies, preferably for the coronary circulation. Thirdly, the predictive value of the vulnerable plaque determinants for a plaque to rupture needs to be investigated first. Until now, no studies have been performed on the predictive value of each of the three markers for a plaque to rupture (e.g. inflammation, large atheroma and thin fibrous cap). This third hurdle brings up another question which may result in a vicious circle: how is plaque rupture assessed retrospectively? Thus, the fourth hurdle is to find a diagnostic measure of plaque rupture. It will not be feasible to determine the occurrence of plaque rupture on a tissue level in a large clinical study. Therefore, surrogate endpoints must be searched for, like rapid acceleration of luminal narrowing on angiography or IVUS. Finally, in a prospective study, the determinants of plaque rupture should be related with the incidence of acute events, taking into account the role of collateral circulation.

Limitations of IVUS and post mortem studies

A major limitation of post mortem and single ultrasound observations is that a single static visualization on a dynamic process is obtained only. Atherosclerosis is a process which is likely to occur over decades. Thus, no inferences regarding causality can be made when a non serial study is performed. *De novo* atherosclerosis, however, is a process which occurs over an extended time period in contrast to, for instance, restenosis which occurs within 6 months. Still, intravascular ultrasound is, by definition, an invasive diagnostic tool, which is a prohibitive factor to study the vascular wall at any given timepoint. The slow progression of *de novo* atherosclerosis virtually excludes serial studies.

In intravascular ultrasound studies the localization of the catheter is of crucial importance. Displacement sensing devices can be used to localize the catheter exactly. The most accepted tool for localization purposes is the automated motorized pull back system that withdraws the catheter at a regular speed of approximately 0.5 mm/s. Starting the pull back at an anatomical landmark facilitates the exact determination of the location of interest. When histopathologic validation needs to be performed shrinkage due to fixation and tissue processing may harm accurate matching with the IVUS image. Also non optimal perfusion pressure may alter plaque morphology.

References

1. Gussenhoven EJ, Essed CE, Lancee CT et al. (1989) Arterial wall characteristics determined by intravascular ultrasound imaging: an in vitro study. J Am Coll Cardiol 1989;14:947-95.
2. NishimuraRA, Edwards WD, Warnes CA, Reeder GS, Holmes DR, Tajik AJ, Yock PG. (1990) Intravascular ultrasound imaging: in vitro validation and pathologic correlation. J Am Coll Cardiol 1990;16:145-154.
3. Lockwood GR, Ryan LK, Gotlieb AI, Lonn E, Hunt JW, Liu P, Foster S. (1992) In vitro high resolution intravascular imaging in muscular and elastic arteries. J Am Coll cardiol 1992;20:153-160.
4. Siegel RJ, Seong Chae J, Maurer G, Berlin M, Fishbein MC. (1993) Histopathologic correlation of the three layered intravascular ultrasound appearance of normal adult human muscular arteries. Am Heart J 1993;126:872-878.
5. Tobis JM, Mallery JA, Gessert J et al. (1989) Intravascular ultrasound cross-sectional arterial imaging before and after balloon angioplasty in vitro. Circulation 1989;80:873 882.
6. Pasterkamp G, Wensing PJW, Hillen B, Post MJ, Mali WPTM, Borst C. (1997) Impact of local atherosclerotic remodeling on the calculation of percent luminal narrowing. Am J Cardiol 1997;79:402-405.
7. Roelandt JRTC, diMario C, Pandian NG et al. (1994) Three dimensional reconstruction of intracoronary ultrasound images. Circulation 1994;90:1044-1055.
8. Slager CJ, Wentzel JJ, Schuurbiers JC et al. (2000) True 3-dimensional reconstruction of coronary arteries in patients by fusion of angiography and IVUS (ANGUS) and its quantitative validation. Circulation 2000;102:511-516.D
9. Davies MJ, Thomas AC. Plaque fissuring-the cause of acute myocardial infarction, sudden ischaemic death, and crescendo angina. Br Heart J 1985;53:363373.
10. van der Wal AC, Becker AE, van der Loos CM, Das PK. Site of intimal rupture or erosion of thrombosed coronary atherosclerotic plaques is characterized by an

inflammatory process irrespective of the dominant plaque morphology. Circulation 1994;89:36-44.

11. Falk E, Shah PK, Fuster V. Coronary plaque disruption. Circulation 1995;92:657-671.

12. Davies MJ, Richardson PD, Woolf N, Katz DR, Mann J. Risk of thrombosis in human atherosclerotic plaques: role of extracellular lipid, macrophage, and smooth muscle cell content. Br Heart J 1993;69:377-381.

13. Salonen JT, Salonen R. Ultrasound B-mode imaging in observational studies of atherosclerotic progression. Circulation 1993;87(suppl II):II-56 - II-65.

14. Burke GL, Evans GW, Riley WA et al. Arterial wall thickness is associated with prevalent cardiovascular disease in middle-aged adults. The Atherosclerosis Risk in Communities (ARIC) study. Stroke 1995;26:386-391.

15. Belcaro G, Nicolaides AN, Laurora G, Cesarone MR, De Sanctis M, Incandela L, Barsotti A. Ultrasound morphology classification of the arterial wall and cardiovascular events in a 6-year follow-up study. Arterioscler Thromb Vasc Biol 1996;16:851-856.

16. Craven TE, Ryu JE, Espeland MA, Kahl FR, McKinney WM, Toole JF, McMahan MR, Thompson CJ Heiss G, Crouse III JR Evaluation of the associations between carotid artery atherosclerosis and coronary artery stenosis. Circulation 1990;82:1230-1242.

17. Adams MR, Nakagomi A, Keech A, Robinson J, McCredie R, Bailey BP, Freedman SB, Celermajer DS. Carotid intima-media thickness is only weakly correlated with the extent and severity of coronary artery disease. Circulation 1995;92:2127-2134.

18. Salonen JT, Salonen R. Ultrasonographically assessed carotid morphology and the risk of coronary heart disease. Arterioscler Thromb 1991;11:1245-1249.

19. Bots ML, Hofman A, Grobbee DE. Common carotid intima-media thickness and lower extremity arterial atherosclerosis. Arterioscler Thromb 1994;14:1885-1891.

20. Wendelhag I, Gustavsson T, Suurkula M, Berglund G, Wikstrand J. Ultrasound measurements of wall thickness in the carotid artery: fundamental principles and description of a computerized image analyzing system. Clin Physiol 1991;11:565-577.

21. Nolsoe CP, Engel U, Karstrup S, Torp-Pedersen S, Garre K, Holm HH. The aortic wall: an in vitro study of the double line pattern in high resolution US. Radiology 1990;175:387-390.

22. Persson J, Formgren J, Israeksson B, Berglund G. Ultrasound-determined intima-media thickness and atherosclerosis. Arterioscl Thromb 1994;14:261-264.

23. Ricotta JJ, Bryan FA, Bond MG et al. Multicenter validation study of real-time (B-mode) ultrasound, arteriography and pathologic examination. J Vasc Surg 1987;6:512-520.

24. Schenk EA, Bond G, Aretz TH et al. Multicenter validation study of real-time ultrasonography, arteriography and pathology: pathologic evaluation of carotid endarteriectomy specimens. Stroke 1988;19:289-296.

25. Gussenhoven EJ, Essed CE, Lancee CT et al. Arterial wall characteristics determined by intravascular ultrasound imaging: an in vitro study. J Am Coll cardiol 1989;14:947-952.

26. Nishimura RA, Edwards WD, Warnes CA, Reeder GS, Holmes DR, Tajik AJ, Yock PG. Intravascular ultrasound imaging: in vitro validation and pathologic correlation. J Am Coll Cardiol 1990;16:145-154.

27. Lockwood GR, Ryan LK, Gotlieb AI, Lonn E, Hunt JW, Liu P, Foster S. In vitro high resolution intravascular ultrasound imaging in muscular and elastic arteries. J Am Coll Cardiol 1992;20:153-160.

28.Fitzgerald PJ, St Goar FG, Conolly AJ, Pinto FJ, Billingham ME, Popp RL, Yock PG. Intravascular ultrasound imaging of coronary arteries: is three layers the norm? Circulation 1992;86:154-158.
29.Tobis JM, Mallery JA, Gessert J et al. Intravascular ultrasound crosssectional imaging before and after balloon angioplasty in vitro. Circulation 1989;80:873-882.
30.Potkin BN, Keren G, Mintz GS, Pichard AD, Satler LF, Kent KM, Leon MB. Arterial responses to balloon angioplasty: an intravascular ultrasound study. J Am Coll Cardiol 1992;20:942-951.
31.Potkin BN, Roberts WC. The effects of percutaneous transluminal coronary angioplasty n atherosclerotic plaque and relation of plaque composition and arterial size to outcome. Am J Cardiol 1988;62:41-50.
32.Pasterkamp G, van der Heiden MS, Post MJ, Borst C, Gussenhoven EJ, Pieterman HJ, van Urk H, Bom N. Discrimination of intravascular ultrasound images using subtraction, conventional averaging and saline flush.Ultrasound Med Biol 1995; 21:149-156.
33.Peters RJG, Kok WEM, Serruys PW et al. Prediction of restenosis following coronary balloon angioplasty. Results of PICTURE. *Circulation* 1997;95:2254-2261.
34.Armstrong ML, Heistad DD, Marcus ML, Megan MB, Piegors DJ. Structural and hemodynamic responses of peripheral arteries of macaque monkeys to atherogenic diet. Arteriosclerosis 1985;5:336-346.
35.Glagov S, Weisenberg E, Zarins CK, Stankunavicius R, Koletis G. Compensatory enlargement of human atherosclerotic coronary arteries.N Engl J Med 1987;316:1371-1375.
36.McPherson DD, Sirna SJ, Hiratzka LF, Thorpe L, Armstrong ML, Marcus ML, Kerber RE. Coronary arterial remodeling studied by high-frequency epicardial echocardiography: an early compensatory mechanism in patients with obstructive coronary atherosclerosis. J Am Coll Cardiol 1991;17:79-86.
37.Hermiller JB, Tenaglia AN, Kisslo KB, Phillips HR, Bahore TM, Stack RS,Davidson CJ. In vivo validation of compensatory enlargement of atherosclerotic coronary arteries. Am J Cardiol 1993;71:665-668.
38.Ge J, Erbel R, Zamorano J, Koch L, Kearney P, Görge G, Gerber TC, Meyer J. Coronary artery remodeling in atherosclerotic disease: an intravascular ultrasonic study in vivo. Coronary artery disease 1993;4:981-986.
39.Gerber TC, Erbel R, Görge G, Ge J, Rupprecht HJ, Meyer J. Extent of atherosclerosis and remodeling of the left main coronary artery determined by intravascular ultr asound. Am J Cardiol 1994; 73:666-671.
40.Pasterkamp G, Wensing PJW, Post MJ, Hillen B, Mali WPTM, Borst C. Paradoxical arterial wall shrinkage contributes to luminal narrowing of human atherosclerotic femoral arteries. Circulation 1995;91:1444-1449.
41.Pasterkamp G, Borst C, Gussenhoven EJ, Mali WPTM, Post MJ, The SHK, Reekers JA, van der Berg FA. Remodeling of de novo atherosclerotic lesions in femoral arteries: impact on the mechanism of balloon angioplasty. J Am Coll Cardiol 1995;26:422-428.
42.Post MJ, Pasterkamp G, Haudenschild CC, Borst C. Arterial remodeling in atherosclerosis and restenosis: a vague concept of a distinct phenomenon. Atherosclerosis 1995;118(suppl):S115-S123.
43.Wong CB, Porter TR, Xie F, Deligonul U. Segmental analysis of coronary arteries with equivalent plaque burden by intravascular ultrasound in patients with and without angiographically significant coronary artery disease. Am J Cardiol 1995;76:598-601.

44.Nishioka T, Luo H, Eigler NL, Berglund H, Kim GJ, Siegel RJ. Contribution of inadequate compensatory enlargement to development of human coronary artery stenosis: an in vivo intravascular ultrasound study. J Am Coll Cardiol 1996;27:1571-1576.
45.Pasterkamp G, Jeremias A, Vink A et al. The contribution of constrictive remodeling and plaque mass on de novo atheroscelrotic luminal narrowing: an intravascular ultrasound study. Circulation 2000;102:3338 (abstract).
46.Pasterkamp G, Jeremias A, Vink A et al. The relative contributions of constrictive remodeling and plaque mass to atherosclerotic luminal narrowing: an intravascular ultrasound study. Eur Heart Journal 2000;21: P2772 (abstract).
47.Ambrose JA, Tannenbaum, Alexopoulos D et al. Angiographic progression of coronary artery disease and the development of myocardial infarction. J Am Coll Cardiol 1988;12:56-62.
48.Little WC, Constantinescu M, Applegate RJ et al. Can coronary angiography predict the site of a subsequent myocardial infarction in patients with mild-to-moderate coronary artery disease? Circulation 1988;78:1157-1166.
49.Pasterkamp G, Schoneveld AH, van der Wal AC, Haudenschild CC, Clarijs RJG, Becker AE, Hillen B, Borst C. The relation of arterial geometry with luminal narrowing and plaque vulnerability: the remodeling paradox. J Am Coll of Cardiol 1998;32:655-662.
50.Pasterkamp G, Schoneveld AH, Hijnen DJ, de Kleijn DPV, Teepen H, van der Wall AC, Borst C. Atherosclerotic arterial remodeling and the localization of macrophages and matrix metalloproteinases in the human coronary artery. Atherosclerosis, 2000;150:245-253.
51.Smits PC, Pasterkamp G, de Jaegere PJ, Eefting FD, Stella PR, Borst C. Coronary artery disease: arterial remodeling and clinical presentation.*Heart* 1999;82:461-464.
52.Schoenhagen P, Ziada KM, Kapadia SR,Crowe TD, Nissen SE, Tuzcu M. Extent and direction of arterial remodeling in stable versus unstable coronary syndromes. Circulation 1999;101:598-603
53.Dangas G, Mintz GS, Mehran R, Lansky AJ, Kornowski R, Pichard AD, Satler LF, Kent KM, Stone GW, Leon MB. Preintervention arterial remodeling as an independent predictor of target lesion revascularization after non stent coronary intervention. Circulation 1999;99:3149-3154.
54.Wexberg P, Gyongyoshi M, Sperker W, Kiss K, Yang P, Hassan A, Pasterkamp G, Glogar D. Pre-existing arterial remodeling is associated with in-hospital and late adverse cardiac events after coronary interventions in patients with stable angina pectoris. J Am Coll Cardiol 2000;36:1860-1869.
55.Mintz GS, Popma JJ, Pichard AD, Kent, KM, Satler LF, Wong SC, Hong MK, Kovach JA, Leon MB. Arterial remodeling after coronary angioplasty. A serial intravascular ultrasound study. Circulation 1996;94:35-43.
56.Post MJ, de Smet BJGL, van der Helm Y, Borst C, Kuntz RE. Arterial remodeling after balloon angioplasty or stenting in the Yucatan atherosclerotic micropig model: an angiographic, serial intravascular ultrasound and histology study. Circulation 1997;96:996-1003.
57.Fuster V. Lewis A Conner Memorial Lecture. Mechanisms leading to myocardial infarction: insights from studies of vascularbiology. Circulation 1994;90:2126-2146.
58.Davies MJ, Thomas A. Thrombosis and acute coronary artery lesions in sudden cardiac ischemic death. N Engl J Med 1984;310:1137-1140.

59.Ge J, Baumgart D, Haude M et al. Role of intravascular ultrasound imaging in identifying vulnerable plaques. Herz 1999;24:32-41.
60.Peters RJG, Kok WEM, Haventih MG, Rijsterborgh H, van der Wal AC Visser CA. Histopathologic validation of intracoronary ultrasound imaging. Echocardiography 1994;7:230-241.
61.de Korte, CL, Pasterkamp G, van der Steen, AFW, Woutman HA, Bom N. Characterization of plaque components with intravascular ultrasound elastography: a validation study in vitro. Circulation 2000;102:617-623.
62.Prati F, Arbustini E, Labellarte A, Dal Bello B, Sommariva L, Mallus MT, Pagano A, A Boccanelli. Correlation between high frequency intravascular ultrasound and histomorphology in human coronary arteries. Heart 2001;85:567-570.
63.Loree HM, Kamm RD, Strigfellow RG et al. Effects of fibrous cap thickness on peak circumferential stress in model atherosclerotic vessels. Circulation Research 1992;71:850-858.
64.Pasterkamp G, Falk E, Woutman H, Borst C. Techniques characterizing the coronary atherosclerotic plaque: influence on clinical decision making. J Am Coll Cardiol 2000;36:13-21.

WHAT VASCULAR SURGEONS WANT FROM VASCULAR ULTRASOUND

Marc R.H.M. van Sambeek MD PhD and Hero van Urk MD PhD

Department of Vascular Surgery, Erasmus University Medical Center Rotterdam, The Netherlands.

Summary: "Endovascular surgery" has emerged from the interest in minimal invasive surgery. The collaboration between cardiologists, interventional radiologists and vascular surgeons has been of eminent importance for the evolution of endovascular techniques. The development of endovascular techniques prompted the need for improved vascular imaging and better diagnostics. Angiography displays only a 'lumenogram' of the vessel. Color flow duplex, computed tomographic angiography and magnetic resonance imaging are important in the pre- and postintervention assessment of vascular disease, however, in smaller peripheral vessels, these techniques do not always give accurate information on the dimensions of the vessel and the extent of the disease. In contrast, intravascular ultrasound (IVUS) provides histology-like cross-sections of the blood vessel, allowing qualitative evaluation of plaque composition and mural thrombus and quantitative assessment of lumen area, vessel area and plaque area.

Although ultrasonic information has distinct advantages over angiographic data, IVUS remains an adjunct to angiography and not a substitute. Future application of IVUS in the peripheral vasculature should be directed at supporting the introduction of interventional devices and procedures such as stents, endovascular grafts and peripheral vascular brachytherapy.

Key words: Intravascular ultrasound, artery, intimal hyperplasia, aneurysm

Endovascular interventions and vascular imaging

In recent years the interest in minimal invasive surgery has been growing and the same trend can be observed in vascular surgery, leading to what is commonly referred to as "endovascular surgery". Although the 1990s represent an era of technical revolution in vascular surgery, it is a misunderstanding to consider endovascular treatment a recent development. In 1947 João Cid dos Santos described the thrombo-endarterectomy[1]; this technique was modified by Vollmar in 1964, to a semi-closed endarterectomy using ring-strippers.[2] In the same year other pioneers, including Dotter and Judkins, published preliminary results on what they called "angioplasty" of the femoropopliteal artery using co-axial catheters.[3] This technique was later modified by Grüntzig in 1974, who replaced the co-axial catheters with dilatation balloons.[4] In the early 1990s, Voldovos and Parodi introduced the endovascular treatment of the abdominal aortic aneurysm with a device composed of a Dacron graft and Palmaz stents.[5,6] In the mid-nineties peripheral endografts were introduced for both aneurysmal and obstructive vascular disease in the iliac and femoropopliteal tract.

The collaboration between cardiologists, interventional radiologists and vascular surgeons has been of eminent importance for further evolution of endovascular techniques. Nowadays a great variety of obstructive and aneurysmal peripheral vascular diseases can be treated with catheter-guided, endovascular and, therefore, less invasive techniques.

Obstructive vascular disease

The development of these endovascular techniques prompted the need for improved vascular imaging and better diagnostics. Angiography displays only a 'lumenogram' of the vessel. This precludes qualitative evaluation of atherosclerotic plaque and quantitative assessment of plaque volume and vessel area. Color flow duplex, computed tomographic angiography and magnetic resonance imaging are important in the pre- and postintervention assessment of vascular disease. However, in smaller peripheral vessels, these techniques do not always give accurate information on the dimensions of the vessel and the extent of the disease, and at the present time cannot yet be used *during* endovascular interventions. In contrast, intravascular ultrasound (IVUS) provides histology-like cross-sections of the blood vessel, allowing qualitative evaluation of plaque composition and mural thrombus and quantitative assessment of lumen area, vessel area and plaque area. Moreover, this technique can be used during intervention. On-line quantitative analyses can be performed using the software that is implemented in most modern equipment.

After extensive validation studies with histology, IVUS was introduced clinically in the early nineties. Initial studies were aimed at guiding balloon angioplasty. From these IVUS studies we learned that lumen area increase was caused predominantly by an increase in vessel area (a media stretch of 70% in average was observed in a series of experiments) and to a lesser extent by plaque area decrease. Overstretching was mostly accompanied by dissection and plaque rupture and, occasionally, by media rupture, detected as rupture of the internal elastic lamina.[7] This study made clear that the quantitative mechanism of balloon angioplasty is unequivocally different than initially proposed by Dotter and Judkins.[3]

Similarly, the mechanism responsible for lumen area change after PTA of the femoropopliteal artery was assessed with IVUS. Multiple IVUS cross-sections derived from the smallest lumen, the treated site and reference site obtained at intervention and at follow-up were matched. The cross-sections were analyzed for change in lumen, plaque and vessel areas. The study revealed that plaque area increase was similar at the most stenotic site, the treated site and reference site. At the most stenotic site lumen narrowing was caused by plaque growth and constrictive remodeling. Both the treated and reference sites showed a significant increase in plaque area, but, the type of vascular remodeling determined the change in lumen area.[8]

Besides guiding angioplasty IVUS has been used also in the operating theatre guiding bypass surgery and endovascular graft surgery.[9,10] From the latter study it could be concluded that compared to angiography, IVUS is more sensitive in demonstrating changes seen at follow-up.

Aneurysmal disease

The use of low frequency IVUS catheters (12.5 MHz) may have important relevance to endovascular aortic surgery. There is an increasing awareness that endovascular stent-graft placement for the treatment of abdominal aortic aneurysm depends on accurate and detailed visualization of the anatomy of the abdominal aorta. From an in vitro study we learned that IVUS provides accurate information on the vessel wall, lesion morphology, and quantitative parameters of the abdominal aorta and iliac arteries.[11] This information can be of benefit, especially when using modular devices for endovascular treatment of abdominal aortic aneurysm. Based on actual measurements at the time of implantation the interventionist can choose the diameter and length of the device. Accurate length and diameter assessment definitely plays an important role in the prevention of endoleaks. On-line actual measurements can be of eminent importance in the assessment of the anatomical configuration of a ruptured abdominal aneurysm when there is an intention to treat this ruptured aneurysm endovascularly. Specific CTA reconstructions are time-consuming, whereas IVUS can provide an almost immediate assessment of diameters and length.

Future perspectives

Ten years experience has confirmed that IVUS imaging for research purposes is highly valuable. However, although the ultrasonic information has some distinct advantages over angiographic data, IVUS remains an adjunct to angiography and not a substitute. Future application of IVUS in the peripheral vasculature should be directed at supporting the introduction of interventional devices and procedures such as stents, endovascular grafts and peripheral vascular brachytherapy. It is likely that IVUS is perfectly suited for precise determination of the size of endovascular devices and for localizing the exact region of stent-graft placement, as well as for evaluating ambiguous angiographic and CT-scan results. In addition, the use of IVUS in longitudinal studies, after treatment of obstructive disease, may disclose the risk factors for restenosis, information that can not be derived radiological. The introduction of a dedicated computer system for automated quantitative assessment of volumes (by 3-D reconstructions) from IVUS images obtained from clinical examination offers the potential for studying progression or regression of atherosclerotic lesions. [12]

It should be acknowledged that the routine use of IVUS for endovascular interventions involves high costs, cumbersome equipment and potential bias regarding image-interpretation. Although the additional value of IVUS has been emphasized in several studies, careful cost-effectiveness analyses should be made for the different areas.

REFERENCES

1. Dos Santos JC. Sur la desobstruction des thrombose arterielles anciennes. Mem Acad Chir 1947;73:409-411.
2. Vollmar J. Rekonstruktive chirurgie der arterien. Stuttgart 1967;24-27:264-270.
3. Dotter CT, Judkins MP. Transluminal treatment of arteriosclerotic obstruction: Description of a new technique and a preliminary report of its application. Circulation 1964;30:654-670.
4. Grüntzig A, Hophoff H. Perkutane Rekanalisation choronischer arterieller Verschlüsse mit einem neuen Dilatationskatheter. Modification der Dotter-technik. Dtsch Med Wochenschr 1974;99:2502-2505.
5. Voldovos N, Karpovich I, Trojan V, Kalashnikova Y, Shekhanin V, Ternyuk N, et al. Clinical experience in the use of self-fixing synthetic prosthesis for remote endoprosthetics of the thoracic and abdominal aorta and iliac arteries through the femoral artery and as intraoperative endoprosthesis for aorta reconstruction. Vasa 1991;33:93-95.
6. Parodi J, Palmaz J, Barone H. Transfemoral intraluminal graft implantation for abdominal aortic aneurysms. Ann Vasc Surg 1991;5:491-499.
7. The SHK, Gussenhoven EJ, Zhong Y, Li W, Egmond van F, Pieterman H, et al. Effect of balloon angioplasty in femoropopliteal arteries assessed by intravascular ultrasound imaging. Circulation 1992;86:483-493.
8. Van Lankeren W, Gussenhoven EJ, van der Lugt A, Kints MJ, du Bois NAJJ van Sambeek MRHM, et al. Serial intravascular ultrasound following balloon angioplasty in the femoropopliteal artery:remodeling or intimal hyperplasia Eur Radiol 1997;7:S203.
9. Van der Lugt A, Gussenhoven EJ, Tai JA, van Sambeek MRHM, Pieterman H, van Urk H. Femoropopliteal venous bypass studied with intravascular ultrasound: a case report. J Endovasc Surg 1996;3:315-320.
10. Van Sambeek MRHM, Hagenaars T, Gussenhoven EJ, Leertouwer TC, van der Lugt A, Hoedt MTC, et al. Vascular response in the femoropopliteal segment after implantation

of an ePTFE balloon-expandable endovascular graft: an intravascular ultrasound study. J Endovasc Ther 2000;7:204-212.

11.Essen van JA, Lugt van der A , Gussenhoven EJ, Leertouwer TC, Zondervan P, Sambeek van MRHM. Intravascular ultrasonography allows accurate assessment of abdominal aortic aneurysm: an in vitro study. J Vasc Surg 1998;27:347-353.

12.Hagenaars T, Gussenhoven EJ, van Essen JA, Seelen J, Honkoop J, van der Lugt A. Reproducibility of volumetric quantification in intravascular ultrasound image. Ultrasound Med Biol 2000;26:367-374

Part 2

The Technical Potential of IVUS

History and Principles

N. Bom[1,2], A.F.W. van der Steen[1,2], C.T. Lancée[1]

[1]Erasmus University Rotterdam and
[2]Interuniversity Cardiology Institute of the Netherlands

Summary: Basic echo principles and various techniques as used in intravascular imaging are explained. Parameters are set in the context of other cardiovascular applications. This is followed by a description of early intraluminal devices and examples of artifacts.

Keywords: Intravascular; echo principles, artifacts, history

Introduction

Intraluminal echography dates back to the very beginning of diagnostic ultrasound. Over the years many fascinating ideas and applications of catheter-tip or gastroscopic-tube-tip-mounted transducers have been described. In the early period of diagnostic ultrasound, one of the compelling reasons for the intraluminal approach was the low sensitivity of the existing echo transducers and, therefore, the need to closely approach the organs to be studied. Yet image quality remained poor and the techniques were considered too invasive to stimulate large scale applications. However, with the introduction of interventional and minimally invasive surgical techniques in the last decade, a new and enthusiastic interest for intraluminal echography has arisen because of the need for better patient selection and management of the interventional procedure.

For intraluminal echo methods the basic echo principles as known from non-invasive diagnostic ultrasound are applicable. These principles include: sound propagation and echo reflection; attenuation; beam formation; the need for time/gain compensation, and relation between probe size, frequency, scan depth and resolution.

Echo principles

Sound waves represent a pressure disturbance that is propagated with a given sound velocity through a medium such as water or human tissue. The sound velocity in blood or water is approximately 1500 m/s. Sound pulses may be reflected from distant structures. With a known sound velocity the time elapsed between transmission of the pulse and reception of the echo can be converted to a distance or depth measurement. It is this principle which is used in SONAR applications and which is also known to be used by, for instance, dolphins.

The principle of IntraVascular UltraSound (IVUS) imaging is illustrated in Figure 1. A transducer T transmits a sound pulse and receives the echoes as created along a narrow sound beam (arrow). Echoes result at changes of acoustic impedance. Thus the plaque, media- and adventitia layer become visible.

(A) represents the transmission pulse effect and four echoes in amplitude mode (A-mode) as a function of time (= depth)

(B) represents these echoes in brightness modulation (B-mode) as used for two-dimensional display purposes.

Electronically or mechanically, the sound beam rotates through 360° to visualise the entire arterial cross-section with IVUS.

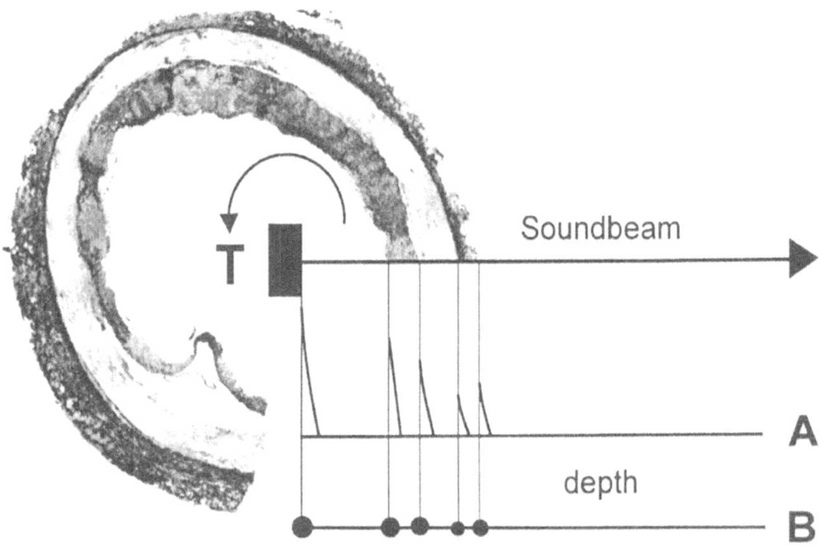

Fig. 1. Echoes are created at boundaries where the acoustic impedance changes. A rotating transducer T transmits a sound pulse and receives the echoes as function of time. With known sound velocity this corresponds to depth. Echoes are indicated in the amplitude mode in A. For the two dimensional display the echoes are used in brightness modulation in B. The rotational transducer position is fully synchronized with echo display direction on the monitor screen.

Attenuation and time/gain compensation

When ultrasound waves travel through a medium, the sound intensity progressively weakens. This process is called attenuation. Attenuation is due to many causes, for instance, the sound energy contained in the transmitted pulse is spread over a larger area as the sound beam widens. Another important contribution to the attenuation is caused by absorption. This process is strongly dependent on the frequency, and increases with frequency. As a result, penetration depth at high frequency is limited.

As a rule of thumb, the attenuation for diagnostic ultrasound waves is 1 dB/MHz/cm. A frequency of 5 MHz can be used in diagnostic areas where penetration of 16 cm is required. In ophthalmologic applications 10 MHz can be applied, since the necessary penetration depth is only a few centimetres. For IVUS imaging, a frequency range between 20 and 40 MHz is used. As will be explained later, resolution can be improved with higher frequencies. Thus there is a tendency to use the highest possible frequency.

In Figure 2, the penetration depth in blood, defined as the -20 dB level measured in reflection, as function of frequency is illustrated. As can be seen, for frequencies around 20 MHz, as used in intravascular imaging, the penetration depth becomes rather small. With the given definition, it is illustrated that a change from 20 to 30 MHz will result in a decrease of penetration depth by 6 mm [1].

Since the intensity of ultrasound decreases as it travels through the medium the echo apparatus incorporates a correction mechanism by amplifying echoes from greater depth. This mechanism is a Time Gain Compensation circuit (TGC). In Figure 3, the gain curve (a) as function of depth (time); the echoes without TGC (b) and with TGC (c) are illustrated [2]. The gain level between t_0 and t_1 as well as the gain ramp function between t_1 and t_2 can be selected by the operator. Some systems

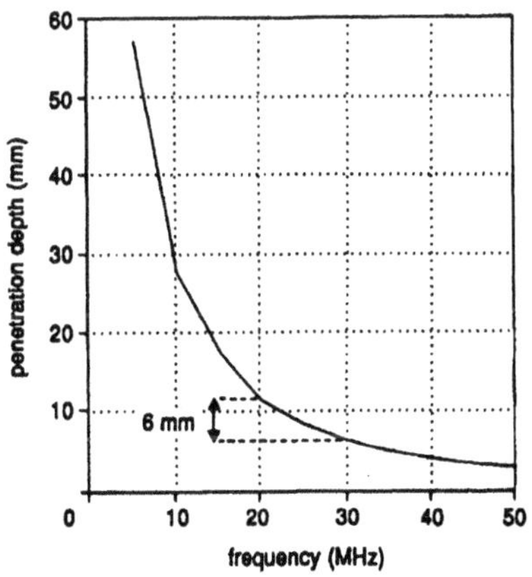

Fig. 2. Penetration depth as function of frequency. Reproduced by permission of the publisher Churchill Livingstone from Bom N, Lancée CT, Gussenhoven WJ, Li W et al. Basic principles of intravascular ultrasound imaging, in Tobis JM, Yock P, eds, Intravascular Ultrasound Imaging. New York 1992, p. 8.

Fig. 3. Principle of Time/Gain Compensation (TGC). The gain can be selected as a function of depth (a). The received echoes without gain compensation are indicated in (b). In order to compensate for attenuation, the TGC is activated; echo results are illustrated in (c).
Reprinted by permission of the publisher Churchill Livingstone from Bom N, Ligtvoet CM, Principles of cardiac ultrasound, in: Roelandt JRTC, Sutherland GR, Iliceto S, Linker D, eds, Cardiac Ultrasound. Edinburgh, 1993, p.13.

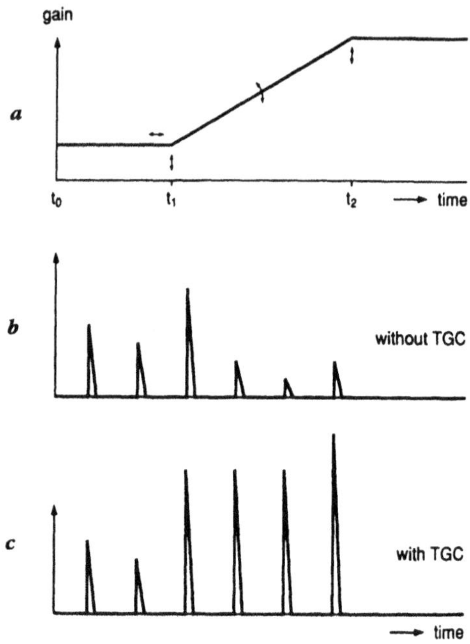

operate with multiple levers for 'near gain' or 'far gain'. One of the reasons for application of TGC is to bring all significant echoes as much as possible within the limited dynamic display range of the monitor.

Relation between probe size, scan depth and resolution versus frequency

As a summary, in Figure 4, the frequency range of various cardiovascular applications is illustrated. In adults (large depth range), low frequencies around 4 MHz must be used. The mid frequency range around 10 MHz is indicated for esophageal echo work in children or intracardiac imaging. At the high frequency, intravascular echography can be used since the distance from echo element to vascular wall to be investigated is very limited here. As can be observed from the figure, the used transducer size decreases with frequency. This is related with the acoustic active transducer surface, which for proper beam forming must encompass a given number of wavelengths (wavelength decreases with frequency).

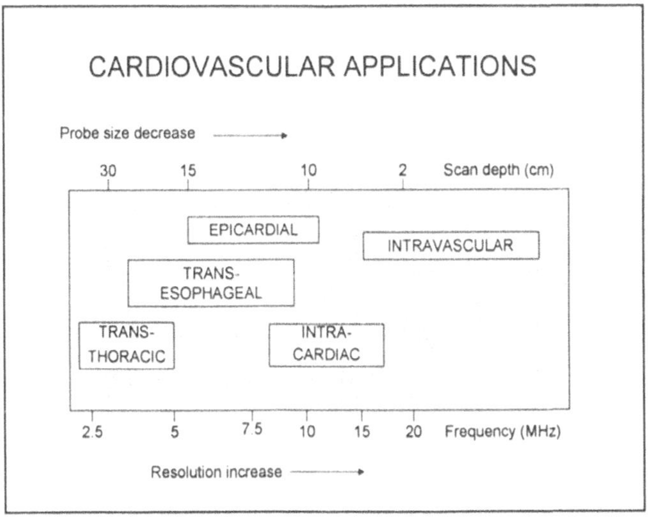

Fig. 4. Frequency range of various cardiovascular echo applications. Since attenuation increases with frequency, possible scan depth decreases from the 20-30 centimeter range for external cardiac application to the millimeter range for intravascular use. The relation with probe size and resolution is indicated as well.

History of intraluminal echo measurements

Already very early in the history of diagnostic ultrasound, information was used to form a cross-sectional image. For this purpose the acoustic beam had to scan through the cross-sectional plane. In order to obtain a realistic display the acoustic beam direction and beam deflection on the display had to be synchronised. Various early systems are illustrated in Figure 5. Mechanically rotated transducers were used such as the one described by Wild [3] in 1955 for rectal tumour location.

Omoto [4] used an intravenous probe with guide wire tip to study cardiac structures, and Ebina [5] described a miniature concave transducer for rotation inside a rubber cuff in the esophagus in 1964. Wells [6] described a rotation device for intravenous echo imaging in 1965. Eggleton [7] approximated a cardiac cross section by rotating a 4-element catheter. Results of his system depended on a stable state of the heart since he accumulated data over many beats for reconstruction of a cross section in a selected steady state. In 1971, Bom [8] initiated a program to develop two-dimensional real-time invasive ultrasonic imaging using state-of-the-art technology.

A 32-element circular phased array with an outer diameter of 3.2 mm mounted at the tip of a No. 9 French catheter as shown in Figure 6 was constructed in 1971. As pointed out in the original paper, the array design was a compromise between the optimal design and the limitations imposed by technological constraints. The final design was chosen to operate at 5.6 MHz with a narrow main beam. With this

INTRALUMINAL IMAGING

Wild	1955	echo-endoscope		rectal tumour location
Omoto	1962	rotating probe C-scan		intracardiac tomography
Ebina	1964	transesophageal P.P.I. scanning	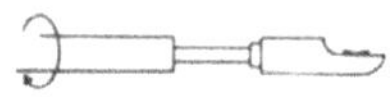	heart and vessels
Wells	1965	rotating mirror		intravenous
Eggleton	1969	4-elements e.c.g. triggered		heart
Bom	1971	32-elements cylindrical phased array		intracardiac tomography

Fig. 5. Examples of early applications of intraluminal imaging with first author, approximate time frame, principle as well as schematic drawing. The application area is indicated as well.

catheter real-time intraluminal images such as left ventricular cross sections were recorded.

Today transesophageal echography has become of major importance in cardiology. Image quality is greatly improved if the attenuating chest structures can be avoided. Most of these systems have gastroscopic pipe mounted transducers. Some of the experience obtained with these systems has been used in development of smaller intraluminal devices. For example, Hisanaga [9] rotates a small transducer

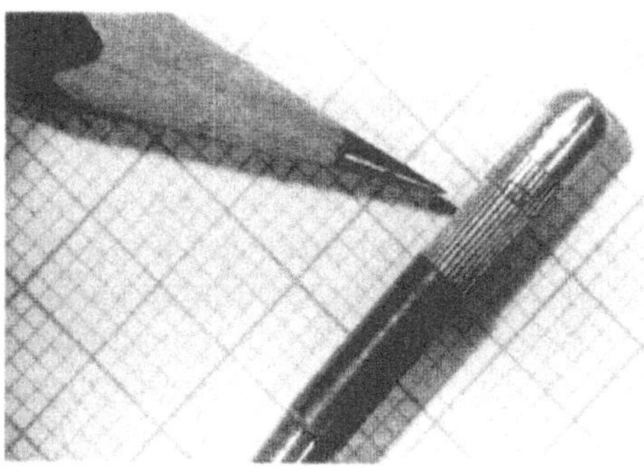

Fig. 6. French 9 diameter 32 element phased array catheter tip constructed for intracardiac and intravascular use in the early seventies as described by Bom [8]. The catheter tip comprises a cylindrical piëzo-electric element, which was subdivided by wire sawing into 32 separate elements.

fixed to a flexible shaft. It rotates at a rate of 15 to 50 cycles/s. DiMagno [10] adapts a 64-element linear array at 10 MHz. The field of view is 3x4 cm. Both Hisanaga and Bertini [9] use a distensible fluid filled bag for contact with the esophageal wall. Souquet introduced the phased array principle for esophageal applications in 1982 [12]. He suggested to rotate the image mechanically through 360°. Further applications of intraluminal imaging were described by Natori [13] and Fukuda [14] who were interested in the diagnosis of intra-abdominal disorders.

Schematic description of IVUS methods

Rotating element

In Figure 7, the rotating tip system is schematically illustrated. The driving motor is connected to the proximal end of the catheter. The shaft (1) must be very flexible and contains the electric wires for the transducer.

The shaft needs to be flexible; yet it must drive the tip in a predictable way. This is not always possible when the catheter must follow a tortuous path. As a result, the beam deflection on the display may not correspond with the acoustic beam deflection. This causes errors in the image. Capacitive and optical feed-back techniques have been suggested to cope with this problem. The effect of Non Uniform Rotation Disorder (NURD) is explained further on. The element (3) is positioned in such a way that no transmission pulse effect appears on the display, since the echo travel time to the dome (2) is of sufficient duration. It allows imaging very close to the catheter outer wall since no 'dead zone' is present. The dome must be acoustically transparent. Results obtained in-vitro are often reported without such a dome, thus increasing image quality and avoiding the liquid filling process necessary for acoustic coupling.

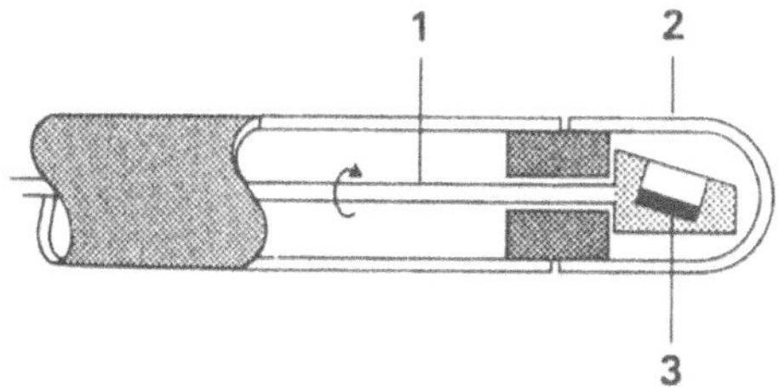

Fig 7. Flex-shaft rotating element principle. This contains a rotating shaft (1); acoustically transparent dome (2); and echo element (3) in a catheter tip. The flex shaft is powered by a proximal motor outside the catheter. In some practical applications an external guide wire is added and can be inserted through an added catheter nose part.

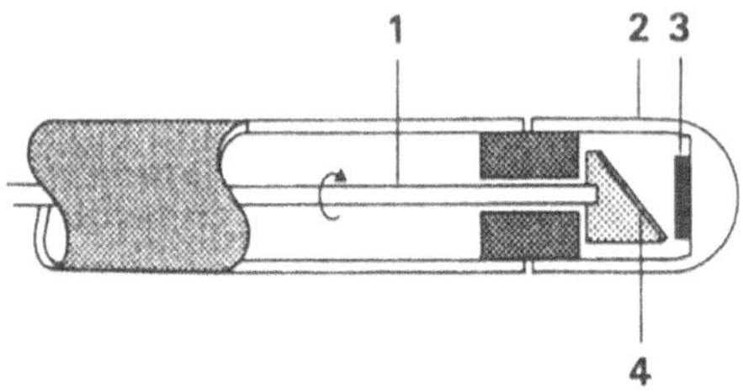

Fig. 8. The rotating mirror principle. This contains the flex shaft (1); The transparent dome (2); the acoustic element (3) and the rotating mirror (4).

Rotating mirror

The rotating mirror technique is similar to the previously described method. Schematically this method is shown in Figure 8. The flexible shaft (1); the transparent dome (2); the echotransducer (3) are complemented with a mirror. The mirror creates an even shorter dead zone outside the catheter due to the longer acoustic pathway inside the catheter. The non-moving transducer avoids the necessity of rotating electric wires. Acoustic lenses and focussing shapes of the mirror have been described.

Electronically switched phased array

The catheter shown in Figure 9 contains many small acoustic elements (2) which are positioned cylindrically around the catheter tip. The number of elements may be any practical number such as 16, 32, 64 or 128. The tip may contain an electronic component to reduce the number of electric wires.

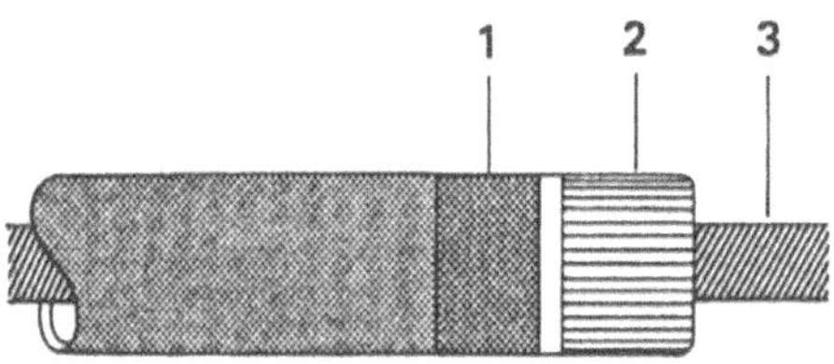

Fig. 9. Electronically switched phased array catheter tip with integrated circuitry for reduction of the number of wires (1); the elements (2) and a guide wire (3).

The construction allows for the introduction of a central guidewire (3). A first principle is identical to the method referred to before [8]. As illustrated for a small number of rotational symmetric elements in Figure 10, by introduction of time delays subgroups of elements may together form a "single larger echo transducer". This process can be repeated with any other subgroup. This allows aperture variation and electronic focussing methods. On the other hand, acoustic element geometry is not optimal and a near-field dead zone may exist (see further). In a second principle, all elements are used individually for transmission and reception. Beam forming is created with computer algorithms using a Synthetic Aperture System (SAS).

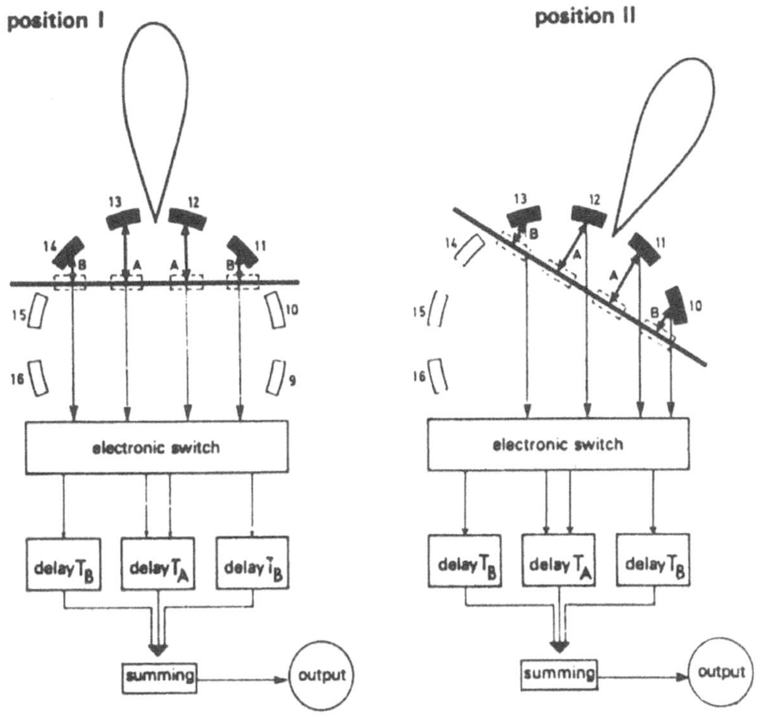

Fig 10. Principle of the phased array catheter tip as described by Bom [8]. Since the individual elements are small in relation to the used frequency, these elements will be omnidirectionally sensitive. By addition of an electronic delay in transmission as well as in reception for each element the elements can be virtually positioned on the solid indicated line as schematically indicated. Used in subgroups of 4 elements (here as a simplified example) a high acoustic directivity can be created. The beam can subsequently be rotated electronically.

Diagnosis of atherosclerosis

Currently diagnosis of vessel obstructions is performed by angiography, which is X-ray imaging of the vessels after injection of a contrast agent through a catheter. The angiogram provides a silhouette of the lumen of the blood vessel of interest and its branching from the position of contrast injection towards down-stream. A stenosis usually can be seen as a restriction of the lumen. Angiography is a powerful and efficient technique to localise blood vessel obstructions and also allows for an estimation of the degree of narrowing [15].

However, the projection information of the angiogram depends on the choice of direction of the X-ray beam, which plays an important roll in revealing a clear sight of the blood vessel narrowing. This is illustrated schematically in Figure 11. The contrast only fills the open lumen, and the angiographic image shows its projection ("shadow"). As indicated for a partly closed lumen (Figures 11B and 11C), the narrowing shown in the image thus depends on the direction of the X-ray.

Even by biplane angiography not enough information can be obtained to exactly reconstruct the 3-dimensional geometry of the lumen of the blood vessel. And apparently, by the nature of the technique, no information can be obtained about the state of the blood vessel wall and the type of the obstructing substances.

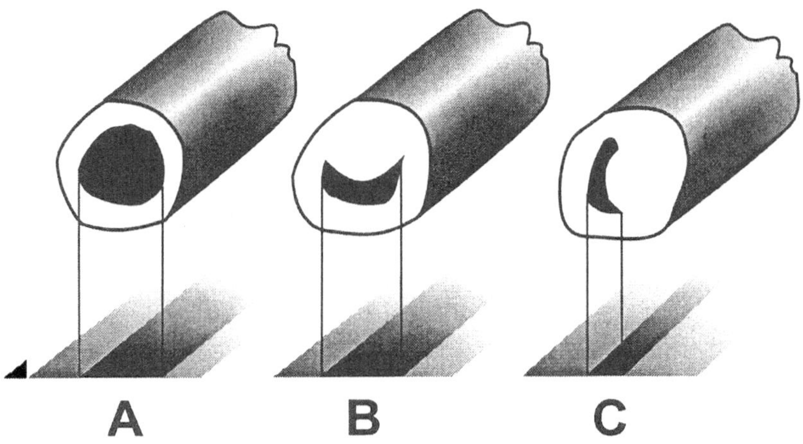

Fig. 11. Simplified explanation of the "shadowgram" as obtained by using X-ray angiography. The projection of the contrast filled lumen is projected on the film as shown in A. When only a narrow open lumen exists the lumen width as indicated on the film will strongly depend on the direction of the X-rays as shown in B and C.

The capabilities of angiography versus IVUS at two locations are illustrated in Figure 12 showing the femoral artery after balloon angioplasty. Dissections are sometimes very obvious on angiography but can be suspected in other cases.

Intravascular ultrasound not only recognises this type of lesion at any sector of the circumference of the vessel wall, it actually visualises the movements of a ruptured plaque inside the lumen. Real-time assessment of vessel wall dynamics offers a great advantage over angiography in these cases.

Present intravascular scanners

A summary of today's IVUS methods is illustrated in Figure 13. Mostly, the intravascular procedure is applied for decision-making when angiographic data are less conclusive. For obvious reasons, there is a strong urge to combine "see and do" in interventional procedures. This leads to catheters in which, for instance, an angioplasty balloon is combined with ultrasonic imaging in or close to the balloon. Another combination may provide guidance during stent implantation (Figure 13d). Since intravascular imaging provides accurate geometrical information within the cross-section, combination with other interventional procedures is likely to expand in future.

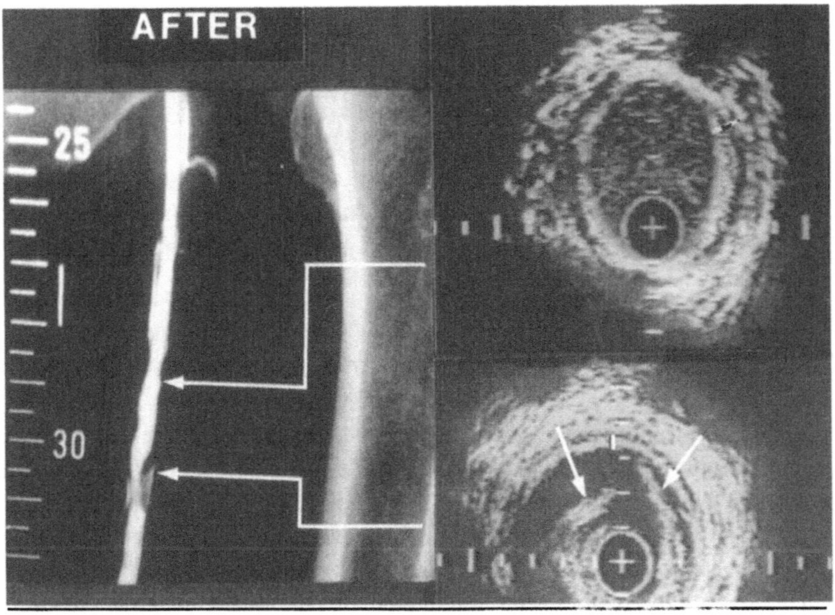

Fig 12. Comparison of the angiographic image of a femoral artery and the intravascular cross sectional images at two levels after an interventional procedure. The arrows indicate plaque and dissection. This figure nicely illustrates the good "overall path finding" capabilities of angiography as well as the excellent "local" information that may be obtained by intravascular echography.

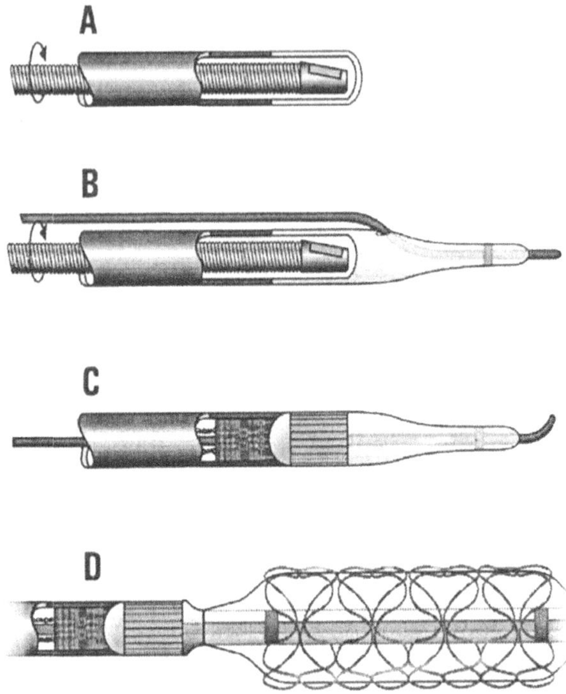

Fig. 13. Overview of presently most frequently used IVUS catheters. A) flex shaft rotating element system. B) Identical to A with added guide wire capabilities. C) Phased array system with guide wire and multiplexing electronics. D) catheter combining imaging with stent delivery capabilities.

Image improvement, distortion and dead zone

Image quality of ultrasonic devices has improved dramatically in recent years. The tendency has been to incorporate higher frequencies and thereby to improve the resolution. This became possible by the use of more sensitive transducers and better signal processing to cope with increased attenuation at raised frequencies. The present frequency range is that from 20-40 MHz. Increase of the frequency will be one of the requirements for further miniaturisation towards an imaging guide-wire in the future. Practical differences between the electronically steered phased array catheter and the flex shaft mechanically rotated technique result from differences in guide wire location; motion of the catheter tip; abnormal rotation of the transducer and differences in length of the so called dead zone.

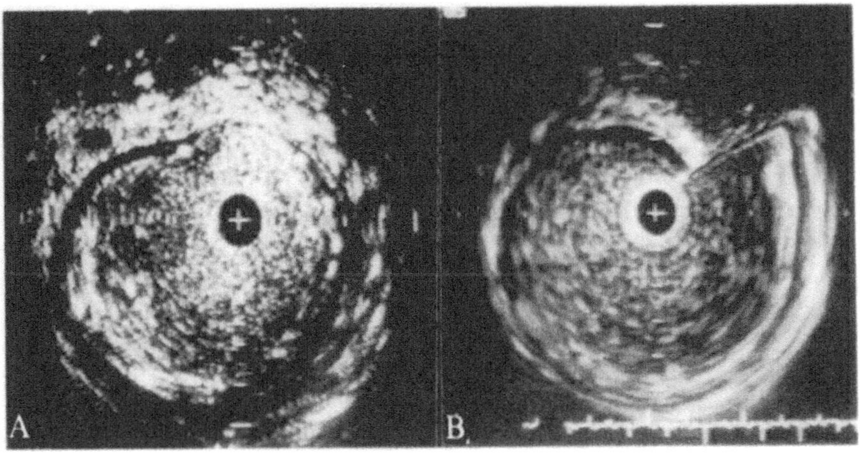

Fig. 14. Examples of image artifacts caused by tip motion in turn caused by the driving shaft (A) and an extreme example of non-linear rotation of the transducer because of partly clamping of the flex shaft somewhere along the tortuous path of the catheter (B).

Most mechanical systems that are presently commercially available are based on rotation of a single element or a mirror in the tip of the catheter. As discussed earlier, the element or mirror is rotated by use of a long, flexible shaft driven by an external motor. With a long catheter through a tortuous path Non-Uniform Rotation Distortion (NURD) may occur. This could cause image artifacts. The artifact in Figure 14A is caused by relative and regular motion of the catheter tip against the arterial wall, which results in the cardioid malformation of the image. In Figure 14B, an extreme example of irregular rotation of the flex shaft as it appears in the image is shown.

The very first echo response is created by the pulse transmission itself and is often described as the transient effect or dead zone. It is caused by saturation of the amplifiers. Very close to the catheter tip, imaging is not possible. If a distance were created between the active transducer elements and the acoustically transparent dome, then the dead zone could be restricted to the internal catheter dimensions. This is impossible for the electronic method where elements are located at the outer circumference of the catheter. As illustrated in Figure 15, the saturation area or dead zone is smallest in a mechanical mirror type IVUS catheter. The largest dead zone results from electronic systems.

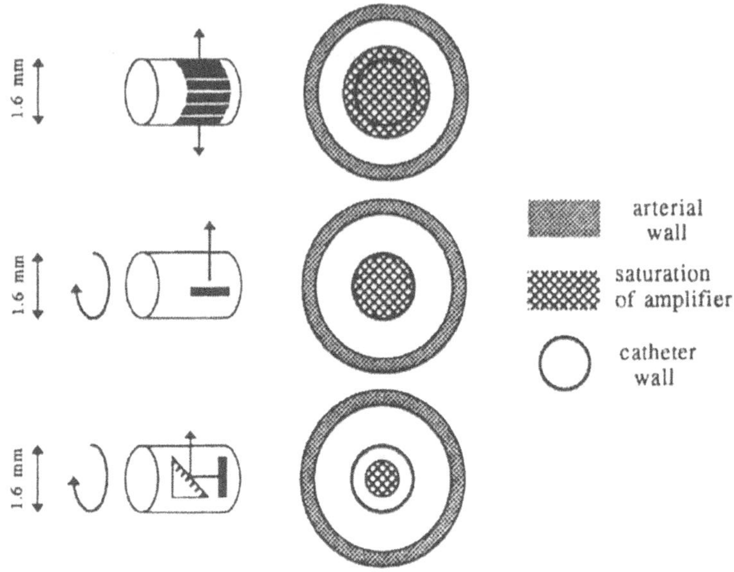

Fig 15. Schematic illustration of the "dead zone" where due to transmission effect (saturation of the amplifier) no return echoes can be detected. As shown, starting from the actual transducer surface, the dead zone is limited inside the catheter for the mirror type IVUS and more extended outside the catheter for the electronic type of IVUS systems.

Conclusion

In this survey, only the principles and early and recent approaches to intraluminal and in particular IVUS imaging have been described. It seems clear that angiography is excellent for "pathfinding", since an excellent overall view is provided. It is subject to X-ray beam direction and only the contrast-filled lumen can be seen. IVUS, on the contrary, is capable of yielding a real-time cross-sectional image. This, however, is much more "local". It allows viewing of plaques as well as wall properties. Since IVUS catheters are expensive, their use is mostly limited to solve unclear situations. The IVUS impact under those conditions, however, is for diagnostic and therapeutic applications of paramount importance.

It is expected that IVUS will become integrated with therapeutic devices and, once IVUS is fully integrated in the catheterisation laboratory, it will be the tool of choice.

Acknowledgement

Figures 6, 7, 8, 9 and 10 have been reprinted from Bom N, ten Hoff H, Lancée C.T. et al. (1989) Early and recent intraluminal ultrasound devices. Int J Cardiac Imag 4:79-88 with kind permission from Kluwer Academic Publishers.

References

1. Bom N, Lancée CT, Gussenhoven WJ, Li W et al. (1992) Basic principles of intravascular ultrasound imaging. In: Tobis JM, Yock P (eds) Intravascular Ultrasound Imaging. Churchill Livingstone Medical Publishers, New York, pp 7-15
2. Bom N, Ligtvoet CM (1993) Principles of cardiac ultrasound. In: Roelandt JRTC, Sutherland GR, Iliceto S, Linker D (eds) Cardiac ultrasound. Churchill Livingsdtone, Edinburgh, pp 9-20
3. Wild JJ, Reid JM (1955) Ultrasonic rectal endoscope for tumor location. Am Inst Ultrasonics Med 4: 59
4. Omoto R (1967) Intracardiac scanning of the heart with the aid of ultrasonic intravenous probe. Jap Heart J 8: 569-581
5. Ebina T, Oka S, Tanaka M, Kosaka S et al. (1965) The diagnostic application of ultrasound to the disease in mediastinal organs. Ultrasono-tomography for the heart and great vessels. Sci Rep Res Inst Tohoku Univ 12: 199-212
6. Wells PNT (1966) Developments in Medical Ultrasonics. World Medical Electronics 66;4:272-277.
7. Eggleton RC, Townsend C, Kossoff G et al. (1969) Computerised ultrasonic visualization of dynamic ventricular configurations. 8th ICMBE, Palmer House, Chicago IL, July 1969, Session 10-3
8. Bom N, Lancée CT, Van Egmond FC (1972) An ultrasonic intracardiac scanner. Ultrasonics 10: 72-
9. Hisanaga K, Hisanaga A, Nagata K et al. (1977) A new transesophageal real-time two-dimensional echocardiographic system using a flexible tube and its clinical application. Proc Jap Soc Ultrasonics Med 32: 43-44
10. DiMagno EP, Regan PT, Wilson DA et al. (1980) Ultrasonic endoscope. Lancet; March: 629-631
11. Bertini A, Masotti L, Zuppiroli A et al. (1984) Rotating probe for transoesophageal cross-sectional echocardiography. J Nucl Med Allied Sci 28: 115-121
12. Souquet J, Hanrath P, Zitelli L et al. (1982) Transesophageal phased array for imaging the heart. IEEE Trans Biomed Engng BME-29: 707-712
13. Natori H, Tamaki S, Izumi S et al. (1983) Clinical application of ultrasound endoscope using linear array transducer for transesophageal ultrasonography of the disease of the mediastinum. In: Lerski A Morley, P (eds) Ultrasound '82, Pergamon Press, Oxford, pp. 339-343
14. Fukuda M (1985) Endoscopic ultrasonography. In: Gill RW, Dadd MJ (eds) WFUMB '85, Pergamon Press, Sydney, pp 13-16
15. Whyting JS, Pfaff JM, Eigler NL (1991) Advantages and limitations of videodensitometry in quantitative coronary angiography. In: Reiber JHC, Serruys PW (eds) Quantitative Coronary arteriography. Kluwer Academic Publishers, Dordrecht, pp 43-54

High Frequency IVUS

Tat -Jin Teo

Boston Scientific
47900 Bayside Parkway Fremont, CA 94538, USA
Email – teot@bsci.com

Summary: A description of IntraVascular UltraSound (IVUS) imaging is provided in this chapter. Section 1 describes and compares the solid state versus the mechanical implementation of IVUS. Section 2 describes the current performance capability attainable in the mechanical implementation. Section 3 provides for some preliminary results with 50 MHz operation. Section 4 provides a detailed description of the theory of operation of an IVUS system. Section 5 describes and compares IVUS to a new intravascular imaging modality called Optical Coherence Tomography (OCT). Section 6 addresses the speed of sound issue in intravascular imaging.

Keywords: IVUS, intravascular ultrasound, OCT, intravascular imaging, minimally invasive imaging.

IVUS technology introduction – mechanical versus solid state

Intra-Vascular UltraSound (IVUS) refers to a minimally invasive ultrasonic imaging modality. It is a catheter based imaging method. There are currently two approaches to IVUS – mechanical versus solid state. We describe the differences between these two types of IVUS systems in terms of the following aspects: (i) Range Dynamic Focusing. (ii) Lateral Scan generation. (iii) Volume scan generation (iv) Imaging at higher frequencies and (v) Active aperture.

Range Dynamic Focusing

This is one of the most fundamental differences between mechanical and solid state IVUS system at present. Along the axis of sound propagation, a typical single channel mechanical IVUS transducer has a fixed focus. At present, this is the only available form of mechanical IVUS system. In principle, however, a mechanical IVUS can also contain several channels corresponding to multiple transducer elements and is capable of range dynamic focusing with the aid of a beam former. A mechanically rotated annular array is one such example.

For solid state IVUS, the multi-channel transducer allows for electronic focusing with the aid of a beam former. Using time and/or phase delays among the different channels, the focus can be changed electronically. This can enlarge the depth of focus since the changing focus can be moved along the path of the beam such that successive ranges in the region of interest are brought into focus. The beam former also performs the lateral scan.

The lack of range dynamic focusing in today's mechanical IVUS system is probably the single most important drawback in situations where large imaging depth is desired.

Lateral scan

In the mechanical approach, a side viewing transducer is mechanically rotated about the long axis of the catheter. A 360 degree cross-sectional image of a blood vessel or a cardiac chamber is therefore generated. In a solid state IVUS, the lateral scan is performed electronically and no mechanical motion is needed. The transducer consists of multiple transducer elements, each with its dedicated electronic channel. In some implementation, the transducer elements have to share a much smaller number of electronic channels through time sharing. These electronic channels make up the beam former. The same beam former that performs the range dynamic focusing is also used for this lateral scan. For solid state IVUS, two scanning geometries are possible. It can either generate a 360 degree cross-sectional image of the vessel or a longitudinal scan containing the long axis of the catheter.

In the 360 degree cross-section scanning geometry, the transducer elements are deployed along the circumference of the catheter. The lateral scan is performed by electronically translating the active aperture around the circumference of the catheter. In the longitudinal scanning geometry, the transducers elements are deployed along the long axis of the catheter. The beam former electronically steers the beam or translates the beam or both. The scanning geometry can then vary between a rectangular shape and a sector shape or some intermediate shape in between. For a given transducer array of a fixed length, the sector shape maximizes the scanned area but limits visibility in the near field. On the other hand, the rectangular shape can provide images with more uniform near/far field visibility while limiting the size of the scanned area.

Volume scan

The mechanical IVUS involves rotation for lateral scanning and a longitudinal motion in the long axis direction to generate a volume scan. This is accomplished by pulling back the transducer/catheter along the sheath or guide catheter. For solid state IVUS, to obtain a volume scan, either a pull back or a rotation is needed depending on whether the lateral scan produces a cross-sectional image or a longitudinal image, respectively.

Presently, on both systems, the lateral scan is typically done at real time or near 30 frames per second while the volume scan is obtained with additional mechanical motion, either with or without ECG gating. There is also a research prototype with a two-dimensional array using which a volume scan can be generated without any mechanical motion[1] for use in intra-cardiac applications.

High Frequency

Compared to a solid state system, a single channel mechanical IVUS system typically can operate at a higher frequency since there is no beam former involved in forming a beam. In the solid state approach, the limitation on employing higher frequencies arises because, in order for the beam former to function properly, the pitch between the elements cannot be too large compared to the wavelength. Higher frequency implies shorter wavelength and that means smaller pitch between the elements. In the solid state cross-sectional mode, to cover the 360 degrees with electronic scanning, more elements will be needed which involves more electronic channels or greater degree of multiplexing with further degradation in the image quality of a dynamic object. In a solid state longitudinal scan mode, small pitch results in smaller active aperture for the same channel count. Currently, the longitudinal scan image format is used typically in intra-cardiac settings where deep penetration is desired and the use of higher frequency is not needed. But for coronary application, greater resolution is still very much desired.

Though not a basic requirement, perhaps due to its higher frequency, mechanical IVUS typically uses analog processing while solid state IVUS typically employs digital processing in its beam former.

Active Aperture

Other parameters being equal, lateral resolution is governed by the size of the active aperture. In the cross-sectional scan mode, the active aperture is constrained by the diameter of the catheter. This is true for either the solid state or the mechanical IVUS system.

A longitudinal mode solid state IVUS, in contrast, deploys the transducer elements along the long axis of the catheter and can achieve much greater aperture. The constraint in this case is primarily the number of micro-coaxial cables that can be fitted within a reasonable size catheter.

Current performance – lateral resolution, axial resolution and penetration.

We provide a list of measures that are currently specified by the United States Food and Drug Administration (FDA) to characterize the axial resolution, lateral resolution and output power, among other parameters. Details on their definition and measurement method can be obtained from the relevant documents available at the FDA's web site and the references therein.

As an example, the Boston Scientific Corporation IVUS instrument is regulated by FDA as a 'Track 1' device for the purpose of acoustic output regulation – these diagnostic ultrasound imaging device do not provide real time display of bio-effect indices. The indices include the Mechanical Index (MI) and the Thermal Index (TI). These track 1 devices are regulated by an acoustic output power limit according to applications. In the Acoustic Output Report stipulated by the FDA Guidance Document[2], the manufacturers are required to measure the following parameters, among others.

1. Pulse Duration – this can be correlated with the axial resolution of the imaging device.
2. 6-dB beam width – this can be correlated with the lateral resolution of the imaging device.

As an example, the Acoustic Output report from Boston Scientific Corporation Atlantis 40 MHz coronary catheter Direction for Use (DFU), lists the following values for these 2 parameters:

Pulse duration – 0.0556 microseconds
6-dB beam width - 0.0417 cm.

With pulse duration, one can obtain a measure of the axial resolution. The 6-dB beam width provides the corresponding measure for lateral resolution. The X-axis beam width refers to the image plane beam width, while the Y-axis refers to the beam width in the thickness direction of the image slice.

All these measurements are transmit-only measurements. For pulsed echo imaging, the 2-way parameters need to be considered. It also reflects the resolution measured at one location specified by the guidance document. To get a measure of how the resolution varies as a function of the imaging depth, several simplifying assumptions are needed. The most drastic simplifying assumption is that the beam is in a homogeneous medium, water, for example. Other simplifying assumptions can include the shape of the active aperture, the attenuation of the medium, the effect of the finite amplitude of the propagation sound waves and the non-linear properties of the propagating medium.

In actual use, the transducer is placed inside a sheath. The sheath material has a propagation speed different from that of surrounding tissue and can cause refraction. The actual axial and lateral resolution measured two way through a sheath is provided below.

Axial resolution – 0.08 mm
Lateral resolution – 0.16 mm.

There is currently no tissue mimicking phantom calibrated at 40 MHz, operating frequency used by this catheter. Hence, penetration measurements, unlike the previously discussed axial and lateral resolution, may some time be subjective and often only relative.

50 MHz performance

Currently, the highest frequency IVUS that is commercially available is the 40 MHz Atlantis catheter from Boston Scientific Corporation. In application where higher resolution is desired, a higher operating frequency will be needed.

The fractional bandwidth of a transducer is defined as the bandwidth divided by the center frequency of the transducer. Bandwidth is inversely proportional to the pulse duration, which governs the axial resolution.

An increase of center frequency from 40 MHz to 50 MHz corresponds to a 25% improvement in axial resolution if the design is similar which yields similar fractional bandwidth.

The lateral resolution is governed by the lateral beam width. For simplicity, consider a circular transducer operating in piston mode. The lateral beam width is described by the following expression:

Lateral beam width = 2.44*λ*focal length/diameter **(1)**

where λ is the wavelength of the transducer.

For a planar transducer, the natural focus corresponds to the near field range given in Equation (2).

Near field range = diameter^2/(4*λ) **(2)**

Note that the lateral resolution is not a function of the bandwidth but the operating frequency of the transducer. As one increases the operating frequency from 40 MHz to 50 MHz, a similar magnitude of improvement in the lateral resolution can also be expected.

Figure 1 compares the signal characteristics of a 40 MHz transducer and a 50 MHz transducer having similar design. The top panels show the time waveform and frequency domain spectrum for the 40 MHz transducer while the bottom two panels show the corresponding 50 MHz waveform and spectrum.

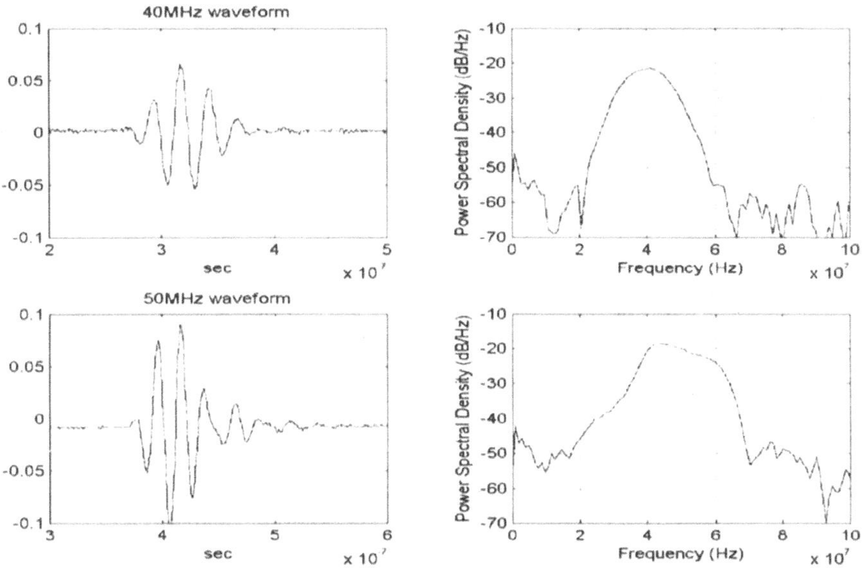

Figure 1: Comparison between 40 MHz and 50 MHz. Left panels – waveform, right panels – spectrum.

A basic IVUS imaging system

The block diagram of a typical mechanical IVUS system is shown in Figure 2. As shown in this basic system block diagram, the pulser in the motor drive provides the electrical excitation to the transducer which launches an ultrasonic waves towards the tissue of interest. The motor drive also rotates the transducer which accomplishes the lateral scan. The T/R switch protects the sensitive receive amplifier from the high voltage excitation. The received echo is the Radio Frequency (RF) signal. This signal is first amplified by the pre-amplification block and followed by band bass filtering and envelope detection. The envelope signal is also sometimes called image signal and is brightness modulated in the display. Currently in the Boston Scientific system, the envelope is digitized for further processing in the digital domain.

It is also common to add a Time Gain Compensation (TGC) to account for the attenuation of the propagation medium so that deeper echoes are amplified more to maintain the signal level.

The block designated as Envelope Detection also performs Logarithmic Compression. This is to accommodate the large dynamic range of the signal onto a display device which typically has a smaller dynamic range.

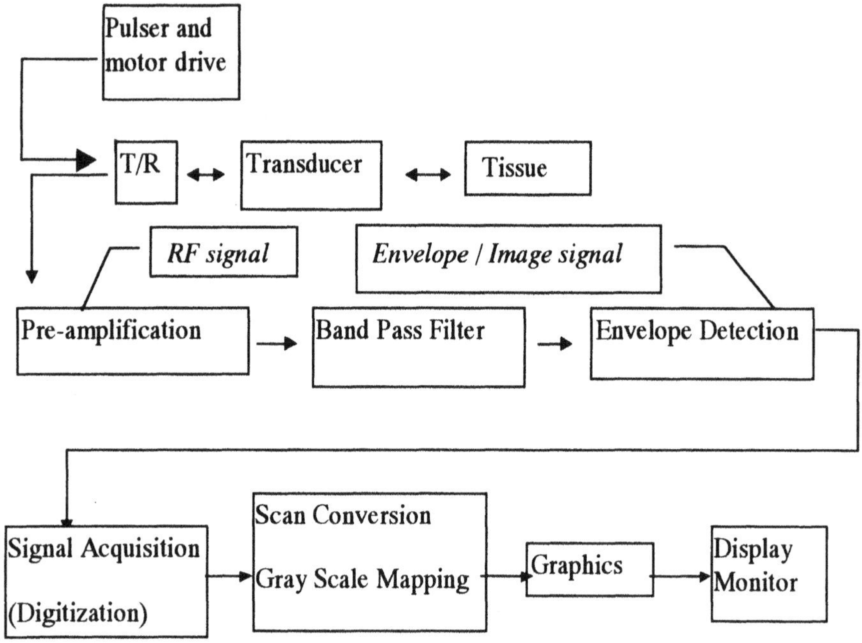

Figure 2: Basic IVUS System Block Diagram

In the block designated as Band Pass Filter, there is currently a set of filters to match the different catheter center frequencies for different applications, ranging from intra-cardiac application, peripheral vessel application to coronary application. This band pass filter reduces the amount of noise outside the signal frequency band of interest and helps to improve the signal to noise ratio of the system.

At the Envelope Detection block, there is a low pass filter to extract the envelope signal. It is this low pass filtered signal that is called the Envelope or Image signal which is digitized by an Analog-Digital-Converter (ADC) into digital samples used for subsequent signal processing.

Figure 3 illustrates the processing steps to transform the RF signal into an Envelope signal appropriate for subsequent image processing in the current mechanical IVUS system.

The top panel in Fig. 3 shows an RF vector. This signal is then amplified and band pass filtered as indicated in the block diagram. The horizontal axis shows the sample number which corresponds to time of arrival, or, equivalently, depth of interest. The vertical axis in the top panel is arbitrary unit in linear scale while that in the second and third panel are in dB. The second panel shows the bipolar RF signal being rectified and log compressed. The third panel shows the low pass filtered version of what was shown in the second panel. Finally, the bottom panel shows the brightness modulated line that makes up the image data.

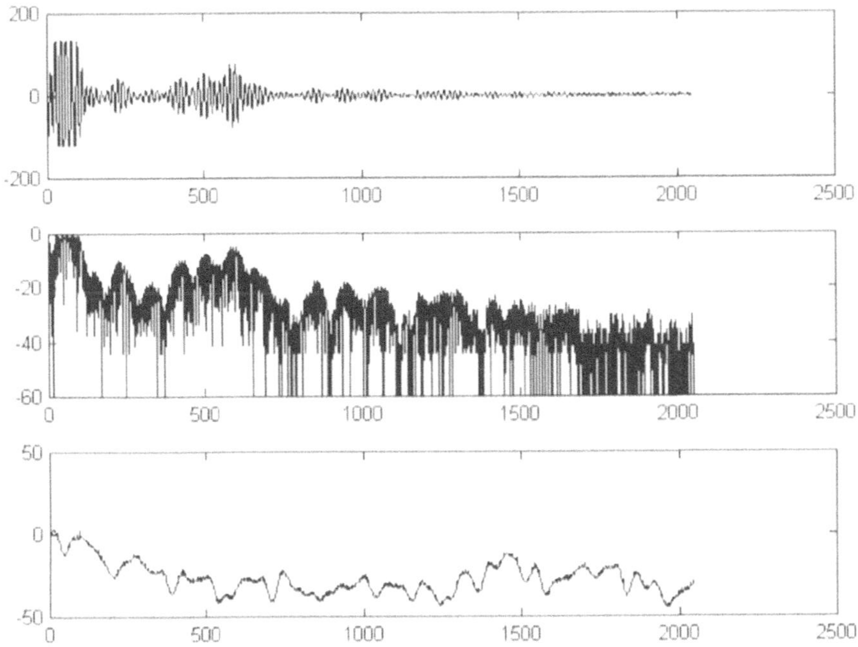

Figure 3: Top to bottom: RF, Rectified and Logarithmic Compressed, Low Pass Filtered, Brightness modulated signal.

The left panel in Fig. 4 shows the Polar Coordinates data. The vertical axis is the azimuth angle θ (Theta) and the horizontal axis is the range axis R. Since the IVUS image is generated by rotating the side-looking transducer about the long axis of the catheter, the collected echoes are in the Polar Coordinates. A scan conversion is typically performed to display the image in the rectilinear format on a display monitor. Even for solid state IVUS that generates a longitudinal image, the image format is also typically of a sector-like display to reduce the rigid length at the foot print of the catheter to facilitate handling. A similar scan conversion is also needed for displaying a rectilinear format image.

Several possible schemes are possible with scan conversion. One quick method is to pick for each (X,Y) pixel the nearest neighbor in the Polar Coordinates domain and used that value for the value at (X,Y) location. A more accurate method will be to pick the four nearest neighbors, two in R, two in θ, and do a bi-linear interpolation to arrive at the (X,Y) pixel value. Still higher order interpolation can be used at the expense of more computation time.

Gray Scale Mapping for the intensity values to the gray scale can also be done at this point. In Figure 5, three gray scale mapping curves are illustrated.

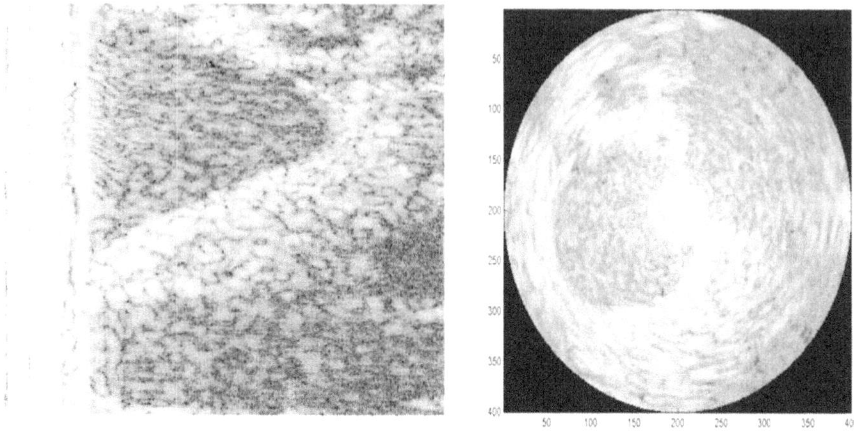

Pre-Scan Conversion Post-Scan Conversion

Figure 4: Scan Conversion

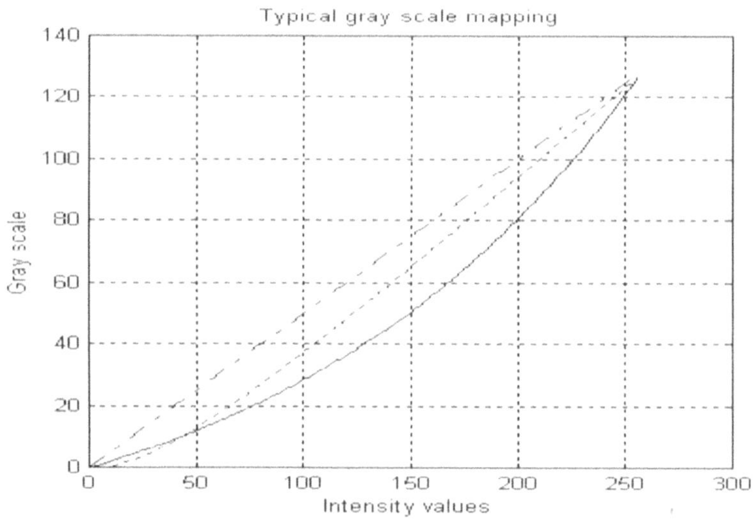

Figure 5: Gray Scale Mapping – 3 examples

Other image processing steps can also be done at this stage. Common methods include border detection, measurement of distance and area, texture analysis and artifact suppression.

Several current IVUS instruments support three dimensional data acquisition via a pull back motion. A volume data set is then acquired which allow further manipulation to display useful information. One common display format is the longitudinal view. This involves defining a cut plane in the original two-dimensional cross-sectional image and collecting all the image data along this cut plane from all the slices in the pull back path. This longitudinal view is convenient for length measurement along the long axis of the vessel. Presently, the tortuousity of the vessel is not taken into account in the display.

Figure 6 shows the screen shot from Boston Scientific Corporation's Galaxy. The top left panel is the cross-sectional IVUS image and the panel next to this is reserved for displaying the fluoroscopic image. The bottom panel shows the longitudinal view described above. On the right hand side is a collapsible window showing the distance and area measurement.

Since the phase information of the RF signal is lost during envelope detection, tissue characterization that exploits the complete amplitude and phase information of the RF signal cannot be performed after the digitization of the envelope signal.

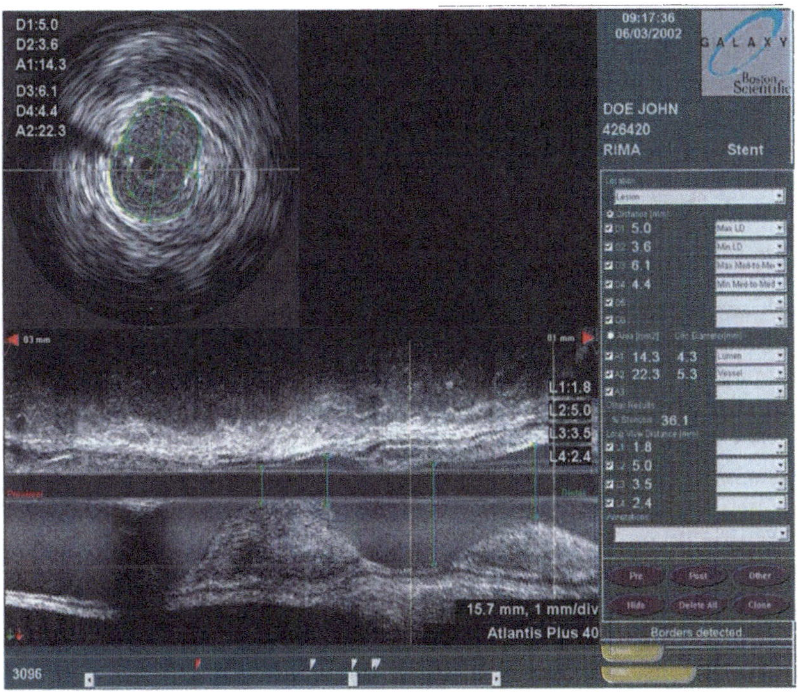

Figure 6: Boston Scientific Corporation's Galaxy screen shot showing longitudinal view

To facilitate RF tissue characterization, digitization will have to occur before the envelope detection. Due to the high frequency of IVUS, several commercially available digitizers are not satisfactory either in terms of sampling frequency or dynamic range or both.

With the rapid advancement in digital electronics and the realization that more information can be obtained with the RF signal, it is natural to expect that all IVUS instrumentation would move towards even greater emphasis on digital signal processing. The field of ultrasonic tissue characterization has a long history in non-invasive diagnostic ultrasound. For IVUS, this approach can yield even greater results since the quality of RF signal is not compromised by a body wall that is highly inhomogenuous.

Comparison with Optical Coherence Tomography (OCT)

Optical Coherence Tomography uses light waves instead of sound waves for forming the image. Unlike with sound waves, the pulse-echo techniques cannot be used with light waves since the speed of light in tissue is too fast for today's electronics. Instead, an interferometry method is used to measure the phase difference between the signal arm and a reference arm. As an illustration, the acoustic echoes from two interfaces separated by 100 microns will be about 130 nano-seconds apart, assuming a sound propagation speed of 1500 m/s, well within the capability of today's electronics for detection. The optical echoes will be within 1 pico-second, assuming a propagation speed for light in tissue of $2.2\text{x}10^8$ m/s.

Figure 7 illustrates the basic OCT system.

There are several similar features between intra-vascular ultrasound and intra-vascular OCT. They are both coherent imaging system and the signal contains both amplitude and phase information. Consequently, the images contain speckle, which is a result of the constructive and destructive interference from the coherent imaging. One implementation of OCT uses the same mechanical rotation format to generate the 360 degree cross-sectional image like its IVUS counterpart. The images from these 2 modalities are both brightness modulated gray scale images obtained using the back-scattered energy, acoustical or optical.

The most striking difference between these two modalities arises due to the difference in wavelength and hence the resolution of the image. Due to the difference in wavelength, the diffraction effect is also dramatically different between these two imaging modalities. In IVUS, the lateral resolution degrades rapidly as a function of depth when there is no dynamic focusing, as is the case with single channel mechanical system. With OCT, the wavelength is about 20 times smaller than that of a 40 MHz IVUS, the diffraction is correspondingly much reduced and the lateral resolution is largely maintained throughout the image depth, albeit much shallower than that of 40MHz IVUS.

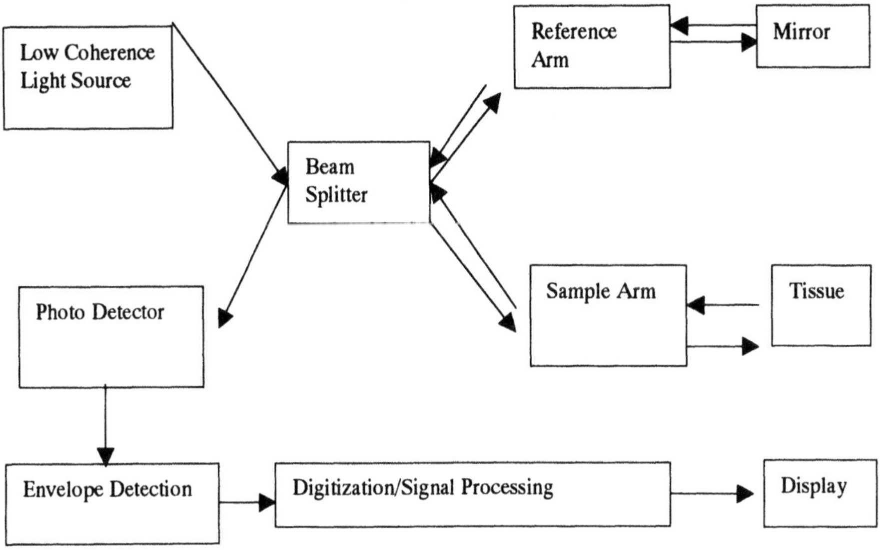

Figure 7: Basic OCT System Block Diagram

One other difference between IVUS and OCT is the fact that light is a transverse wave and is capable of supporting polarization. This can provide additional information via birefringence, a property of medium having polarization dependent propagation speed, something not possible with longitudinal acoustical waves.

The following table summarizes the imaging performance comparison by abstracting from clinical literature[3] and discussion with OCT practitioners in the field:

Table 1: Resolution comparison between IVUS and OCT

	IVUS	OCT
Penetration	4 mm*	1.25 mm with Saline flush* otherwise 0.5mm in blood
Axial Resolution	80 μm*	15 μm*
Lateral Resolution	200 μm	20 μm

* The reference used 30 MHz IVUS. Reference did not mention lateral resolution.

From the above table, it is clear that the penetration in blood is currently the main limitation of OCT. Presently, one needs to use an occluding balloon or a saline flush to displace the blood. Such additional step represents the main barrier to the adoption of this imaging modality. There are catheter designs that can reduce the amount of saline needed for obtaining good images. The efficacy of this approach remains to be seen.

Speed of sound for use in Intra-vascular imaging

Since pulsed echo imaging converts all time intervals into distances for image display, the choice of the value of propagation speed becomes an important one. In the author's experience, a value of 1.54 mm/µsec is a common value among U.S. manufacturers of non-invasive diagnostic ultrasound imaging equipment. In Japan, the author is given to understand that 1.53 mm/µsec may be the preferred value[4].

Boston Scientific Corporation mechanical IVUS instrument uses 1.5625 mm/µsec for speed of sound in blood. The different choice of the value of speed of sound among different IVUS manufacturers makes comparison across different equipment difficult and need to be undertaken with great care.

For mechanical IVUS which has a transducer rotating inside a sheath, the presence of the sheath with its own speed of sound needs to be taken into account. The parameters needed to correct for this are the thickness of the sheath and the speed of sound in the sheath. By taking these two parameters into account, the measurement accuracy is improved. This new algorithm was first implemented by Boston Scientific for its 40 MHz catheters on its ClearView IVUS system. Consequently, when comparing a 30 MHz catheter and a 40 MHz catheter on a ClearView IVUS instrument, one needs to add a factor of 175.2 µm to the 30 MHz catheter diameter measurement to be equivalent to that of the 40 MHz catheter diameter measurement. On the newer Galaxy IVUS instrument, this adjustment factor is no longer needed across all catheters.

References

1. "Intracardiac Catheter 2-D Arrays on a Silicon Substrate" Warren Lee et.al. IEEE Transactions on Ultrasonics, Ferroelectrics, and Frequency Control vol. 49, no. 4, April 2002, pp. 415-425.
2. "Information for Manufacturers Seeking Marketing Clearance of Diagnostic Ultrasound Systems and Transducers" Issued September 30, 1997, US Department of Health and Human Services, FDA, Center for Devices and Radiological Health, Office of Device Evaluation
3. I.K. Jang et. al. JACC 39(4) 2002
4. Mr. Yagami, Terumo Corporation. Private communication. Also Japan Standards Institute publication.

Quantitative IVUS Flow Estimation

A.F.W. van der Steen[1,2], F.A. Lupotti[1,2] F. Mastik[1], E.I. Céspedes[1], S.G. Carlier[1], W. Li[1,2], P.W. Serruys[1,2], N. Bom[1,2]

[1]Thorax Centre Erasmus University Rotterdam The Netherlands,
[2]Interuniversity Cardiology Institute of the Netherlands.

Summary: Coronary flow assessment can be useful for determining the heamodynamic severity of a stenosis and evaluate the outcome of interventional therapy. We developed a method for measuring the transverse flow through the imaging plane of an intravascular ultrasound catheter. This permits simultaneous assessment of vessel geometry and function with the same device. Furthermore it should give more accurate information than combination devices, because lumen diameter and velocity are determined at the same location.

Flow velocity is estimated, based on decorrelation estimation from sequences of radiofrequency (RF)-traces acquired at (nearly) the same position. Signal gating yields a local estimate of the velocity. Integrating the local velocity over the lumen gives the quantitative flow. This chapter describes quantitative flow measurement using a single element transducer, as well as a solid state phased array catheter. Specific advantages and problems that come with both catheters have been identified and have been solved.

The methods have been tested through computer modeling, in vitro experiments using a flow phantom and in vivo experiments in a porcine animal model, and validated against a Doppler flowire in humans.

Keywords: Blood Flow, Intravascular Ultrasound, decorrelation, rf data processing

Introduction

Currently, IVUS is the only clinically available technique, capable of providing real time cross-sectional images *in vivo.* It delivers information that is not available from X-ray angiography. Since the number of available interventional techniques for treatment of atherosclerotic luminal narrowing increases, specific diagnostic information that can aid in the selection becomes increasingly important. For this reason, IVUS is more and more routinely used for guiding interventional procedures [1] and for studying the mechanisms for restenosis [2].

In the last couple of years the possibilities to access a sufficient amount of RF data at adequate speed have improved significantly. This has given the possibility to study IVUS data in a totally different way [3, 4]. In the traditional way the emphasis was always on imaging itself while RF-analysis also allows tissue characterization [5-7], elasticity imaging [8-12], flow estimation [13, 14]and enhanced boundary detection[15].

Assessment of blood flow has long been recognized as an important approach to evaluate the functional status of a diseased artery. Intracoronary Doppler is a conventional ultrasonic method for measurement of blood velocities and has been applied for coronary flow reserve assessment and to evaluate the severity of coronary lesion [16, 17]. The use of Doppler velocimetry alone, however, does not allow direct measurement of flow volume. For quantitative measurement of flow volume, the vessel cross-sectional area and the blood velocity must be assessed coincidentally in time and space. Intravascular ultrasound (IVUS) is a tomographic imaging technique that allows accurate measurement of the vessel cross-sectional area, irrespective of vessel geometry [18]. Several studies have demonstrated the feasibility of simultaneous use of intracoronary Doppler for velocity measurement and IVUS imaging for area measurement to estimate blood flow [19]. One potential limitation of this approach is that the flow volume is estimated from the mean Doppler velocity with an assumed velocity profile. Additionally, the simultaneous use of multiple ultrasound transducers could cause possible interference between devices, for instance, cross-talk between the Doppler and IVUS transducer [20].

A unique feature of IVUS techniques is that the direction of blood flow is almost normal to the imaging plane. As blood particles move across the imaging plane, the received echo signals decorrelate at a rate proportional to flow velocities. The temporal correlation properties of the radio frequency (RF) blood scattering signals in vivo have been documented extensively [15]. In this paper, we review the principles of a method that allows a direct measure of local velocity as well as volume flow by means of analysis of the intravascular RF echo signals. The flow measurements obtained with the decorrelation method (IVUS-flow) are compared to the data of an electromagnetic flow meter (EM-flow) in vivo under various flow conditions in an animal model. Furthermore clinical data are presented.

Since a rotating single element transducer suffers from some inherent motion artifacts we are currently developing the method for an array transducer system. These transducers get to their inherent resolution and image quality by synthetic aperture focussing techniques. Since this is based on reconstructing of the image from multiple pulses from different directions towards the same “static” object, this technique is not directly suited for flow estimation, because the blood particles are moving in the time course of the acquisition. The scanning protocol of the array had to be changed. In this paper another scanning protocol is described, based on a computer model of the array and included in an intravascular ultrasound system. Preliminary patient data using this catheter are presented.

Furthermore possibilities to use this method for automatic boundary detection in IVUS are discussed.

Methods

Decorrelation of blood echoes

The displacement of an ultrasound-scattering material such as blood moving through the beam of an ultrasound transducer results in concomitant changes (decorrelation) in the received echo signal. The decorrelation phenomenon can be clearly observed in the intravascular environment in which the ultrasound beam is transmitted perpendicularly to the arterial tissue and blood flow (Fig. 1). In IVUS, it is commonly observed from real-time display that the scattering pattern of flowing blood varies rapidly over time. This is due to the fact that the flow stream drives the randomly distributed blood particles constantly washing in/out of the image plane, causing the received echo signals to decorrelate as a function of time delay.

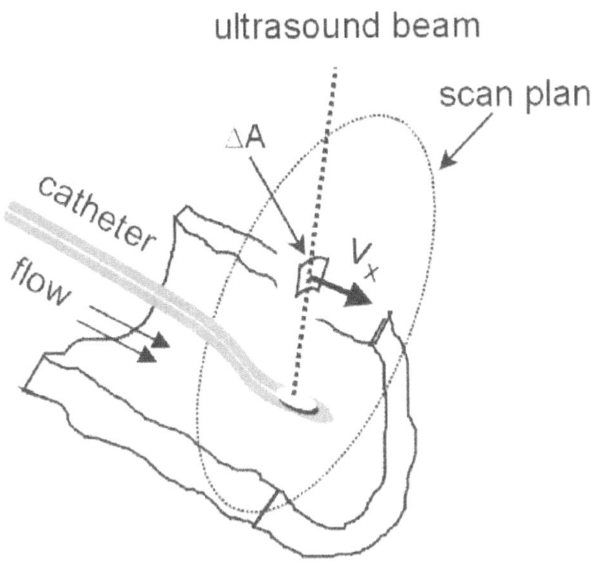

Fig1. The coordinates of the IVUS imaging set up. The scanning plane is perpendicular to the direction of blood flow.

The speed of the blood decorrelation process has proven to be related to the flow velocity. In other words, the faster blood particles move across the ultrasound beam, the higher the decorrelation rate of the received signals will be. The relationship between echo decorrelation and flow velocity is approximately linear. This result suggests that if the velocity-correlation relation is pre-quantified for the type of transducers used, the measured decorrelation value from blood scattering signals can be converted into flow velocity.

Decorrelation-to-flow quantification

The method is based on decorrelation processing of a series of RF signals (time RF sequence) acquired approximately at the same transmission angle.

Spatial decorrelation model: Unlike most other medical imaging techniques, majority of IVUS signals is acquired in the near field of the ultrasound beam (0.1~4 mm). The relationship between the correlation of echo signals and the scatterer motion across the ultrasound beam (lateral decorrelation) can be approximated with a linear model in the near field [13, 21]. The slope termed the displacement decorrelation slope (mm^{-1}), is characterized by the beam pattern. The linear dependence of signal correlation upon the scatterer lateral displacement has been validated in both computer simulation and in vitro with a flow phantom. Our results show that, because the beam pattern may change significantly in the near field, the slope is a beam dependent factor that requires to be calibrated for the type of transducer used [13, 21].

Time decorrelation measurement: At each transmission angle, a sequence of RF traces separated by a fixed time interval is acquired at a high pulse repetition rate. The decorrelation rate of the range-gated echo is evaluated in two steps:

1. Shifting over the time delay where the cross-correlation function reaches maximum first aligns the RF-traces. This is a phase matching procedure that removes the decorrelation due to the axial displacement (along the ultrasound beam) of blood scatterers.
2. Then the correlation coefficient between pairs of traces in the aligned RF windows are calculated. A slope is obtained by fitting a linear model through the correlation versus displacement. The slope α_i^t, termed the time decorrelation slope (s^{-1}), is related to the transverse blood velocity (normal to the ultrasound beam) for a given beam position.

Velocity estimation: Thus, the transverse velocity can be estimated at each window position from the ratio between the *calibrated* decorrelation function and the *measured* decorrelation slope from a time RF sequence. Note that the calibration procedure needs to be carried out only once for a specific type of transducer.

Volume flow integration: The local velocity measurement in a number of small, consecutive range windows can be repeated over the lumen cross-section to provide two-dimensional flow mapping. As illustrated in Fig. 1, the volume flow is calculated by integrating the local transverse velocity with its corresponding area element over the complete imaging plane Because the lateral motion of wall tissues has a much lower velocity th an blood, the contribution of tissue velocities can be automatically removed by setting a threshold. Thus, no contour tracing of the arterial lumen is needed.

Additional decorrelation factors

Other than blood flow itself several potential decorrelation factors can be identified that may affect the velocity estimation using the method described above. These factors can be measurement system dependent, or dependent on physiological processes.

System dependent factors are the change of the velocity profile by the catheter, secondary flow that is caused by the presence of the catheter, and impediment of flow. Other than that, decorrelation caused by rotation of the element (in case of the single element catheter) and noise (especially a factor or the array cath eters). Physiological factors are the influence of aggregation and various flow profiles.

1) The presence of the catheter does change the velocity profile, but this does not affect the flow estimation, since the values of the transverse velocity profile are integrated over the full cross section and the exact profile is not important.

2) The method is able to discriminate between axial and transverse flow, but incapable of discriminating transverse motion through the plane from in plane transverse motion. The latter one is referred to as secondary flow and a significant presence of this could cause a bias on the flow estimation. This secondary flow has been studied using Finite Element Modelling [22]. The flow through a curved vessel with a catheter inside at a variety of positions is compared to the flow without a catheter present. The main conclusion of the study was that the secondary flow with a catheter inside is of minor influence on the flow estimation [23]. A peculiar fact is that secondary flow with a catheter inside is even less than without catheter. The catheter seems to have a regulating effect on the secondary flow.

3) The impediment of the catheter on the flow is an issue. Obviously, when the free lumen has been narrowed to the size of the catheter it is of significant influence. At this moment it is unclear at which level of narrowing catheter size becomes a significant factor. After balloon angioplasty or stent placement in a coronary artery the catheter will only occupy roughly 10 % of the lumen, so it will not be a significant factor for assessing the coronary flow reserve to judge whether the procedure has been performed right.

4) When the flow is large enough the rotation of the element will not affect the decorrelation of the ultrasound signals. Furthermore the rotation creates a baseline decorrelation that can be corrected for. However the rotation does create a minimal threshold in the velocity estimation. When the blood flow velocity reaches values on the order of the angular velocity of the element signal analysis will not be able to discriminate the two. Especially when the rotation is non

uniform the rotation can be a factor the phase where the angular velocity is highest.

5) The correlation-based method is highly influenced by the presence of noise in the RF signals. Correction for noise is crucial for measuring intravascular transverse blood-flow. If no correction for noise is applied, an overestimation of the blood velocity will occur. Information on the noise level is already present in the signals themselves[24]. When a burst of ultrasound signals is transmitted towards flowing blood the signals will be decorrelated as function time. In a noise free environment this decorrelation function has the value of one at time leg zero. By fitting a curve through the data, the effect of noise on the decorrelation is determined. In presence of noise the value at leg zero will be slightly lower than one. By correcting the correlation function for this difference and then applying the methods above the influence of noise can be eliminated. Since the decorrelation due to noise is estimated and corrected per window, the noise correction method is dynamic and not dependent on previous measurements.

6) The potential influence of aggregation on decorrelation has been studied using computer models[25] and in vitro measurements on blood in a couette system[26]. First, the decorrelation patterns for a scattering medium formed by single red blood cells (RBCs) and aggregates of RBCs were studied and compared with the auto-convolution of the acoustical beam. The RF decorrelation patterns were in agreement with the reference curve when the scattering medium was formed by single RBCs. When aggregates of RBCs were aligned with the flow, the decorrelation pattern showed to be influenced by the length of the aggregate s. The longer the aggregates of RBCs the lower the rate of decorrelation; this would result in an underestimation of the blood flow velocity. However by a more realistic situation, aggregates of RBCs at random angles and lengths. *In vitro* measurements using a couette system on aggregation of RBCs also showed close agreements between the mean decorrelation pattern and the theoretical values at all shear rates, for the normal and disease blood groups, except during very low shear rates (1 and 5 s^{-1}) and for high aggregating blood (i.e. large aggregates of RBCs)

7) The influence on decorrelation by blood velocity profiles such as random spread of flow, linear velocity gradients and parabolic flow was also studied by means of computer simulations[27]. This study was performed for the RF and envelope signals and for transverse and axial flow. The envelope mean decorrelation pattern for axial spread of blood velocity was faster than the one for axial plug flow. For the RF and envelope signals, transverse blood-flow is not significantly influenced by random spread of flow velocity, linear velocity gradient or parabolic flow. Similar results were found for axial flow and for the RF signals when random spread flow was considered.

In summary: all decorrelation factor described above can be controlled in a post procedural situation.

Implementation

The methods described above have been implemented on a rotating single element IVUS system (DuMed/Endosonics Inc) and on an array system (JOMED Inc). In both cases the systems have been adapted for this purpose. The rotational speed of the Dumed catheter has been lowered slightly and the repetition rate of the pulse had been increased. The flow of the array catheter has been acquired using the catheter in linear array mode. In this mode four elements are excited simultaneously and the same elements are used for receiving. In both cases the RF signals are acquired from the Ultrasound machine, and the data processing is performed on an external workstation.

Results

The methods have been validated in a number of ways.

Single element catheter

In vitro data showed good agreement with a calibrated electromagnetic flow meter [13]. In the carotid artery of a pig the IVUS flow showed good agreement with Electromagnetic flow estimation (Fig 2) [19, 20].

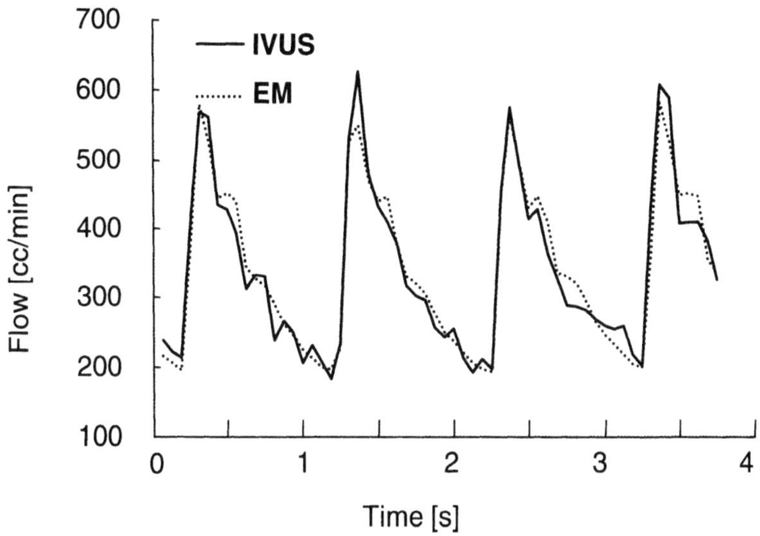

Fig 2. Flow curves of phasic EM-flow (dotted line) and IVUS-flow (solid line) in a pig carotid artery, showing an almost identical beat-to-beat change in blood volume flow. [17,18]

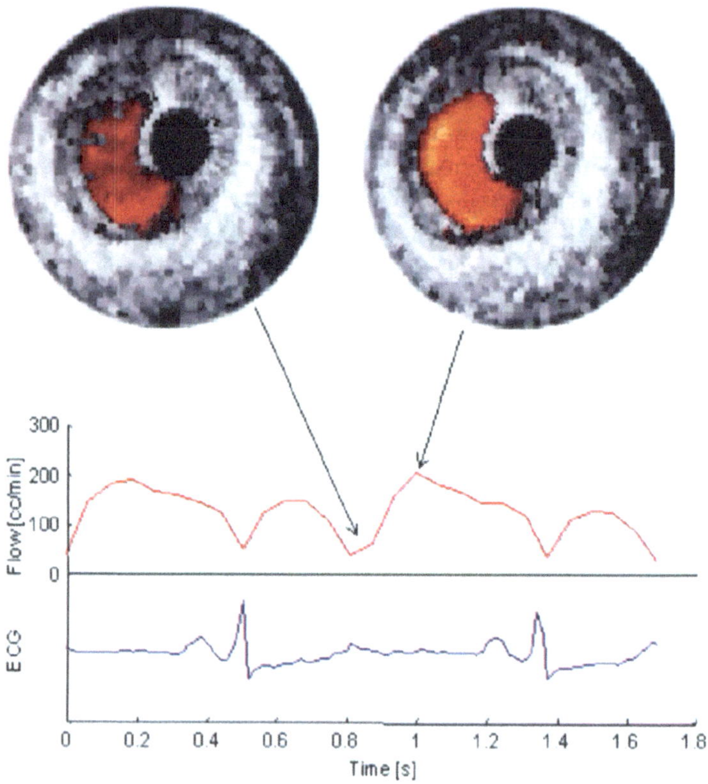

Fig 3. Upper panel: Coronary flow mapping obtained in patient for low (left) and high (right) flow rates during the cardiac cycle. Lower panel: IVUS flow curve showing a typical 2-phasic diastolic flow pattern of a coronary artery.

The relation under a variety of flow conditions approached unity. In patient studies the IVUS flow is compared to the values obtained by a Doppler flow wire [28, 29]. Again a good agreement was found. This method has been evaluated extensively in the Catheterization laboratory [19, 23] (Fig 3).

Array catheter

IVUS quantitative blood flow measurements using an array were also performed in Yorkshire pigs. These experiments showed the correlation-based method capable to measure quantitative volume blood flow *in vivo*, which means blood flow measurements performed in an artery, under pulsatile flow, and the inevitable motion of the catheter due to breathing and heart beating (from 100 to 120 beats/min) (Fig. 4)[30].

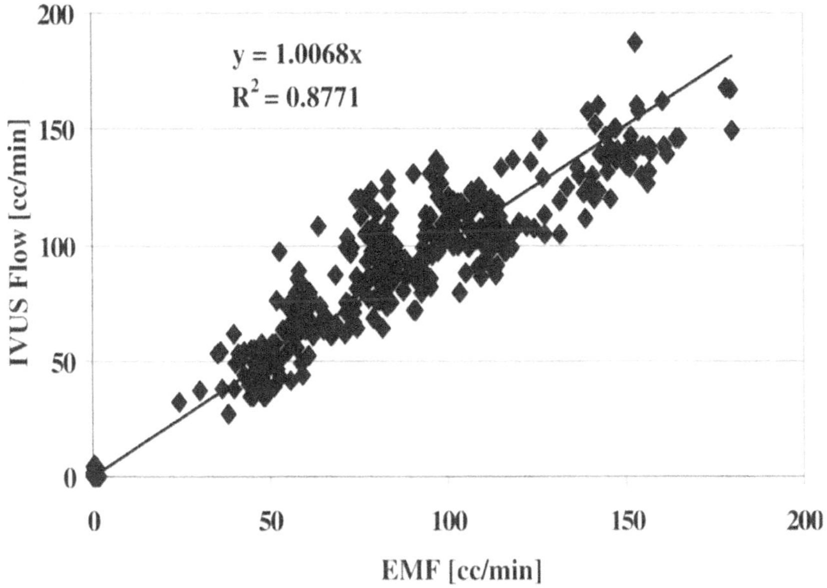

Fig 4. IVUS flow measurements in a Yorkshire pig compared to a calibrated Electromagnetic Flowmeter (EMF).

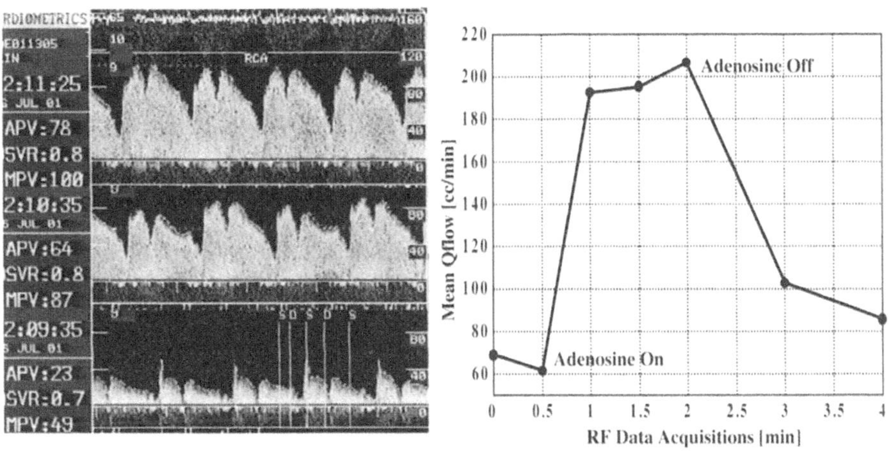

Fig 5. Coronary Flow Reserve (CFR) in a patient during a hyperemia cycle, left panel: Doppler velocity measurements (CFR of 3.3), right panel: IVUS blood flow measurements (CFR of 3.0).

Finally, the technique was advanced to coronary application in patients. Coronary Flow Reserve (CFR) is a useful index to evaluate the outcome of an intervention such as a stent implantation. CFR is estimated from the ratio of the maximum flow at peak hyperaemia to the baseline flow. The hyperaemia is induced by the injection of adenosine. Results for CFR practiced in patients based on IVUS blood flow measurements showed a good relationship when compared to those based on Doppler blood velocity measurements (Fig. 5). The results presented demonstrate the feasibility of IVUS blood flow using an array catheter to be applied in a clinical environment.

Acknowledgement: This project was financially supported by The Dutch Technology Foundation (STW), NWO, and a research grant from Endosonics/JOMED Inc.

Reference

1. J. Baptista, C. Di Mario, Y. Ozaki, J. Escaned, R. Gil, P. J. de Feyter, J. R. T. C. Roelandt, and P. W. Serruys, "Impact of plaque morphology and composition on the mechanisms of lumen enlargement using intracoronary ultrasound and quantitative angiography after balloon angioplasty," *American Journal of Cardiology*, vol. 77, pp. 115-121, 1996.
2. G. S. Mintz, J. J. Popma, K. M. Kent, L. F. Satler, S. C. Wong, M. K. Hong, J. A. Kovach, and M. B. Leon, "Arterial remodeling after coronary angioplasty. A serial intravascular ultrasound study," *Circulation*, vol. 94, pp. 35-43, 1996.
3. N. Bom, W. Li, A. F. W. van der Steen, C. T. Lancée, E. I. Céspedes, C. J. Slager, and C. L. de Korte, "Novel developments in intravascular imaging,"*European Journal of Ultrasound*, vol. 7, pp. 9-14, 1998.
4. A. F. W. van der Steen, E. I. Cespedes, C. L. de Korte, S. G. Carlier, W. Li, F. Mastik, C. T. Lancee, J. Borsboom, F. A. Lupotti, R. Krams, P. W. Serruys, and N. Bom, "Novel developments in intravascular imaging," *IEEE ultrasonics symposium*, pp. 1733-1742, 1998.
5. T. Spencer, M. P. Ramo, D. M. Salter, T. Anderson, P. P. Kearney, G. R. Sutherland, K. A. A. Fox, and W. N. McDicken, "Assessment of regional vascular distensibility in deceased iliofemoral arteries by intravascular ultrasound," *Ultrasound in Medicine and Biology*, vol. 20, pp. 529-542, 1997.
6. S. L. Bridal, P. Fornes, P. Bruneval, and G. Berger, "Correlation of ultrasonic attenuation (30 to 50 MHz) and constituents of atherosclerotic plaque," *Ultrasound in Medicine and Biology*, vol. 23, pp. 691-703, 1997.
7. A Nair, BD Kuban, EM Tuzcu, P Schoenhagen, SE Nissen, and D. Vince., "Coronary plaque classification with intravascular ultrasound radiofrequency data analysis," *Circulation*, vol. 106, pp. 2200-2206, 2002.
8. C. L. de Korte, S. G. Carlier, F. Mastik, M. M. Doyley, A. F. W. van der Steen, P. W. Serruys, and N. Bom, "Morphological and mechanical information of coronary arteries obtained with Intravascular elastography: *a feasibility study in vivo.*," *European Heart Journal*, vol. 23, pp. 405-413, 2002.

9. E. I. Céspedes, C. L. de Korte, and A. F. W. van der Steen, "Intraluminal ultrasonic palpation: assessment of local and cross-sectional tissue stiffness," *Ultrasound in Medicine and Biology*, vol. 26, pp. 385-396, 2000.
10. A. F. W. van der Steen, C. L. de Korte, and E. I. Cespedes, "Intravascular Ultrasound Elastography," *Ultraschal in Med.*, vol. 19, pp. 196-201, 1998.
11. E. Brusseau, C. Perrey, P. Delachartre, M. Vogt, D. Vray, and H. Ermert, "Axial strain imaging using a local estimation of the scaling factor from RF ultrasound signals.," *Ultrasonic Imaging*, vol. 22, pp. 95-107, 2000.
12. C. de Korte, "intravascular ultrasound elastography," *PhD Thesis Erasamus University Rotterdam*, 1999.
13. W. Li, A. F. W. van der Steen, C. T. Lancee, E. I. Cespedes, and N. Bom, "Blood flow imaging and volume flow quantitation with intravascular ultrasound," *Ultrasound in Medicine and Biology*, vol. 24, pp. 203-214, 1998.
14. A. F. W. van der Steen, E. I. Céspedes, S. G. Carlier, F. Mastik, F. A. Lupotti, J. M. G. Borsboom, W. Li, P. W. Serruys, and N. Bom, "Flow estimation using an intravascular imaging catheter," *Ultrasonics*, vol. 38, pp. 363-368, 2000.
15. W. Li, A. F. W. van der Steen, and C. T. Lancee, "Temporal correlation of blood scattering signals in vivo from radio frequency intravascular ultrasound," *Ultrasound in Medicine and Biology*, vol. 22, pp. 583-90, 1996.
16. J. W. Doucette, P. D. Corl, H. M. Payne, A. E. Flynn, M. Goto, M. Nassi, and J. Segal, "Validation of a Doppler guide wire for intravascular measurements of coronary artery flow velocity," *Circulation*, vol. 85, pp. 1899-1911, 1992.
17. S. G. Carlier, C. Di Mario, M. Kern, and P. W. Serruys, "Intracoronary Doppler and pressure monitoring," in *Textbook of Interventional Cardiology (3rd ed)*, E. J. Topol, Ed. Philadelphia: W.B.Saunders Company, 1998, pp. (in press).
18. C. di Mario, S. H. K. The, S. Madrestma, R. J. van Suylen, R. A. Wilson, N. Bom, P. W. Serruys, E. J. Gussenhoven, and J. R. T. C. Roelandt, "Detection and characterization of vascular lesions by intravascular ultrasound: an in vitro study correlated with histology," *Journal of American Society Echocardiography*, vol. 5, pp. 135-146, 1992.
19. T. M. Chou, K. Sudhir, S. Iwanaga, K. Chatterjee, and P. G. Yock, "Measurement of volumetric coronary blood flow by simultaneous intravascular twodimensional and Doppler ultrasound: validation in an animal model.," *Am Heart J*, vol. 128, pp. 237-43, 1994.
20. J. M. Isner, J. Kaufman, K. Rosenfield, A. Pieczek, R. Schainfeld, K. Ramaswamy, and B. D. Kosowsky, "Combined physiologic and anatomic assessment of percutaneous revascularization using a Doppler guidewire and ultrasound catheter.," *Am J Cardiol*, vol. 71, pp. 70D-86D, 1993.
21. W. Li, C. T. Lancée, E. I. Céspedes, A. F. W. van der Steen, and N. Bom, "Decorrelation of intravascular echo signals: Potentials for blood velocity estimation," *Journal of the Acoustical Society of America*, vol. 102, pp. 3785-3794, 1997.
22. J. J. Wentzel, R. Krams, A. F. W. van der Steen, W. Li, E. I. Céspedes, N. Bom, and C. J. Slager, "Disturbance of 3D velocity profiles induced by an IVUS catheter: evaluation with computational fluid dynamics.," *Computers in Cardiology*, vol. 24, pp. 597-600, 1997.
23. R. Krams, J. Wentzel, E. I. Cespedes, R. Vinke, S. G. Carleir, A. F. W. van der Steen, C. T. Lancee, and C. J. Slager, "Effect of catheter placement on 3D velocity profiles in curved tubes," *ultrasound in med and biol*, vol. 25, pp. 803-810, 1999.

24. F. A. Lupotti, C. L. de Korte, F. Mastik, and A. F. W. van der Steen, "Dynamic Noise Correction for IVUS Quantitative Volume Blood Flow: Methods and numerical validation," *Ultrasound in Medicine and Biology*, vol. 28, pp. 1053-1060, 2002.
25. F. A. Lupotti, E. I. Céspedes, and A. F. W. van der Steen, "Decorrelation Characteristics of Transverse Blood Flow along an Intravascular Array Catheter: Effects of Aggregation of Red Blood Cells," *Ultrasound in Medicine and Biology*, vol. 27, pp. 409-417, 2001.
26. F. A. Lupotti, A. Zimmer, M. Daronat, F. S. Foster, A. F. W. van der Steen, and G. Cloutier, "Effects of Aggregation of Red Cells and Linear Velocity Gradients on the Correlation-based method for Quantitative IVUS Blood Flow at 20 MHz," *submitted in Ultrasound in Medicine and Biology*, 2002.
27. F. A. Lupotti, A. F. W. van der Steen, F. Mastik, and C. L. de Korte, "Decorrelation based blood flow velocity estimation: Spread of Flow Velocity, Linear Flow Velocity Gradients and Parabolic Flow," *IEEE Transactions on Ultrasonics, Ferroelectrics, and Frequency Control*, vol. 49, pp. 705-714, 2002.
28. S. G. Carlier, W. Li, E. I. Cespedes, A. F. W. van der Steen, J. N. Hamburger, N. Bom, and P. W. Serruys, "Simultaneous morphological and functional assessment of a renal artery stent intervention with intravascular ultrasound," *Circulation*, vol. 97, pp. 2575-6, 1998.
29. S. G. Carlier, E. I. Céspedes, W. Li, A. F. W. van der Steen, J. R. van Meegen, F. Mastik, J. Honkoop, P. W. Serruys, and K. Bom, "In vivo validation of a new intracoronary volumetric blood flow meaurement method with intravascular ultrasound," *Circulation*, vol. 96, pp. I-703 (abst), 1997.
30. F. A. Lupotti, F. Mastik, C. L. de Korte, S. G. Carlier, W. J. van der Giessen, and A. F. W. van der Steen, "Quantitative IVUS blood flow: varidation in vitro, in animals and in patients." *Ultrasound Med Biol*, 2003 Apr: 29(4): 507-15

Intravascular Elastography: from idea to clinical tool

C.L. de Korte[1,2], PhD; F. Mastik[1,2]; J.A. Schaar[1,2], MD; P.W. Serruys[1], MD, PhD; A.F.W. van der Steen[1,2], PhD

[1] Thoraxcenter, Erasmus Medical Center, Rotterdam, The Netherlands
[2] Interuniversity Cardiology Institute of the Netherlands, Utrecht, The Netherlands

Abstract: The composition and morphology of the atherosclerotic lesion are currently considered more important determinants of acute coronary ischemic syndromes than the degree of stenosis. Especially identification of lipid pools remains limited using conventional intravascular ultrasound (IVUS). IVUS elastography, a new technique to assess the local mechanical properties of the vessel wall and plaque is discussed in this paper.

The underlying principle of elastography is that the deformation of tissue by a mechanical excitation is a function of its mechanical properties. The deformation of the tissue is determined using ultrasound. For intravascular purposes, the intraluminal pressure is used as the excitation force. The radial strain in the tissue is obtained by cross-correlation techniques on the radio frequency (rf) signal. The strain is color-coded and plotted as a complimentary image to the IVUS echogram.

IVUS elastography was validated *in vitro* using diseased human coronary and femoral arteries. These experiments demonstrate that the strain in fibrous plaques is different from the strain in fatty plaques ($p<0.001$). *In vivo,* in an atherosclerotic Yucatan minipig animal model also higher strain values were found in fatty material ($p<0.001$). Preliminary patient data revealed low strain values in calcified material and stented lesions. High strain values were found at the shoulders of plaques.

In conclusion, intravascular elastography appears to be a unique tool to determine local mechanical properties in atherosclerotic lesions to identify fibrous and fatty tissue. Experiments have demonstrated the feasibility of this technique to be applied *in vivo.*

Keywords: Intravascular, Ultrasound, Elastography, Vulnerable, Plaque, Stress, Strain, Tissue characterisation.

Introduction

IntraVascular UltraSound (IVUS) is the only commercially available clinical technique providing real-time cross-sectional images of the coronary artery in patients [1]. Using the morphology of the coronary wall and plaque are obtained. Furthermore, calcified and non-calcified plaque components can be identified. However, the sensitivity to identify fatty plaque components remains low [2, 3]. Recent rf-based tissue identification strategies appear to have better performance [3, 4].

Identification of different plaque components is of crucial importance to detect the rupture prone plaque. Typical parameters of these plaques are an eccentric plaque with a large atheroma covered by a thin fibrous cap [5, 6]. The presence of macrophages in the cap further increases the vulnerability of these plaques [7]. It is known that the mechanical properties of fibrous and fatty plaque components are different [8-10]. Furthermore, fibrous caps with inflammation by macrophages are locally weakened [11]. All the stress that is applied on the plaque by the blood is concentrated in the cap since the lipid pool is unable to withstand forces on it [12, 13]. The cap will rupture if it is unable to withstand the stress applied on it.

In 1991, Ophir et al. [14] proposed a method to measure the strain in tissue using ultrasound. The tissue was strained by applying an external force on it. Different strain values were found in tissues with different mechanical properties. Implementing this method for intravascular purposes has potential to:

1. Identify different plaque components.
2. Identify high strain regions due to increased circumferential stress.

In this chapter, different strategies to image the elastic properties of tissue are discussed. Different methods to apply intravascular elastography are discussed. The work of different groups to apply the technique on phantoms, in vitro, and in animals and patients are described.

Methods for elasticity imaging using ultrasound

Several techniques have been proposed to study the mechanical properties of tissues using ultrasound. The main differences are in the way of excitation (vibration at low or high frequency or (quasi-)static deformation).

Elastography

Ophir and colleagues [14, 15] have developed an imaging technique called elastography, which is based on (quasi-)static deformation of a linear isotropic elastic material (fig. 1). The tissue under inspection is deformed and the strain (direct or indirect using displacement) between pairs of ultrasound signals with and without deformation is determined. [16].

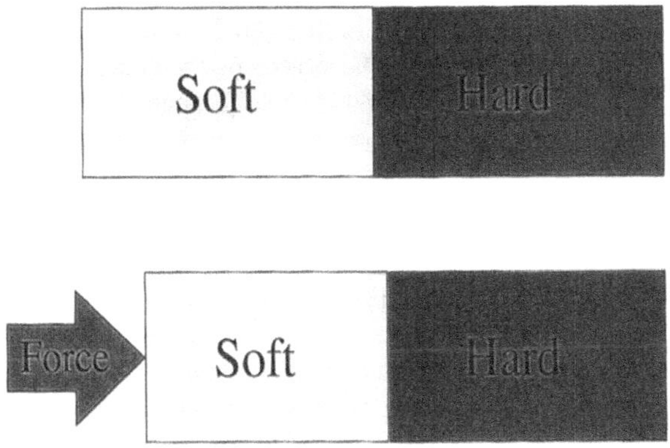

Figure 1. Principle of elastography. The deformation of tissue is a function of its mechanical properties. When a force is applied, the soft tissue deforms more than the hard tissue.

Sonoelasticity

Lerner and Parker [17] presented preliminary work on vibration amplitude 'sonoelasticity imaging'. With this method, a low-frequency vibration (20–500 Hz) is applied externally to excite internal vibrations within the tissue under study. A stiff inhomogeneity surrounded by soft tissue produces a disturbance in the normal vibration eigenmode patterns. Doppler detection techniques are employed to make a real-time vibration image. In some organs, modal patterns can be created, revealing additional information to the shear wave speed of sound of tissue [18]. Recently, Sandrin et al. have reported the results obtained in phantoms of a new method based on such a low-frequency shear wave (200 Hz) visualized by means of an ultrafast imaging system (up to 10,000 frames/s) composed of a linear array of transducers at 3.5 MHz [19].

Vibro-acoustography

Fatemi et al. [20, 21] described a method using acoustics as a mechanical excitation. An oscillatory radiation force is produced by interfering focused ultrasound beams. The resulting acoustic emission is directly related to the mechanical frequency response of the object in the focus of the beams. This response is measured using a hydrophone. A 2D image is formed using raster-scanning. This technique offers high resolution, speckle free images of superficial structures.

Intravascular Elastography

Although Ophir *et al* never explored the quasi-static approach for intravascular purposes this approach seems to be the most fruitful concept. The compression can be obtained from the systemic pressure difference that is already available in intravascular applications. Additionally, well-controlled deformation is possible by using a compliant intravascular balloon [22]. Sonoelasticity imaging has not been evaluated for vascular tissue, but given the large wavelengths and the level of inhomogeneity in a diseased vessel, it seems not to be the technique of first choice. Vibro-acoustography has limited suitability for intravascular applications since it requires a relative long acquisition time of a non-moving target.

The principle of intravascular elastography is illustrated in figure 2. An ultrasound image of a vessel-phantom with a hard vessel wall and a soft eccentric plaque is acquired at a low pressure. In this case, there is no difference in echogenicity between the vessel wall and the plaque resulting in a homogeneous IVUS echogram. A second acquistion at a higher intraluminal pressure (approx 5 mmHg) is performed. The elastogram (image of the radial strain) is plotted as a complimentary image to the IVUS echogram. The elastogram reveals the presence of an eccentric region with increased strain values thus identifying the soft eccentric plaque. The different strategies to perform intravascular elastography are in the way of detecting the strain and the source of deforming the vascular tissue.

Detection methods

Envelope based methods.

Two groups worked on intravascular elastography using the envelop of the ultrasound signal. Talhami *et al.* [23] introduced a technique to assess the strain that is based on the Fourier scaling property of the signals and uses the chirp Z-transform. The scaling property directly represents the strain in the tissue. The chirp-Z transform was determined from the envelope signal to overcome decorrelation of the rf-signal due to deformation of the tissue data are used for this analysis. The result is displayed as a colour-coded ring around the image of the vessel. The technique seems relatively easy to implement. Initial results on vascular tissue *in vitro* and *in vivo* are described. The technique was not further developed and validated.

Ryan and Foster [24] developed a technique based on speckle tracking in video signals. The strain can be determined from these displacement estimates. The technique was tested using vessel mimicking phantoms. It was shown that in a phantom that was partly made of soft and partly made of hard material the displacement in soft material was larger than the displacement in hard material.

An advantage of envelope-based methods is the fact that the correlation function is smoother than the rf-based correlation function. This prevents 'peak hopping', meaning that the correlation function is maximized around the wrong peak. This makes the method less noise sensitive. Furthermore, the video signal is commonly available from any commercial echosystem. A disadvantage is the limited resolution and the low sensitivity of the method for low strain values. Since small tissue strains are expected for intravascular applications, and the arterial wall is relatively thin, it is expected that the use of the high frequency rf signal will seriously improve the resolution. Based on work of Varghese *et al.* [25], a smaller variance of the strain estimate is expected using RF data instead of envelope data.

RF based methods

Shapo *et al* [26, 27] developed a technique based on cross-correlation of A-lines. The group proposes a large deformation to maximize the signal-to-noise ratio of the displacement and strain estimation. Since the deformation that is needed is larger than the deformation that occurs in arteries in vivo, the tissue is deformed using a non-compliant balloon that is inflated up to 8 atm. Large displacement will decorrelate the ultrasound signals to such an extent that correlation detection is unreliable. For this reason, the cross-correlation is calculated in several intermediate steps of intraluminal pressure. For detection, they use a phase-sensitive speckle tracking technique. The technique was demonstrated in simulations and tissue mimicking phantoms. Recent this group [28] presented data on a compliant balloon containing an intravascular catheter. The compliant balloon is inflated to 2 atm. to obtain strain values up to 40%. This method was tested in phantoms and in vitro. The phantom (with one half soft and one half hard material) revealed strains from 20-40% in the soft part and strains lower than 20% in the hard part. These results were corroborated by finite element analysis. In vitro, low strain values (7%) were found for fibrous tissue in the human femoral artery and high strain values (35%) were found in thrombus in a rabbit aorta.

De Korte et al [29, 30] incorporated the 'classical' way of doing elastography for intravascular purposes. The vascular tissue is strained by different levels of the intraluminal pressure. The processing is visualised in figure 3. The local displacement of the tissue is determined using cross-correlation analysis of the

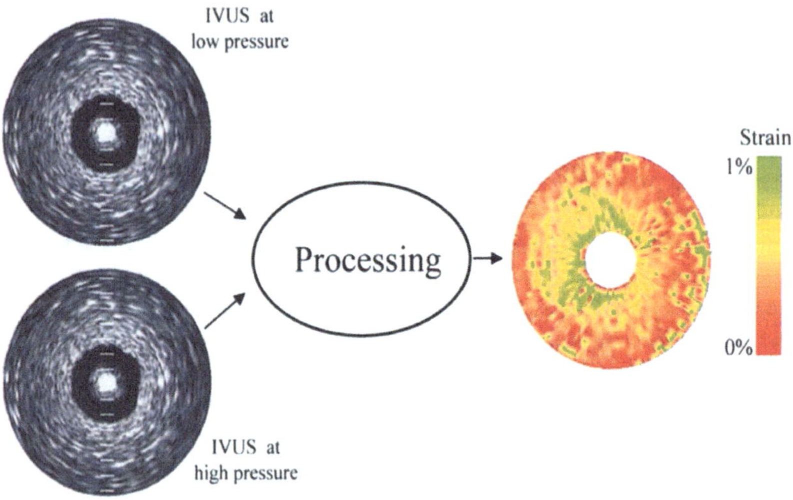

Figure 2. Schematic representation of elastographic acquisition. Two echo images are acquired with different intraluminal pressures: one at low pressure (left bottom) and one at high pressure (left top). The incremental deformation of the tissue is determined using cross-correlation analysis (see figure 3). The strain is plotted in a colour coded image.

gated rf-signals. The finite difference is applied to determine the strain. The strain is determined for the full vessel wall and plaque and plotted as a complimentary image to the IVUS echogram or the strain is determined in a significant part of the vessel wall and plotted as a colour-coded ring (palpogram) at the lumen vessel-wall boundary. [31]. The method was validated using vessel phantoms with the morphology of an artery with an eccentric soft or hard plaque [32]. The plaque could be clearly identified from the vessel wall using the elastogram, independently of the echogenicity contrast between vesselwall and plaque (fig 2). The technique was validated in vitro and tested in atherosclerotic animal models and during interventions in patients.

This cross-correlation based technique is especially suited for strain values smaller than 2.5%. These strain values are present during in vivo acquisitions when only a part of the heart cycle is used to strain the tissue [33]. The maximum strain that will be present between the systolic and diastolic pressure is in the order of 10 %. For these strain rates, another approach that takes into account the change in shape of the signals can be applied. This 'local scaling factor estimation' technique [34] has been recently described for intravascular purposes [35] and has proven to be more robust to large deformations. The signal after compression is processed as a delayed and scaled replica of the signal before deformation. An adaptative strain estimation method based on the computation of local scaling

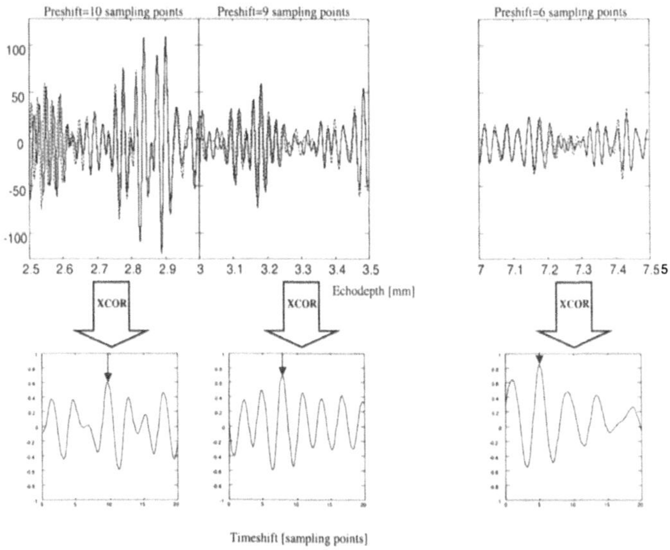

Figure 3. Principle of time delay estimation using the peak of the crosscorrelation coefficient function. In the upper part, both the rf-traces (with the 2nd trace preshifted for better visual comparison) are shown for windows with increasing echodepth. In the lower part, the corresponding crosscorrelation coefficient function for each window is plotted, showing a decreasing position of the peak with increasing echodepth (From de Korte et al. Ultraound Med. Biol. 23(5): 735-746; 1997 with permission).

factors has been applied to compute elastograms of cryogel phantoms mimicking vessels and of a freshly excised human carotid artery using a 30 MHz mechanical rotating single element ultrasound scanner (ClearView, CVIS, Boston Scientific Corp.) [36].

In vitro validation

De Korte et al [37] performed a validation study on excised human coronary (4) and femoral (9) arteries. Data were acquired at room temperature at intraluminal pressures of 80 and 100mmHg. Coronary arteries were measured using a solid state 20 MHz array catheter (EndoSonics, Rancho Cordova, CA, USA). Femoral arteries were investigated using a single element 30 MHZ catheter (DuMed/ EndoSonics, Rijswijk, The Netherlands) that was connected to a modified motorunit (containing the pulser and receiver and a stepper-motor to rotate the catheter). The rf-data was stored and processed off-line. The visualized segments were stained on the presence of collagen, smooth muscle cells and macrophages. Matching of elastographic data and histology was performed using the IVUS echogram. The cross-sections were segmented in regions (n=125) based on the strain value on the elastogram. The dominant plaque types in these regions (fibrous, fibro-fatty or fatty) were obtained from histology and correlated with the average strain and echo-intensity.

A typical example of a coronary artery is shown in figure 4. The strain in the plaque is low indicating hard material. The histology corroborates this finding since the plaque is composed of fibrous material (collagen and smooth muscle cells). Increased strain values at the shoulder indicate soft material. This shoulder has an increased macrophage concentration and presence of foam cells.

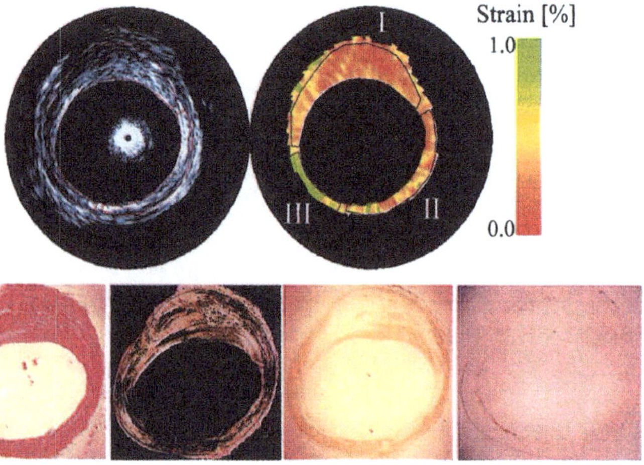

Figure 4. Intravascular echogram (a) and elastogram (b) of a diseased human femoral artery with the corresponding histology: (c)Picro-sirius-red, (d) Anti-alpha actin and (e) anti cd-68 antibody. The echogram reveals an eccentric plaque between 9 and 3 o'clock. The elastogram reveals low strain in the plaque (0.24%), a similar strain in the non-diseased vessel wall between 3 and 7 o'clock (0.32%) and a high strain in the vessel wall between 7 and 9 o'clock (0.96%). The histology reveals a fibrous composition of the plaque (c & d). The region with high strain contains fatty foam cells at the lumen-vessel wall boundary and an increased macrophage activity (e) indicating this region as an erosive plaque region (from de Korte et al. Circulation 102(6): 617-623; 2001 with permission).

Mean strain values of 0.27%, 0.45% and 0.60% were found for fibrous, fibro/fatty and fatty plaque components. The strain for the three plaque types as determined by histology differed significantly (p=0.0002). This difference was independent on the type of artery (coronary or femoral) and was mainly evident between fibrous and fatty tissue (p=0.0004). The plaque types did not reveal echo-intensity differences in the IVUS echogram (p=0.992). Conversion of the strain into Young's modulus values resulted in 493kPa, 296 kPa and 222 kPa for fibrous, fibro/fatty and fatty plaques. Although these values are higher than values measured by Lee et al [10], the ratio between fibrous and fatty material is similar. Since fibrous and fatty tissue demonstrated a different strain value and high strain values were often co-localised with increased concentrations of macrophages, these results reveal the potential of identification of the vulnerable plaque.

Animal studies

Recently, IVUS elastography was validated in vivo using an atherosclerotic Yucatan minipig [38, 39]. External iliac and femoral arteries were made atherosclerotic by endothelial Fogarty denudation and subsequent atherosclerotic diet for the duration of 7 months. Balloon dilation was performed in the femoral arteries and the diet was discontinued. Before termination, 6 weeks after balloon dilation and discontinuation of the diet, data were acquired in the external iliac and femoral artery in 6 yucatan pigs. In total, 20 cross-sections were investigated with a 20 MHz Visions® catheter (Jomed, Rancho Cordova, CA). The tissue was strained by the pulsatile blood pressure. Two frames acquired at end diastole with a pressure differential of approx. 4 mmHg were taken to determine the elastograms.

After the ultrasound experiments and before dissection, X-ray was used to identify the arterial segments that had been investigated by ultrasound. The specimen were frozen in liquid nitrogen. The cross-sections (7μm) were stained for collagen (picro Sirius red and polarized light), fat (oil red O) and macrophages (alcalic phosphatase). Plaques were classified as absent, as early fibrous lesion, as early fatty lesion or as advanced fibrous plaque. The mean strain in these plaques and normal cross-sections was determined.

Strains were similar in the plaque free arterial wall and the early and advanced fibrous plaques. Univariate Analysis of Variance revealed significantly higher strain values in cross-sections with early fatty lesions than in fibrous plaques (p=0.02) independently on the presence of macrophages. Although a higher strain value was found in plaques with macrophages than in plaques without macrophages, this difference was not significant after correction for fatty components.

Patient studies

Preliminary acquisitions were performed in patients during percutaneous transluminal coronary angioplasty (PTCA) procedures [33, 40]. Data were acquired in patients (n=12) with an EndoSonics InVision echoapparatus equipped with an rf-output. For obtaining the rf-data, the machine was working in ChromaFlo mode resulting in images of 64 angles with unfocussed ultrasound data. The systemic pressure was used to strain the tissue. This strain was determined using crosscorrelation analysis of sequential frames. A likelihood function was determined to obtain the frames with minimal motion of the catheter in the lumen, since motion of the catheter prevents reliable strain estimation. Minimal motion was observed near end-diastole. Reproducible strain estimates were obtained within one pressure cycle and over several pressure cycles. Validation of the results was limited to the information provided by the echogram. Strain in calcified material (0.20%) was lower ($p<0.001$) than in non-calcified tissue (0.51%).

Recently, high-resolution elastograms were acquired using an EndoSonics InVision echoapparatus [41]. The beam-formed image mode (512 angles) ultrasound data (F_c = 20 MHz) was acquired with a PC based acquisition system. Frames acquired at end-diastole with a pressure difference of approx. 5 mmHg were taken to determine the elastograms.

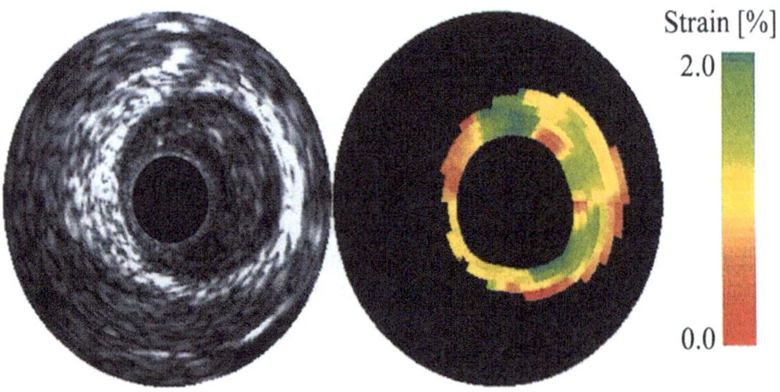

Figure 5. IVUS echogram and elastogram of coronary artery with eccentric plaque obtained in a patient. The elastogram reveals low strain in the calcified region (as identified from the echogram by the bright echo with distal shadowing) and moderate to high strain in the shoulders of the plaque.

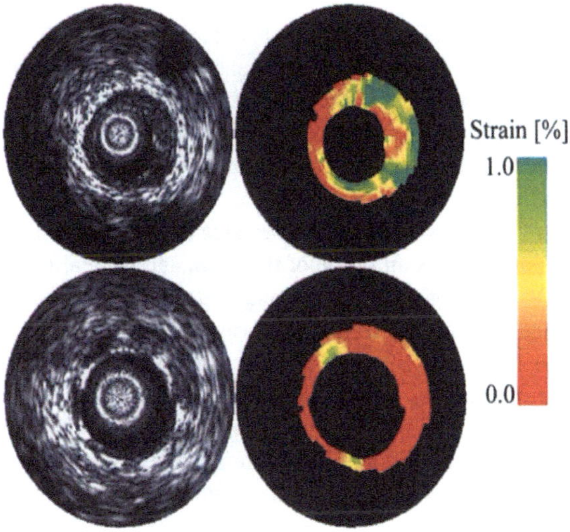

Figure 6. Elastogram and echograms of a mixed plaque pre and post intervention. The elastogram pre-intervention (upper row) reveals a mixed composition of the plaque. Low and high strain values are observed in the core of the plaque. Especially at the shoulders increased strain values were found. After stenting (bottom row), low strain values were observed. Due to the deformation of the material and the struts of the stent, the plaque is not strained by the pulsatile intracoronary pressure. At the shoulders of the plaque and looking between two struts, high strain spots were still present.

Typical examples of elastograms obtained in patients are presented in the figures 5 and 6. The elastogram of a mixed plaque reveals low strain values in the centre of the plaque. The echogram reveals a large calcification in this area. Increased strain values were observed at the shoulders of the plaque. In figure 6, an elastogram obtained pre-intervention is shown. The plaque contains soft and harder material identifying the plaque as being composed of soft tissue. The elastogram obtained after placing a stent revealed very low strain values, except for two regions: these are between the stent struts and at the shoulders of the plaque.

Discussion

Identification of plaque components and the proneness of a lesion to rupture is a major issue in interventional cardiology. Intravascular ultrasound echography is a real-time, clinical available technique capable of providing cross-sectional images and identifying calcified plaque components. Since elastography only requires ultrasound data sets that are acquired at different levels of intraluminal pressure, it can be realised using conventional catheters. It has been shown by several groups that elastograms of vessel-like phantoms and arteries in vitro can be produced. Furthermore, the feasibility of IVUS elastography in vivo in animals and patients was demonstrated.

The question still not answered is the relevance of the information given by the elastogram. An elastogram is an image of the strain and is therefore an artefactual representation of the Young's modulus. Using finite element analysis, an image of the Young's modulus can be reconstructed using the strain and/or displacement information (also known as solving the inverse problem) [42, 43]. Analysis of more complex geometries has to be performed to identify the differences between and the similarities of strain and modulus images. Currently, the inverse problem is solved for strain images acquired from human artery specimen in vitro. The resulting Young's modulus images and the strain images will be related to the histology.

There is no clinical available technique capable of identifying the rupture prone plaque. Identification of these plaques is of paramount importance to investigate the underlying principle of plaque rupture, the effectiveness of pharmaceutical treatments and on the long-term preventing sudden cardiac deaths. IVUS elastography has proven to be able to identify the rupture prone plaque in vitro with high sensitivity and specificity and in vivo experiments demonstrate the power to identify fibrous and fatty plaque components. Additionally, in animals a high sensitivity and specificity to identify fatty tissue and macrophages was found. Therefore, IVUS elastography may be one of the first techniques that can be applied in patients to assess the vulnerability of plaques.

Conclusion

Intravascular elastography has proven to be technique capable of providing information on the plaque composition. IVUS elastography was tested in vitro in several institutes using different IVUS systems, different sources of mechanical stimulus and different processing techniques. Experiments in vivo in animals and patients demonstrate that IVUS elastography may develop into a clinical available tool to identify the rupture prone plaque.

References

1. Mintz GS, Nissen SE, Anderson WD, et al (2001) ACC Clinical Expert Consensus Document on Standards for Acquisition, Measurement and Reporting of Intravascular Ultrasound Studies (IVUS). A report of the American College of Cardiology Task Force

on Clinical Expert Consensus Documents. Journal of the American College of Cardiology 37(5):1478-92.

2. Prati F, Arbustini E, Labellarte A, et al (2001) Correlation between high frequency intravascular ultrasound and histomorphology in human coronary arteries. Heart 85:567-570.
3. Komiyama N, Berry G, Kolz M, et al (2000) Tissue characterization of atherosclerotic plaques by intravascular ultrasound radiofrequency signal analysis: An in vitro study of human coronary arteries. American Heart Journal 140:565-574.
4. Hiro T, Fujii T, Yasumoto K, et al (2001) Detection of Fibrous Cap in Atherosclerotic Plaque by Intravascular Ultrasound by Use of Color Mapping of Angle-Dependent Echo-Intensity Variation. Circulation 103(9):1206-1211.
5. Falk E, Shah P,Fuster V (1995) Coronary plaque disruption. Circulation 92(3):657-671.
6. Fuster V, Stein B, Ambrose J, et al (1990) Atherosclerotic plaque rupture and thrombosis. Evolving concepts. Circulation 82(suppl II):II.47-II.59.
7. Moreno PR, Falk E, Palacios IF, et al (1994) Macrophage infiltration in acute coronary syndromes: implcations for plaque rupture. Circulation 90(2):775-778.
8. Loree HM, Tobias BJ, Gibson LJ, et al (1994) Mechanical properties of model atherosclerotic lesion lipid pools. Arteriosclerosis and Thrombosis 14(2):230-234.
9. Loree HM, Grodzinsky AJ, Park SY, et al (1994) Static circumferential tangential modulus of human atherosclerotic tissue. Journal of Biomechanics 27(2):195-204.
10. Lee RT, Richardson G, Loree HM, et al (1992) Prediction of mechanical properties of human atherosclerotic tissue by high-frequency intravascular ultrasound imaging. Arteriosclerosis and Thrombosis 12:1-5.
11. Lendon CL, Davies MJ, Born GVR, et al (1991) Atherosclerotic plaque caps are locally weakened when macrophage density is increased. Atherosclerosis 87:87-90.
12. Loree HM, Kamm RD, Stringfellow RG, et al (1992) Effects of fibrous cap thickness on peak circumferential stress in model atherosclerotic vessels. Circulation Research 71:850-858.
13. Richardson PD, Davies MJ,Born GVR (1989) Influence of plaque configuration and stress distribution on fissuring of coronary atherosclerotic plaques. Lancet 21:941-944.
14. Ophir J, Céspedes EI, Ponnekanti H, et al (1991) Elastography: a method for imaging the elasticity in biological tissues. Ultrasonic Imaging 13:111-134.
15. Céspedes EI, Ophir J, Ponnekanti H, et al (1993) Elastography: elasticity imaging using ultrasound with application to muscle and breast in vivo. Ultrasonic Imaging 17:73-88.
16. Céspedes EI, Huang Y, Ophir J, et al (1995) Methods for estimation of subsample time delays of digitized echo signals. Ultrasonic Imaging 17:142-171.
17. Lerner RM,Parker KJ, eds. Sonoelasticity images, ultrasonic tissue charactertization and echographic imaging. 7th European Communities Workshop, ed. J. Thijssen. 1987: Nijmegen, The Netherlands.
18. Parker KJ,Lerner RM (1992) Sonoelasticity of organs, shear waves ring a bell. Journal of Ultrasound in Medicine 11:387-392.
19. Sandrin L, Catheline S, Tanter M, et al (1999) Time-resolved pulsed elastography with ultrafast ultrasonic imaging. Ultrasonic Imaging 21:259-271.
20. Fatemi M,Greenleaf JF (2000) Probing the dynamics of tissue at low frequencies with the radiation force of ultrasound. Phys Med Biol 45(6):1449-1464.
21. Fatemi M,Greenleaf JF (1998) Ultrasound-stimulated vibro-acoustic spectrography. Science 280(5360):82-85.

22.Sarvazyan AP, Emelianov SY,Skovorada AR, (1993), Intracavity device for elasticity imaging: US patent.
23.Talhami HE, Wilson LS,Neale ML (1994) Spectral tissue strain: a new technique for imaging tissue strain using intravascular ultrasound. Ultrasound in Medicine and Biology 20(8):759-772.
24.Ryan LK,Foster FS (1997) Ultrasonic measurement of differential displacement and strain in a vascular model. Ultrasonic Imaging 19:19-38.
25.Varghese T, Ophir J, Konofagou E, et al (2001) Tradeoffs in elastographic imaging. Ultrasonic Imaging 23:216-248.
26.Shapo BM, Crowe JR, Erkamp R, et al (1996) Strain imaging of coronary arteries with intraluminal ultrasound: experiments on an inhomogeneous phantom. Ultrasonic Imaging 18:173-191.
27.Shapo BM, Crowe JR, Skovoroda AR, et al (1996) Displacement and strain imaging of coronary arteries with intraluminal ultrasound. IEEE Transactions on Ultrasonics, Ferroelectrics, and Frequency Control 43(2):234-246.
28.Choi CD, Skovoroda A, Emelianov S, et al (2000) Strain imaging of vascular pathologies using a compliant balloon catheter. IEEE Ultrasonics Symposium. Puerto Rico, USA, pp 1771-1774
29.de Korte CL, Céspedes EI, van der Steen AFW, et al (1998) Intravascular ultrasound elastography: Assessment and imaging of elastic properties of diseased arteries and vulnerable plaque. European Journal of Ultrasound 7(3):219-224.
30.de Korte C, van der Steen A, Céspedes EI, et al (2000) Characterisation of plaque components and vulnerability with intravascular ultrasound elastography. Physics in Medicine and biology 45(6):1465-1475.
31.Céspedes EI, de Korte CL,van der Steen AFW (1997) Intravascular ultrasonic palpation: assessment of local wall compliance. IEEE Ultrasonics Symposium. Toronto, Canada, pp 1079-1082
32.de Korte CL, Céspedes EI, van der Steen AFW, et al (1997) Intravascular elasticity imaging using ultrasound: feasibility studies in phantoms. Ultrasound in Medicine and Biology 23(5):735-746.
33.de Korte CL, Carlier SG, Mastik F, et al (1999) Intracoronary elastography in the catheterisation laboratory: preliminary patient results. IEEE Ultrasonics Symposium. Lake Tahoe, CA, USA, pp 1649-1652
34.Alam S, Ophir J,Konofagou E (1998) An adaptive strain estimator for elastography. IEEE Transactions on Ultrasonics, Ferroelectrics, and Frequency Control 45:461-472.
35.Brusseau E, Perrey C, Delachartre P, et al (2000) Axial strain imaging using a local estimation of the scaling factor from RF ultrasound signals. Ultrasonic Imaging 22:95-107.
36.Brusseau E, Fromageau J, Finet G, et al (2001) Axial strain imaging of intravascular data: results on polyvinyl alcohol cryogel phantoms and carotid artery. Ultrasound in Medicine and Biology 27(12): 1631-1642.
37.de Korte CL, Pasterkamp G, van der Steen AFW, et al (2000) Characterization of plaque components using intravascular ultrasound elastography in human femoral and coronary arteries in vitro. Circulation 102(6):617-623.
38.de Korte CL, Sierevogel M, Mastik F, et al (2001) Intravascular Elastography in Yucatan pigs: validation in vivo. European Heart Journal 22(abst suppl):251.

39. de Korte CL, Sierevogel M, Mastik F, et al (2002) Identification of Atherosclerotic Plaque Components with Intravascular Ultrasound Elastography in vivo: a Yucatan pig study. Circulation 105(14):1627-1630.
40. de Korte CL, Carlier SG, Mastik F, et al (2002) Morphological and mechanical information of coronary arteries obtained with Intravascular elastography: *a feasibility study in vivo.* European Heart Journal 23(5):405-413.
41. de Korte CL, Doyley MM, Carlier SG, et al (2000) High resolution IVUS elastography in patients. IEEE Ultrasonics Symposium. Puerto Rico, USA, pp 1767-1770
42. de Korte CL, Céspedes EI, van der Steen AFW, et al (1996) Image artifacts in intravascular elastography. IEEE EMBS. Amsterdam, The Netherlands, pp paper no. 685
43. Soualmi L, Bertrand M, Mongrain R, et al, (1997) Forward and inverse problems in endovascular elastography, In: S. Lees and L.A. Ferrari, (Eds) Acoustical Imaging. Plenum: New York, USA. pp 203-209.

3D ICUS

Nico Bruining, Ronald Hamers, Patrick W. Serruys and Jos R.T.C. Roelandt

Thoraxcenter, Department of Cardiology, Erasmus Medical Center and Erasmus University, Rotterdam, The Netherlands

Summary: Two-dimensional intravascular (IVUS), and intracoronary, ultrasound (ICUS) allows the study of the extent and distribution of atherosclerotic plaques by a unique visualization of both the vascular lumen and vessel wall, which is well described in other chapters in this book. However, it is difficult to acquire a three-dimensional (3D) concept of the spatial vascular structure based on a series of two-dimensional (2D) parallel slices of ICUS images (e.g. a tomographic data set). A 3D reconstruction of a tomographic data set may overcome this limitation[1-5]. Furthermore, these tomographic data sets can be used for volumetric quantification of coronary and other vessels. This can be useful to evaluate new pharmaceutical agents (progression/regression studies) and new interventional techniques (restenosis studies). This chapter mainly deals with 3D reconstruction and quantification of coronary vessels; however, the applied methodologies are used for other vessels too.

Keywords: Intracoronary ultrasound, image processing, stents, computer-assisted methods, coronary vessels, three-dimensional reconstruction

Rationale of 3D ICUS

Why 3D reconstruction of ICUS images? In clinical practice an ICUS catheter can be used to examine and interrogate the area of interest of a coronary vessel by moving the ICUS catheter forward and backwards. Since ICUS images only showing a very thin (e.g. local) cross-sectional views of the vessels, it is difficult for the observer to build up in the mind a 3D representation of the investigated segment. Digitisation of a stack of 2D views (a tomographic image data set) and producing a 3D reconstruction with computer software can help in this respect. The role of ICUS for serial studies and consequently quantitative measurements is still gaining interest. A longitudinal 3D reconstructed image plane makes it easier to find treated segments back when comparing baseline and follow-up studies. Trying to do the same limited, to 2D cross-sectional images only, is difficult.

Furthermore, measurements of the target area (lesion and stent lengths) are easily obtained from 3D reconstructed longitudinal views.

How to acquire 3D ICUS image data sets?

In recent years several methods have been developed to acquire tomographic ICUS data sets for 3D reconstruction:

- Manual pullback.
- Manual pullback with the ICUS catheter shaft placed in a catheter displacement-sensing device.
- Motorised pullback by a continuous speed velocity motor, applying a pullback speed of 1 or 0.5mm/sec[2].
- Motorised pullback by a continuous velocity motor and use of an ECG-labelling device so that only one phase during a R-R interval is acquired at a continuous speed of 0.5mm/sec, off-line ("Pseudo-Gating")[6].
- Motorised ECG-gated and triggered pullback applying a dedicated workstation and using a pullback device equipped with a stepping motor, on-line[7-12].

At present most 3D systems are based on an image acquisition with a pullback device that withdraws the ICUS catheter at a continuous speed (0.5mm/s). However, artifacts and inaccurate measurements of the 3D reconstruction may result from the systolic-diastolic changes of the vascular dimensions, diminished the cardiac motion; the respiration and the cyclic movement of the catheter[3, 9]. This problem can be solved by the use of a pullback device driven by an accurate stepping motor which is controlled by a steering logic of a 3D acquisition station, ensuring an ECG-gated acquisition of a 3D ICUS image data set[10] (Fig. 1). For peripheral vessels this is not necessary.

Fig. 1. Pullback device, incorporating a stepping motor, containing in this case a hand-held CVIS motor unit.

Catheterization laboratory "hardware" set-up

A scheme of the equipment set-up of an ECG-gated 3D acquisition in the catheterisation laboratory is shown in Fig. 2. If you think away, in Fig. 2, the 3D and quantitative analysis workstation and imagine that the pullback device is battery powered and designed to pullback the catheter with a continuous speed (e.g. non-gated acquisition), the two most used set-ups for 3D ICUS, today, are presented. At present, there are two major ICUS ultrasound scanner and catheter manufacturers (CVIS, Sunnyvale, CA, USA, and Jomed (better known as Endosonics, Middlesex, United Kingdom).The CVIS catheters are 30- or 40MHz mechanical rotating element catheters; Endosonics uses an electrical 20MHz phased array catheter. Both manufacturers are also delivering, with their ultrasound scanners, battery powered continuous speed pullback devices, in which in case of the CVIS the hand-held motor driving unit of the catheter can be clamped and by Endosonics the shaft of the catheter.

To perform an ECG-gated 3D acquisition an additional workstation is necessary at this moment. At the Thoraxcenter a custom-made pullback device (Fig. 1) was developed which can be controlled by the EchoScan system of TomTec (TomTec GmbH, Unterschleissheim, Germany, Fig. 2). At the moment the mechanical element catheters of CVIS are mostly used for ECG-gated studies. The catheter

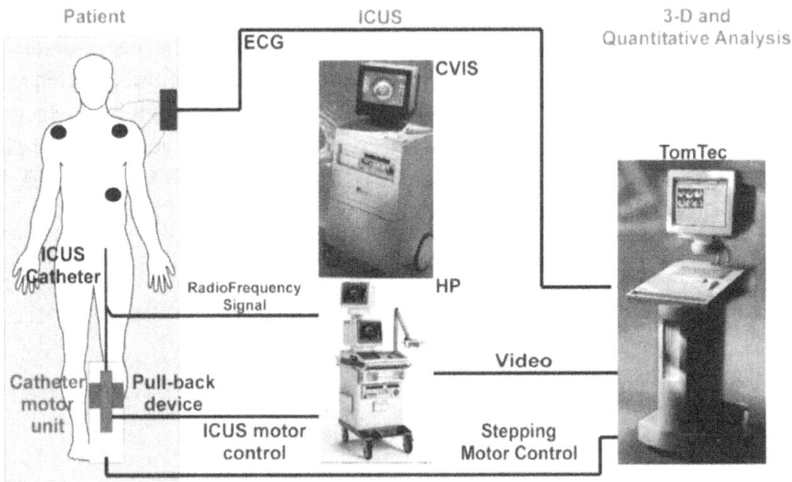

Fig.2. The setup in the catheterization laboratory. Besides the basic intracoronary ultrasound (ICUS) system, a 3D image acquisition workstation (Echoscan, TomTec GmbH) was used (displayed at the right-hand side. The 3D reconstruction station received the video output of the ICUS system and the patient's electrocardiographic (ECG) signal to control the stepping motor.

configuration (external echo-transparent sheath with an independent imaging cable inside) guarantees that pulling a defined distance at the proximal end of the catheter results in an equivalent movement of the tip of the catheter. However, the pullback device was constructed so, that also the phased array catheters of Endosonics can be used. The stepping pulses to move the motor are generated by the steering logic in the acquisition station where the scan distance, the longitudinal step resolution, the ECG- and optional the respiration intervals can be set.

Image Acquisition

The process of ECG-gated image acquisition is given in Fig. 3. The mean acquisition time to acquire 1cm of axial vessel length requires on average 1 minute, applying longitudinal intervals of 0.2 mm. The ICUS images are digitized in real-time in the catheterization laboratory. After the pullback is completed; the images are saved in a 4D (3D + time) tomographic data set. The image resolution

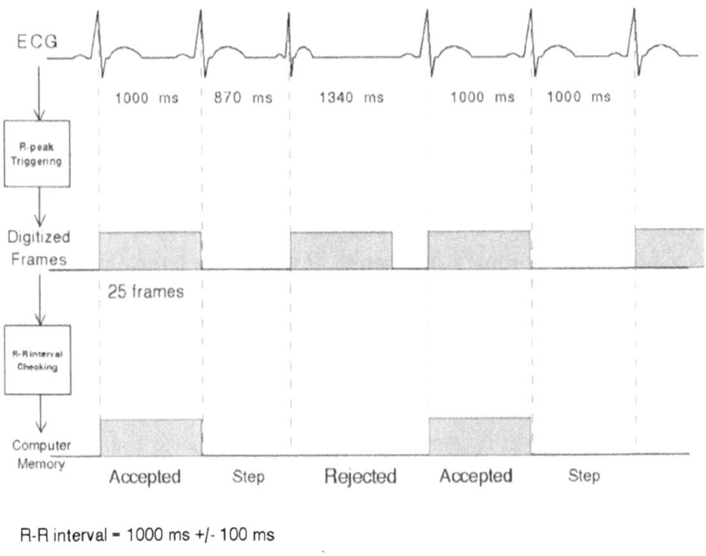

Fig. 3. Diagram showing the image acquisition schematics. The peak of the R-wave triggered the gated image acquisition. After acquisition of each cardiac cycle, the software checked if the R-R interval met the preset range. If this was the case, the images were transferred to the computer main memory, and the stepping motor received pulses to move the catheter axial by 0.2mm. Otherwise, the images were removed from the memory of the frame grabber, and a new cardiac cycle was acquired with the catheter remaining at the same site.

is 256*256 pixels with 8 bits per pixel, and usually 6 images per R-R cycle are acquired. However, newer 3D workstations are capable of acquiring images at a higher resolution and with more images per cardiac cycle without prolonging pullback time. In practice it is possible to acquire almost every heartbeat, if the heart rate is by example 60 BPM an acquisition of 60 * 0.2 mm = 12 mm/min is reached.

The non-ECG-gated pullbacks applying the continuous speed motorized devices are requiring per cm pullback a time of 20 seconds (10mm / 0.5mm/s = 20s). The images are mostly stored on S-VHS videotape for off-line processing. Newer generations ultrasound scanners are also capable of storing the ICUS images on CD-ROM, this will increase the quality, since the process of translating the digital video images to analog videotape and subsequently digitizing them again for off-line quantification can deteriorate image quality. However, the industry has not reached consensus how to store 3D (or 4D) tomographic ICUS image data sets in the DICOM standard, yet.

The new generation ultrasound scanners are also capable of building up in real-time one non-ECG-gated longitudinal 3D reconstruction for on-line analysis in the catheterisation laboratory.

3D processing

The acquired 3D ICUS image data set can when digitised be used for 3D visualisation (volume rendering) and for quantitative analysis, which we like to refer to as quantitative coronary ultrasound (QCU). The term 3D reconstruction sometimes leads to verbal confusion; it is often being used to identify the different modalities, visualisation and quantification.

3D volume rendering

Image processing

ECG-gated data sets: During the post-processing phase the images are formatted in the correct sequence according to their ECG phase in volumetric data sets (256*256*256 pixels/each 8 bits). A gray level threshold range is used to separate and subtract the blood pool and background from the coronary wall structures in each cross-section (ea. segmentation), followed by the application of several algorithms to reduce noise, enhance edges, and reduce spatial artifacts (ROSA filter). Volume rendering techniques are then applied for *dynamic* 3D reconstruction.

For non-ECG-gated sets mostly the same procedure is followed, only the step of formatting the images according to their correct place in the phase of the cardiac cycle is naturally skipped.

3D Image Display

Rendering of the volumetric data sets is now applied to perform 3D reconstructions. Rendering is a technique to produce 2-D images from a 3D data set in order to make it visible in our 2-D display world (e.g. computer monitor or printout). The observer must indicate a viewing point (viewing point: indicating the segment of interest from distal to proximal) to and a cut-plane (cut-plane: intersection through a coronary artery segment through sigittal or even oblique planes) through the 3D data set. Currently the most used viewing point is at the end of the pullback (proximal in the coronary vessel) and a cut through the data set is made so that the coronary artery is sliced into halve. In the 3D rendering mode the coronary segment is tilted to create depth perception. Another viewing approach is to actually follow the path of the imaging catheter and to create a "fish-eye" view of the luminal vessel wall on different positions in the coronary vessel of the tomographic data set. When showing the stack of those "fish-eye" 3D rendered images like a movie on the computer, a black-and-white endoscopic examination can be simulated.

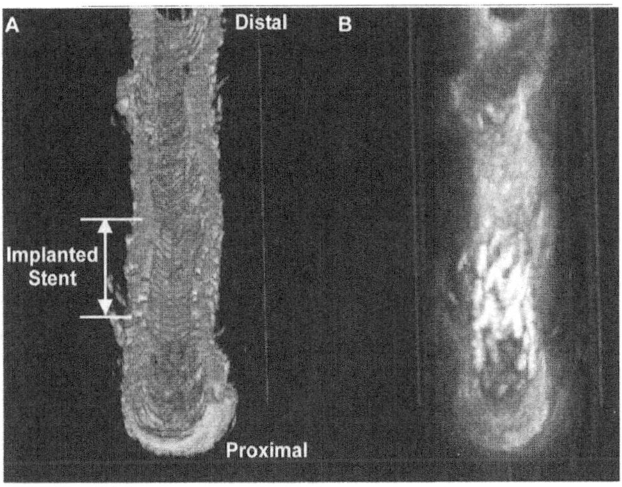

Fig. 4. (A) Gradient shading technique, (B) maximum rendering mode of a non-ECG-gated study. This causes the ribbon appearance of the vessel wall at the site of the implanted stent. In panel A the site of the stent can be identified by its smoother appearance (covered stent surface). Applying the maximum rendering mode it is possible to "peel" off the cover of the stent and to make the individual stent struts visible, panel B.

There are different rendering modes available. Two of the ones mostly used are:

- Gradient shading: In this rendering technique a realistic illu mination model is used. The light source is assumed to be mounted on the observer's head. Light emitted from this light source is reflected at the surface of the object. This type of rendering provides very realistic and detailed views of the object, but is highly susceptible to artifacts in the voxel data (Fig. 4A).
- Maximum mode: In this representation an operator defined maximal gray value along each ray through the data volume is displayed so that only bright structures (such as calcifications and stent struts) in the volume are represented. This "maximum mode display" allows creating a "transparent" echographic view of the coronary vessel (Fig. 4B).

It is possible to "remove disturbing objects" (e.g. cut) from the data set which may sometimes be necessary since in some cases the blood pool and/or the near-field artifact of the catheter can not be sufficiently removed by use of the gray level threshold without destroying actual tissue data.

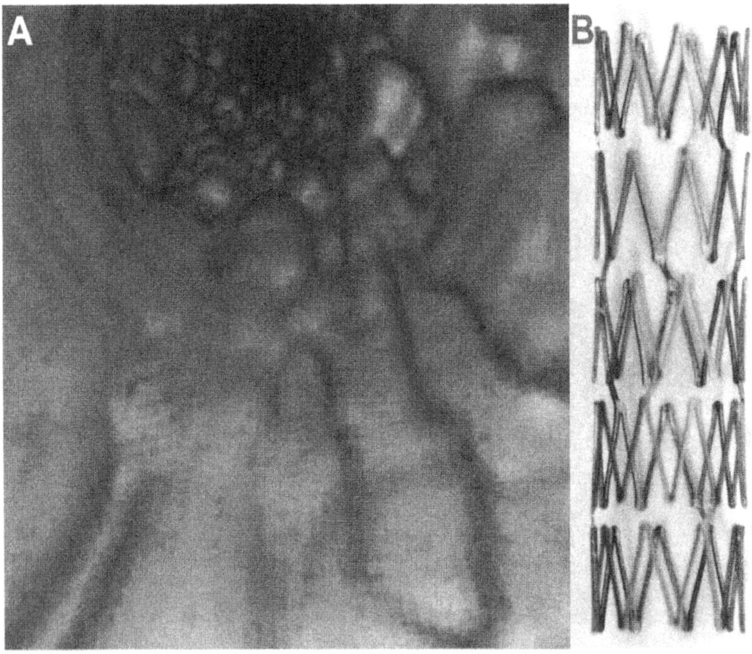

Fig. 5. In panel A an endoscopic 3D reconstruction of a Sci-med Radius implanted stent in-vivo is presented. From three to six o'clock the struts of the stents are clearly visible. For comparison, in panel B, a photo ofthis stent is presented.

Clinical example: 3D reconstruction of stents

3D reconstruction of stents identifying individual struts can be achieved. But it is possible that there are strut dropouts caused by: the architecture of the stent, the thickness of the struts, the angle of the ultrasound beam[13], and ICUS catheter resolution[14, 15]. Stents, which have a predominantly longitudin al layout of their stent struts, with thick strut "wires" can be well visualised (Fig. 4 and 5). The most impressive 3D reconstructions are the "Endoscopic" views (Fig. 5A), mimicking a "black-and-white" low-resolution angioscopic catheter[16]. However, struts do have a blurred ultrasound appearance caused by the physical nature of the ultrasound beam and the sound reflections of the used metal alloys of the stents. Measurements indicated that the size of the stent struts here are two to three times larger than original.

3D Quantitative Coronary Ultrasound

In addition to the above-described 3D volume rendered reconstructions, the possibility to perform quantitative measurements is clinically considered far more important. The 3D acquired ICUS image data sets can conveniently be used for both purposes. For quantitative analysis the "raw" images are being used, no image filtering, smoothing or interpolation is being applied.

Non-ECG-gated approach

The most simple way of performing volumetric analysis is by analyzing n-number of 2D cross-sectional images (e.g. to draw on the ICUS console of the ultrasound scanner the lumen and vessel area), to note the outcome and forward the video-tape, by example with two seconds assuming the catheter moved in this case with 2 * 0.5mm/s = 1 mm, and accumulating all the found volumes for a final end-result.

Another approach is to digitize the pullback from videotape with a computer equipped with a frame grabber and using computer assisted methods to detect the lumen, stent and vessel contours, also known as image segmentation. Two major different software approaches are being used momentarily for segmentation of ICUS images: 1) acoustic quantification; 2) contour detection.

In the early 90's Indec (Indec Systems, Capitola, CA, USA) was the first company commercially selling a 3D workstation incorporating 3D visualization and QCU based on pullbacks performed by hand[17]. Not long after this introduction the continuous speed motorized devices, operated with 1 mm/s later followed by 0.5 mm/s, were introduced. This is often referred to as "automated-pullbacks" and is the most used technology. Actually, the application of these devices was the big break-through for, and initialized, the development of QCU

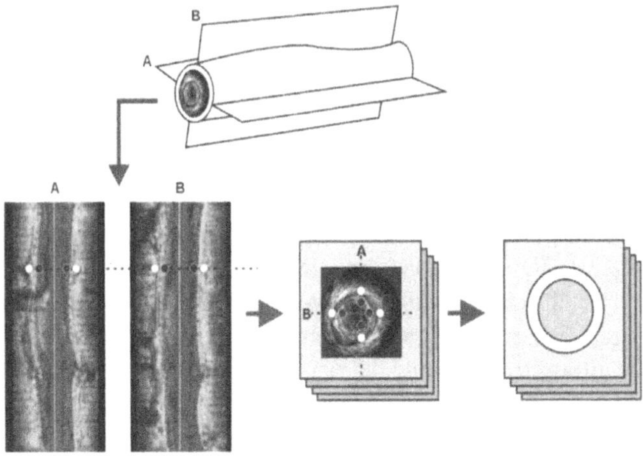

Fig. 6. Contour detection by Li. et al. The contour information of the two longitudinal contours is used to guide the detection in the individual crosssections. Firstly, two perpendicular longitudinal planes are reconstructed from the tomographic data set. In these planes the intima leading edge and the external contour of the vessel are interactively detected (minimum-cost algorithm). Secondly, contour detection is performed in the cross-sectional images using the four-edge points derived from the longitudinal contours. The result of the final contour detection is finally checked and may be corrected by the user.

software. The Echoquant software of Indec was based on the application of an acoustic quantification method[2, 18, 19]. The acoustic quantification method makes a distinction between the blood pool and luminal wall applying a statistical pattern recognition algorithm. Comparing the pattern of flowing blood to the pattern of the vessel wall, assuming that there is much more variation in the blood pool. The method is able to distinguish between these different patterns and thus detect the interface between the vessel and blood. Finally, segmentation and contour detection can be performed.

The QCU approach developed by Li et al. [20], applying a two-step approach is in modified approaches still used. After digitization of the videotape, as first step two perpendicular 3D longitudinal reconstructions of the coronary vessel are displayed (Fig. 6). With help of a semi-automated lumen and media contour detection algorithm, the observer must identify in these two longitudinal views the lumen and vessel wall and make corrections were necessary. The second step is the detection of these contours in each individual cross-section. For this the minimum-cost algorithm is applied. The contours are forced to go through the 4-contour points, which were derived from the longitudinal 3D reconstructions. After the detection, the observer can inspect all the individual cross-sections and is able to manually correct them in case of error. A maximum of 200 individual cross-sectional images can be analyzed.

ECG-gated off-line approach

As earlier mentioned, the non -ECG-gated approach can suffer from artifacts and subsequently possible inaccurate measurements. The motion of the catheter with respect to its longitudinal position to the coronary vessel wall (the z-axis) results in typical saw-tooth (or parabolic) shaped appearances of the vessel wall in 3D reconstructions (Fig. 7A). This makes it difficult to develop a contour-detecting algorithm.

In Fig. 7B, the differences in longitudinal 3D reconstruction of both acquisition methods (gated vs. non-gated) are presented. The methodology for contour detection is more or less similar to the non-gated approach.

Besides the more pleasant display of the smooth contours of the vessel wall this approach leads to significantly reduced analysis times, and it has also been shown that there are significant differences in quantitative results[7, 8, 11]. The use of ECG-gated ICUS image data sets is thus of particular interest when potentially

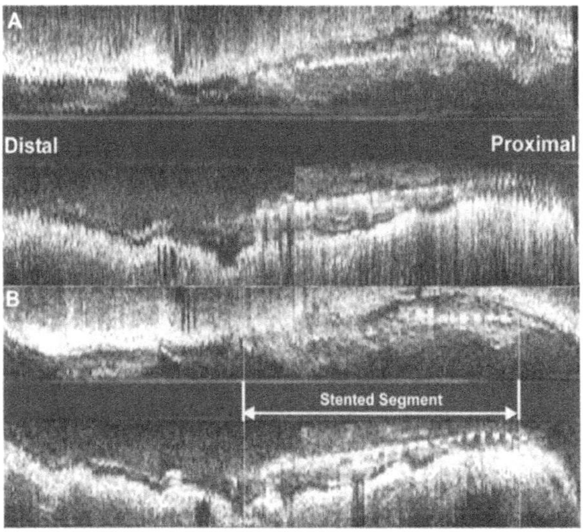

Fig. 7. In panel A, a longitudinal 3D reconstruction is presented of a non-ECG-gated pullback. In panel B, the longitudinal reconstruction of the same segment is presented, but now the image acquisition was performed ECG-gated. It can be appreciated that distal of the stented segment and proximal stent, the contours are saw-tooth shaped. This makes it difficult to perform an automated detection.

small volumetric changes, such as expected in progression-regression studies of atherosclerosis, are being investigated.

Commercially there are at the moment three vendors selling QCU soft- and hardware:

- Indec for add-on equipment for "on-line" analysis of non-ECG-gated data sets plus off-line software. The add-on equipment can be of interest for institutes who do not want to buy a complete new ICUS scanner but want to refurbish their older equipment. The system shows an on-line 3D non-ECG-gated longitudinal reconstruction, for analysis in the cathlab, and can store analyzed data digitally.
- Medis (Medis, Leiden, The Netherlands) for off-line analysis of non-ECG-gated data sets.
- Tomtec, a 3D workstation for acquisition of ECG-gated data sets and for off- and on-line analysis of both gated and non-gated sets.

New QCU developments

Development of quantitative software addressing new clinical needs is at the moment limited by a small amount of companies, mentioned earlier, and a few research institutes. At the Thoraxcenter validation of a new quantitative software package, developed by Curad, is well under way (Curad B.V., Wijk bij duurstede, Netherlands). This software package is based to translate both videotape-based images via a digitisation station and digital acquired images (such as the ECG-gated sets) to the DICOM medical image standard (Fig. 8). The tomographic data sets are stored on a server, which is connected to a hospital network. On workstations throughout the hospital the QCU software can import these data sets for further analysis. The contour detection is based on *n*-number of longitudinal 3D reconstructions (free of choice how many with intervals of 5 degrees, thus a maximum of 72). Lumen, stent and vessel wall contours can be detected (Fig. 9).

In practice 4-longitudinal reconstructions, and thus 8 longitudinal contours (2 per 3D longitudinal reconstruction) seems to be sufficient to automatically draw the contours in the individual cross-sections. In case of eccentric vessel structures a ninth contour can be added. The contour detection algorithm uses a modified Derich digital filter. Through the found contour points in the longitudinal 3D reconstructions, in the individual cross-sectional images a Bspline is fitted.

This new QCU software package is at the moment under validation. First results are showing that with no loss in accuracy, the speed of analysis is greatly improved. A time less than 20 minutes to complete an ECG-gated analysis was found, independent of the number of included individual cross-sections. This is mainly due to the fact that the analysis only takes place in the longitudinal planes.

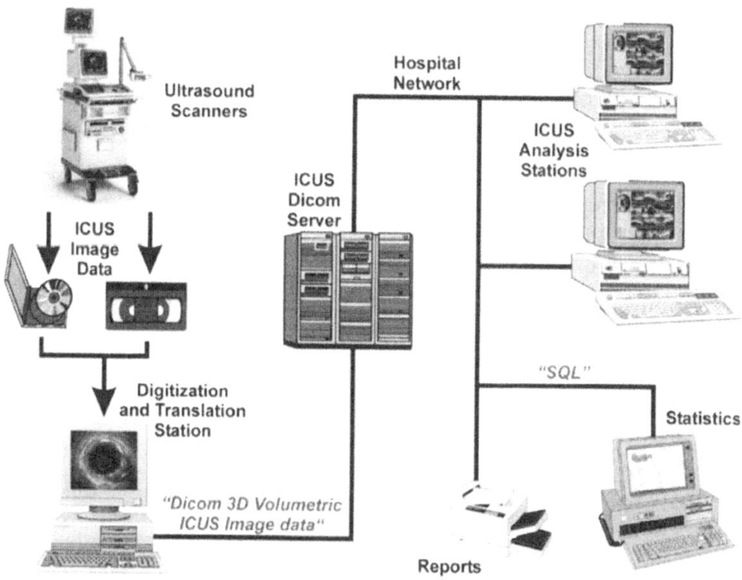

Fig. 8. This is a schematic overview of the ICUS image data flow. The pullback information is digitised or translated into the Dicom image standard and stored on a Dicom server connected to the hospital network. Analysis stations can retrieve the data from this server for further analysis. Statistics can be performed by retrievingnumerical data via SQL queries.

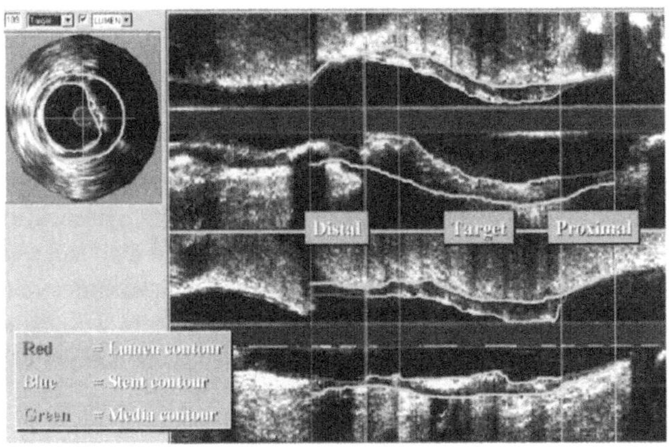

Fig. 9. This is a screen-shot of the new QCU program. The investigated segment can be divided in multiple parts to make further statistical analysis more convenient. The contour detection in the cross-sectional images can be checked in the panel at the top left.

Limitations

The described QCU software has some limitations:

- The curvature of the vessel is not taken into account so that an artificially straightened reconstruction is produced. However, the resulting error in the final results seems to be rather small[21]. Visualization of the true curvature of the coronary vessel can only be accomplished by combining bi-plane angiography with the 3D ICUS image data set[22].
- The resolution of the ICUS transducers determines the upper resolution limit of 3D reconstructions.
- The ECG-gated acquisition time is longer compared with continuous velocity pullbacks.

Future developments

The industry is constantly working to improve their catheters (higher frequencies, >40MHz), ultrasound scanners (digital imaging, storage and image transfer), add-on equipment (improved and smaller pullback devices) and software for imaging, image acquisition, analysis and three-dimensional reconstruction. Disadvantages such as occluding the vessel by means of the catheter could maybe be prevented by use of an imaging guide-wire or by use of a forward scanning catheter[23, 24]. In this case the site of the stenosis is not occluded during an acquisition.

Further improvements could be found in improved on-line quantification and incorporation of ECG-gating functionality in the ultrasound scanners.

Conclusions

Quantitative coronary ultrasound is a technique that has undoubdtfully claimed its place in research of coronary vessels being treated with new pharmaceutical agents or interventional techniques. The feasibility of ECG-gated pullbacks for producing dynamic 3D volume rendered reconstructions has been demonstrated and even more important allowing high quality quantitative results with rather smooth contours, not showing cyclic saw-shaped artifacts often seen in non-ECG-gated approaches.

References

1. Rosenfield K, Losordo DW, Ramaswamy K, et al. (1991) Threedimensional reconstruction of human coronary and peripheral arteries from images recorded during two-dimensional intravascular ultrasound examination.Circulation 84:19381956

2. Di Mario C, von Birgelen C, Prati F, et al. (1995) Three-dimensional reconstruction of intracoronary ultrasound : clinical of research tool? Br Heart J:26-32
3. Roelandt JR, di Mario C, Pandian NG, et al. (1994) Three-dimensional reconstruction of intracoronary ultrasound images. Rationale, approaches, problems, and directions.Circulation 90:1044-1055
4. von Birgelen C, Erbel R, Di Mario C, et al. (1995) Three-dimensional reconstruction of coronary arteries with intravascular ultrasound.Herz 20:277-289
5. Mintz GS, Pichard AD, Satler LF, et al. (1993) Three-dimensional intravascular ultrasonography: reconstruction of endovascular stents in vitro and in vivo.J Clin Ultrasound 21:609-615
6. von Birgelen C, Di Mario C, Prati F, et al.: Intracoronary ultrasound: Three-dimensional reconstruction techniques. In: de Feyter PJ, Di Mario C, Serruys PW, (eds): "Quantitative coronary imaging", Rotterdam: Barjesteh, Meeuwse & Co., 1995: pp 181-197.
7. von Birgelen C, de Vrey EA, Mintz GS, et al. (1997) ECG-gated three-dimensional intravascular ultrasound: feasibility and reproducibility of the automated analysis of coronary lumen and atherosclerotic plaque dimensions in humans.Circulation 96:2944-2952
8. von Birgelen C, Mintz GS, Nicosia A, et al. (1997) Electrocardiogram-gated intravascular ultrasound image acquisition after coronary stent deployment facilitates on-line three-dimensional reconstruction and automated lumen quantification.J Am Coll Cardiol 30:436-443
9. Bruining N, von Birgelen C, Mallus MT, et al. 1996 ECG-gated ICUS image acquisition combined with a semi-automated contour detection provides accurate analysis of vessel dimensions, IEEE Computer Society Press, Computers In Cardiology, 53-56
10. Bruining N, von Birgelen C, Di Mario C, et al. 1995 Dynamic three-dimensional reconstruction of ICUS images based on an ecg-gated pull-back device, IEEE Computer Society Press, Computers In Cardiology, 633-636
11. Bruining N, von Birgelen C, de Feyter PJ, et al. (1998) ECG-gated versus nongated three-dimensional intracoronary ultrasound analysis: implications for volumetric measurements.Cathet Cardiovasc Diagn 43:254-260
12. Bruining N, Sabate M, de Feyter PJ, et al. (1999) Quantitative measurements of in-stent restenosis: A comparison between quantitative coronary ultrasound and quantitative coronary angiography.Catheter Cardiovasc Interv 48:133-142.
13. Di Mario C, Madretsma S, Linker D, et al. (1993) The angle of incidence of the ultrasonic beam: a critical factor for the image quality in intravascular ultrasonography.Am Heart J 125:442-448
14. Benkeser PJ, Churchwell AL, Lee C, et al. (1993) Resolution limitations in intravascular ultrasound imaging.J Am Soc Echocardiogr 6:158-165
15. Engeler CE, Ritenour ER, Amplatz K (1995) Axial and lateral resolution of rotational intravascular ultrasound: in vitro observations and diagnostic implications.Cardiovasc Intervent Radiol 18:239-242
16. Bruining N, von Birgelen C, de Feyter PJ, et al. (1998) Ultrasound Appearances of Coronary Stents as Obtained by Three- Dimensional Intracoronary Ultrasound Imaging In Vitro.J Invasive Cardiol 10:332-338.
17. Hausmann D, Sudhir K, Mullen WL, et al. (1994) Contrast-enhanced intravascular ultrasound: validation of a new technique for delineation of the vessel wall boundary.J Am Coll Cardiol 23:981-987.

18.Prati F, Mallus MT, Lioy E (1998) Three-dimensional reconstruction techniques applied to intracoronary images.G Ital Cardiol 28:460-467.
19.Di Mario C, von Birgelen C, Prati F, et al. (1995) Three dimensional reconstruction of cross sectional intracoronary ultrasound: clinical or research tool?Br Heart J 73:2632
20.Li W, von Birgelen C, Di Mario C, et al. 1994 Semi-automated contour detection for volumetric quantification of intracoronary ultrasound, IEEE computer society press, Computers in cardiology, 277-280
21.Schuurbiers JC, von Birgelen C, Wentzel JJ, et al.(2000) On the IVUS plaque volume error in coronary arteries when neglecting curvature.Ultrasound Med Biol 26:1403-1411.
22.Slager CJ, Wentzel JJ, Schuurbiers JC, et al. (2000) True 3-dimensional reconstruction of coronary arteries in patients by fusion of angiography and IVUS (ANGUS) and its quantitative validation.Circulation 102:511-516.
23.Evans JL, Ng KH, Vonesh MJ, et al. (1994) Arterial imaging with a new forward-viewing intravascular ultrasound catheter, I. Initial studies.Circulation 89:712-717
24.Ng KH, Evans JL, Vonesh MJ, et al. (1994) Arterial imaging with a new forward-viewing intravascular ultrasound catheter, II. Three-dimensional reconstruction and display of data.Circulation 89:718-723

Coronary 3-D angiography, 3-D ultrasound and their fusion

C.J. Slager, J.J. Wentzel, J.C.H. Schuurbiers, J.A.F. Oomen, F.J.H. Gijsen, R. Krams, W.J. van der Giessen, P.W. Serruys, P.J. de Feyter.

Thoraxcenter, Cardiology Department, Hemodynamics Laboratory, Erasmus Medical Center, Rotterdam, The Netherlands.

Summary: Coronary contrast angiography is an extremely useful X-ray lumen imaging technique with high spatial and temporal resolution. By nature, its shadow imaging character permits visualization of a complete arterial branching system. Main restriction is that it provides only lumen information. Extending its use to biplane imaging allows reconstruction of 3-D location of points, 3-D shape of curves and with some approximation also 3-D reconstruction of the coronary lumen. Intravascular ultrasound imaging (IVUS) presents lumen and also wall information with even higher detail, but does so in a fragmented tomographic way. Simple 3-D reconstruction of an artery from IVUS, neglecting vessel curvature, is obtained by combining serial data, acquired during a controlled catheter pullback. Biplane imaging allows prediction of the true 3-D path passed by an IVUS catheter tip during pullback. Fusing IVUS data with this 3-D path allows true 3-D reconstruction of coronary lumen and wall (ANGUS). ANGUS imaging has opened the wide field of image based fluid mechanical modeling in the coronary arteries in the patient. With proper boundary conditions, the shear stress of the flowing blood induced on the endothelium can be calculated. Shear stress has a profound influence on vascular biology. Coronary artery wall data related to shear stress in patients are presented.

Keywords: angiography, X-ray, IVUS, 3-D reconstruction, image fusion, shear stress

Coronary X–ray imaging

In the coronary catheterization laboratory, X-ray imaging has become the method of choice to study coronary luminal narrowing and to guide catheter interventions. Many technical developments have contributed to this status. To mention are the application of short high intensity pulses in the generation of X-rays, of sensitive

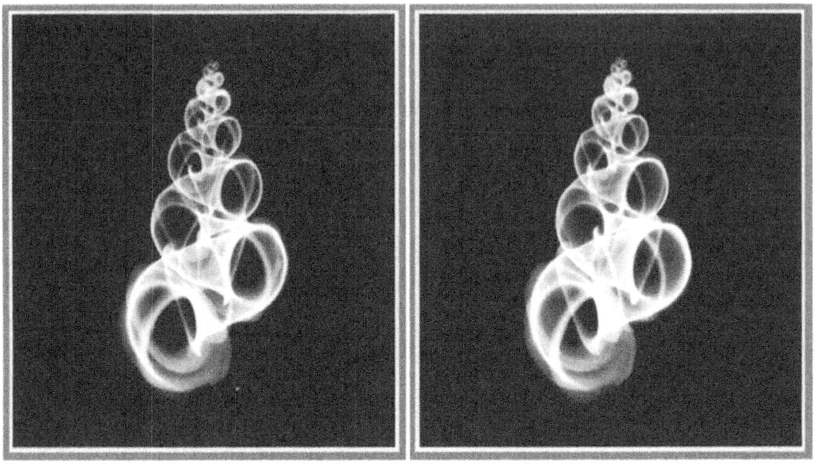

Figure 1. Stereo radiograph of Precious Wentletrap by Peter Abrahams, Portland, Oregon. Tip for viewing: Hold images close to face, move the figure slowly away to normal viewing distance, keep the left eye aiming at the left and the right eye at the right (parallel viewing). Note that 3 images will be perceived, the middle should deliver 3-D. Do not fixate to focus at this first stage of viewing, at some normal viewing distance focusing happens to occur automatically.

and accurate image intensifiers and of capturing X-ray images by high resolution and high-speed digital video cameras. Complemented with digital image storage and on line enhancement techniques, the coronaries can be observed and recorded at high frame rates during long lasting periods, while keeping radiation exposure within acceptable limits. For the trained operator the X-ray shadow images provide a highly useful 2-D overview of a myriad of 3-D branching patterns enabling the guidance of catheters and wires to the site of application. Particularly, X-ray imaging in its modality of contrast coronary angiography, i.e. by injecting a radio-opaque contrast medium to the blood flow enables fast and high quality visualization of a selected coronary lumen. However, despite the major advantages of coronary angiography and its common use for lumen size estimation, important limits exist to gather quantitative morphological data from the contrast angiogram. Indeed, the dark side of the shadow imaging technique is its masking of 3-D information and its inability to reveal with sufficient contrast the dimensions and condition of the diseased arterial wall. Both of these shortcomings and ways for improvement will be discussed in this chapter.

3-D stereo acquisition

Since long, stereographic imaging is known to be a potential extension of the X-ray imaging technique from 2-D into 3-D. This modality was invented already within a year after Roentgen's discovery of X-rays in 1895[1]. The technique has been widely used, for example to observe fractures of bones with great spatial detail. For those having the ability of stereo viewing, Figure 1 shows an interesting picture of a sea shell, clearly illustrating the huge amount of 3-D information available in such a static pair of stereo X-ray images. A single X-ray unit, which sequentially records the object from two slightly differing viewing angles (6-7 degrees), can produce such a set of images.

Quantitative 3-D reconstruction of marker positions

For quantitative purposes, such a pair of images, taken from different perspectives, allows reconstruction of identifiable image structures and points in 3-D space. Reconstruction of marker positions, or points will be discussed first, as this will be easiest to comprehend. A current application for example is to 3-D reconstruct marker positions attached to implanted artificial prostheses[2]. Deformation and remodeling of the implant and its environment can thus be monitored accurately. The first application in the cardiac field, addressing the reconstruction of the 3-D location of coronary branch points on the epicardium and their motion was published a few decades ago[3]. Indeed for these quantitative 3-D applications, increasing the viewing angle between both images to a maximum of 90 degrees will maximize accuracy. For 3-D study of moving cardiac structures a rapid filming biplane X-ray unit is required, and to minimize matching errors the recording should be done at exactly the same moment. However, in practice the two independent viewing systems of such a biplane installation generate images in an alternating way. Therefore, good matching may require combining image data of one of the viewing directions with that obtained by interpolation from two successive images of the other direction. Only if the motion of an object is truly periodic and thus predictable, one may consider 3-D reconstruction from sequential monoplane recordings with different views. Then afterwards, images collected at a similar cardiac phase can be combined to get a proper set of stereo pairs.

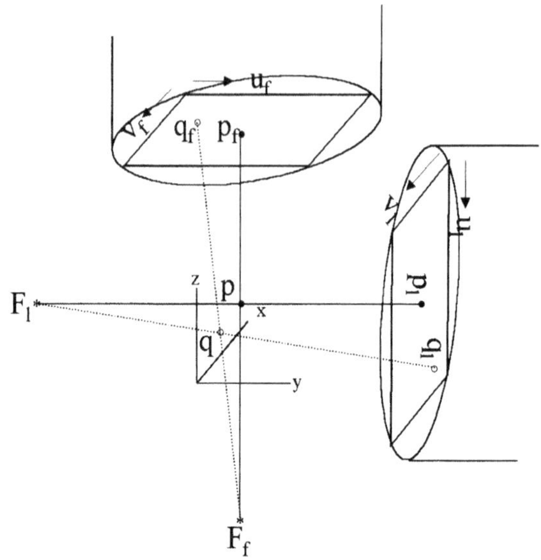

Figure 2. Two 3-D points p and q are projected from a frontal (f) and a lateral (l) X-ray focal source F_f and F_l unto their shadow images p_f , q_f and p_l , q_l . Locations in the respective frontal and lateral images are described in a u,v coordinate system, while the location of p and q is expressed in a 3-D x,y,z coordinate system. Mathematical equations can be formulated to transform the point coordinates from one coordinate system into the other.

3-D frame of reference and calibration

As an essential requirement for quantitative 3-D reconstruction, a 3-D calibrated frame of reference is needed in which projection sources, 3-D object points and 2-D projection planes can be defined. This enables the description of the mathematical transformation[3] from 2-D to 3-D and the reverse. Figure 2 illustrates the spatial interrelationship of two points p and q in 3-D and their biplane projections.

Several solutions have been implemented for the purpose of calibration and reference definition. As the most common solution, all modern cardiac X-ray units assume a locally defined 3-D coordinate system in the catheterization laboratory and provide a calibrated digital storage and read out of the imaging geometry. Data stored for each view are the distances between image intensifier, X-ray focus and isocenter as well as the viewing direction expressed in two angles. These

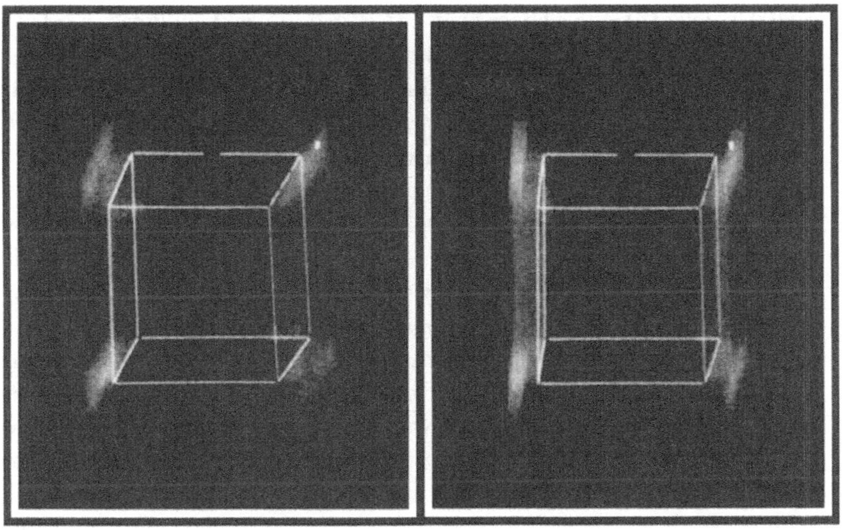

Figure 3. Stereo pair (see also figure 1) of calibration cube, used to define a calibrated 3-D reference system. Dot: origin of x,y,z coordinate system; 1 and 2 (barely visible) interruptions in the edge indicate y and z axis respectively. Viewing direction: backside of the investigation cradle, caudal from the left.

angles describe the cranio-caudal (angulation) and left-right (rotation) deviation from the frontal viewing direction. The isocenter is the 3-D spatial location which, independent of the viewing direction, will be projected in the center of the X-ray image. Usually the isocenter is taken as the origin of the 3-D coordinate system. An object positioned in the isocenter can be imaged from different perspectives without object translocation. To derive for each particular view the different 2-D/3-D transformation mathematics from the geometry read out it must be noted that brand dependent differences in the mechanical ways to reach a certain perspective view[4] need to be taken into account. For example, an X-ray source and image intensifier combination mounted to the ceiling moves quite different from the one mounted at the floor and this has consequences for the finally obtained view. As one may have noticed in practice, an identical geometry read out, for example 45° cranial angulation and 45° rotation will show a (rotational) difference in the obtained images between such differently moving units.

Unfortunately, in practice discrepancies exist between the X-ray systems geometry read out and the actual 3-D position and orientation of the image determining components[5], which cause problems in a point's 3-D reconstruction. One of the reasons for this is the weight of the image intensifier and X-ray source. When laterally orientated, this causes the beam connecting the X-ray source and central point of the image intensifier to sag by several millimeters below the isocenter. In that case the transformation mathematics obtained for the ideal case, will show that the beams connecting the foci and the corresponding 2 projections of a point (Figure 2) will not intersect in 3-D. Therefore, solutions have been found to calculate in 3-D an estimated point's location[6,7] that shows the best mathematical fit for the observed 2-D point's projections. Also corrections for the X-ray geometry read out may be applied, but after some experience with such methods, we have chosen to apply another 3-D frame of reference defining method by using a 3-D calibration object[3], thereby neglecting the on line provided geometry read out. Reasons for this choice were the variations in geometry read out over time and its dependency on the actual viewing direction, which required a frequent and complex adjustment of correction factors. The 3-D calibration object usually contains a series of markers[3] with known 3-D positions that determine the location and the dimensions of the 3-D coordinate system defining the spatial frame of reference. Our design of a simple cube with clearly identifiable 4-cm long edges as is shown in Figure 3 also suffices this purpose. Different interruptions in the edge line patterns aid in recognizing the respective x, y and z-axes.

As a practical consequence of this approach, each X-ray geometry setting applied during an investigation needs to be identically reproduced when imaging the calibration cube. Fortunately, using the digitally stored information, the modern X-ray units automatically reproduce the earlier applied geometry in a very accurate way. A practical limitation of this method is that on line 3-D reconstruction of the object under study may not be possible because most applications will require determination of the optimal imaging geometry during the study. Therefore, applying a predetermined calibrated geometry for an investigation may provide unsatisfactory views. One should also notice that positioning of the cube in the isocenter fully determines the 3-D coordinate frame of reference. Therefore, the 3-D reconstructed marker locations of an object expressed in x,y,z coordinates will vary depending on the cube's positioning. This, however, has no implications for the reconstruction of distances between markers and their relative positions within the object, i.e. the 3-D characteristics of an object will be reconstructed correctly, only its position and rotation with respect to the reference system may vary. Of course, when investigating an object at a second occasion using the digitally stored geometry to reproduce the frame of reference, not exact repositioning of the object in this absolute frame of reference leads to a similar type of variability.

A third innovative method has been proposed to even circumvent the necessity of recording the imaging geometry. Indeed, if an object contains a variety of well recognizable unique markers, sufficiently distributed over 3-D space, two different perspective images of the object suffice to derive the unique relative orientation of both perspective views[8]. Only one combination of viewing directions will fit with both observed projections of all markers. Because this method poses strict requirements to the minimum number and spatial distribution of markers in order to get sufficient accuracy its practical applicability will be limited. Also, an additional size calibration of the object will be needed because the absolute calibration to real dimensions cannot be derived from determining the viewing directions only.

Quantitative 3-D reconstruction of curves

Quite a difference exists between the 3-D reconstruction of single points and that of entire curves from a set of stereo images. The reason for that is that the projections of curves do not contain identification marks indicating which point on one curve should be matched with the one on the other curve. Particularly, when the mathematical transformations between 2-D and 3-D do not deliver exact point reconstructions because of non intersecting beams, ambiguities in the 3-D curve reconstruction will occur. This also holds when a point of one curve matches with a number of points on the other curve. To find the 'best' 3-D curve from 2-D projections of the searched curve, an overview in matrix format of all possible, best fit solutions of any pair of points defined on the curves can be constructed[7,5]. Degree of fitting may be measured for example by determining the average distance between the point's projections in 2-D with the original points. In Figure 4 this procedure is visualized. For example, when sampling each 2-D curve by 64 equidistant points a total of 64x64 possibilities for reconstructing a 3-D point exist. In Figure 4 such a resulting matrix is shown in which for any pair of 2-D curve points the error of the corresponding 3-D point's projections is indicated in gray scale coding. This example is obtained from a curve calculation applied to the biplane views of the centerline of a catheter that was positioned in a coronary artery.

An optimal solution for the best fitting 3-D curve can be derived from this matrix by searching for the course through the matrix that delivers the lowest total error[7,5]. In previous work[9,10] we have proposed another way to reconstruct a curve from its two projections, making use of the relatively simple 3-D course of a catheter located in a coronary artery. First, we determined a planar 3-D curve, being part of an ellipse, which 2-D projections roughly fitted the actual 2-D catheter projections. Subsequently, similar as above, a distance error was determined in the 2-D projections. Using this information, the spatial curve was further adjusted by iterative techniques to improve 2-D fitting. Then, by subsequently adding more degrees of freedom the spatial curve was further detailed and refined until an optimal fitting in 2-D was achieved. Although this method worked well and the results were very accurate[10], the time required for the iterative procedure was rather long and in some rare cases no convergence of the iteration could be obtained.

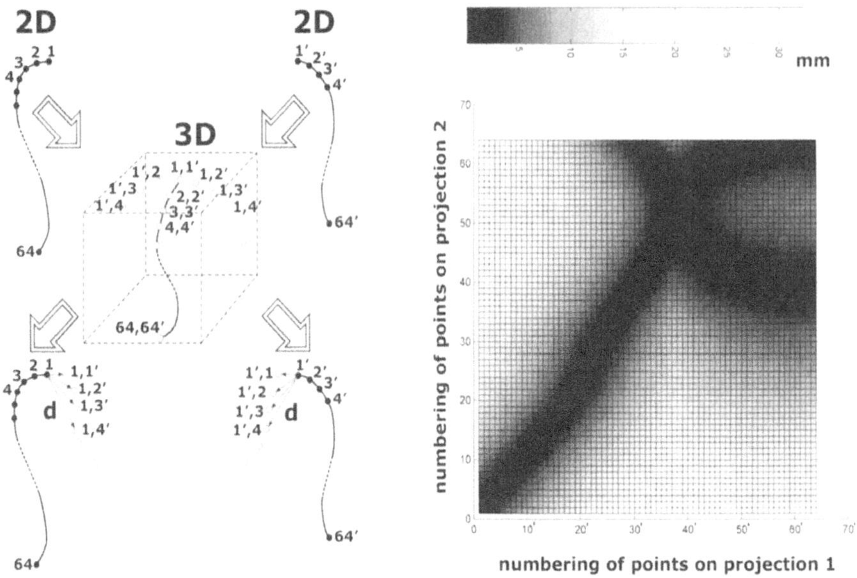

Figure 4. Left panel: two 2-D projections of a curve are equidistantly sampled by 64 points. Pairing each possible point combination (1-1' to 1-64', 2-1' to 2-64' etc.) to achieve a 3-D point gives 64x64 best fit solutions. Projection of each of the obtained 3-D points to 2-D and comparing their locations with the corresponding originals, gives error distances d. Right panel: Error distance (mm) averaged for both projections (i.e.1 to 1-1 is combined with 1' to 1'-1 etc.) is indicated in gray scale in this matrix presentation. The optimal curve must be found along the lowest error path through this matrix.

Improved new 3-D curve reconstruction method

Recently, we found an interesting new solution for the problem of 3-D curve reconstruction, being very fast and accurate. Briefly this method operates in the following way and is illustrated in Figure 5.

Having two projections of a curve of which the beginning and end points are by definition matching points, we first assign 64 equidistant points to each projection, numbered 1 to 64 from proximal to distal. Then pairing the identically numbered points and using the 2-D/3-D mathematical transformation rules (obtained from the cube's calibration and reference frame definition method), 64 best-fit points

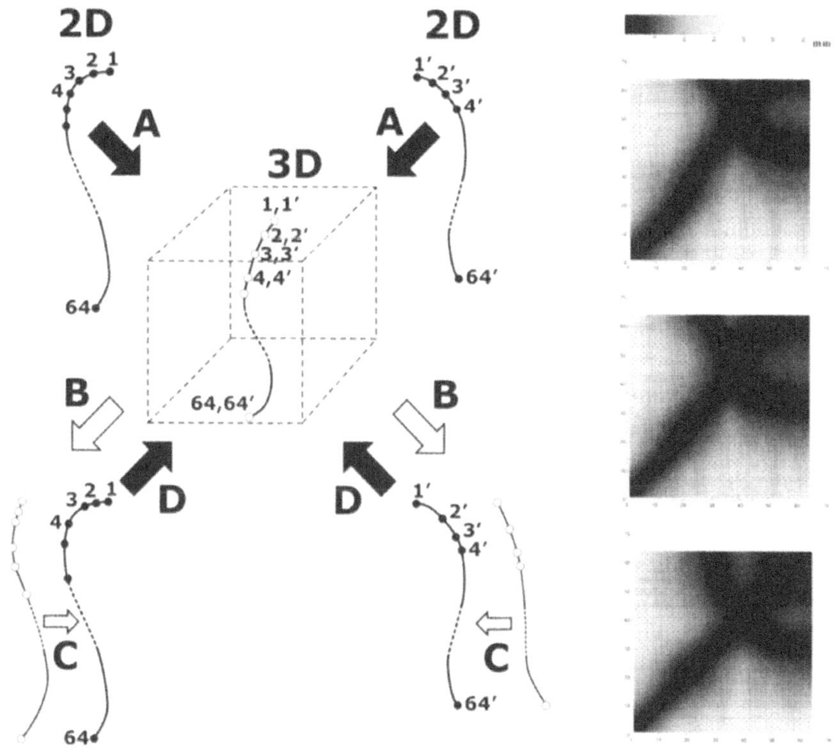

Figure 5. Left panel: the original 2-D curves are sampled equidistant by 64 points. **Arrow A,** first step: Only corresponding numbers are paired, this renders a first approximation of the 3-D curve reconstruction. This 3-D curve is then **equidistantly resampled in space** and as a second step (**arrow B**) the thus obtained 64 points projected in 2-D. This results in a non-equidistant distribution of the points on the 2-D curves because of different degrees of foreshortening. **Arrow C.** The obtained 2D foreshortening pattern is next applied to sample the original 2-D curves accordingly. **Arrow D.** Again, corresponding numbers are paired which delivers the next 3-D solution. Steps B,C,D are repeated several times which delivers the eventual 3-D curve reconstruction. **Right panel:** Distribution of errors for the matrix example shown in Figure 4 before and after the first and eighth resampling round. The diagonal (0,0' to 64,64') by definition shows the error at each stage of the solution.

are reconstructed in 3-D space. Note that in the approach described before, 64x64 points would need to be determined. Indeed, it is unlikely that this initial assignment of 64 pairs would deliver optimally matching pairs. As a result the first derived 3-D curve through these points will deviate from the eventual solution. For reasons mentioned later, a moving average filter is applied to the 3-D curve points that sequentially covers three successive points. This filtering step is applied 3 times. Subsequently, using a 3-D spline interpolation technique

(MATLAB®, The MathWorks, USA) a 3-D continuous curve description is obtained. This curve is then *equidistantly resampled in space* by assigning a new set of 64 points. Subsequently, using the 3-D to 2-D transformation rules these 64 points are projected in 2-D. In general the 2-D projections will not show an equidistant point distribution, because most 3-D curves contain parts with spatially varying directions, which lead to different degrees of foreshortening in the projection. Subsequently, the thus obtained 2-D pattern of non-equidistant point distribution is utilized in a second round of sampling to distribute in an identical pattern 64 points on the original 2-D curve projections. This appears to greatly improve the 3-D matching of the newly obtained 64 pairs, which again provide a spatial curve. After this the procedure of 3-D filtering, spatial equidistant resampling,2-D projection and redistribution, etc. is repeated. In general, less than 10 iterations, requiring just a few seconds on a standard PC, suffice to reach an accurate curve reconstruction. Accuracy, defined as the average distance between the original 2-D curve drawings and the projections of the reconstructed 3-D curve, is generally well below a few tenths of a mm. In this respect, scaling of the drawings is corrected for the geometric magnification introduced by the shadow imaging. Similarly, reconstruction length is accurate to within a few percent. In our experience, the method is very robust and can handle very complex spatial curves, also those showing crossings in their projections.

Blind planes

The 3-D moving average filter, mentioned above, tends to shorten local curve length of parts with high curvature. We intentionally implemented this "tension" feature in the 3-D curve in order to reduce fundamental problems with curve reconstruction near what we call blind planes. If a local planar part of a curve is projected by the biplane X-ray sources so that both shadows of a bended part project as straight lines, all local curve information is lost. In this situation the viewing planes, which are the planes between the X-ray foci and the 3-D curve points, align with the local curve plane. For a situation like the one depicted in figure 2, this would imply that p; q and both foci are in one plane. Any curve between p and q located in this so called blind plane cannot be reconstructed. Obviously, at image acquisition, projections should be selected such that this situation is avoided as much as possible. However, this cannot always be achieved and most long curves, e.g. any half circle, will have some small part passing through a blind plane. Figure 4B also indicates the curve to be tangent to a blind plane i.e. at coordinates (38', 52) where the two different low error curves cross. Near those blind plane locations, robustness of the reconstruction method is required to avoid local spatial excursions of the reconstructed curve near the blind plane, as these will not be observed in the projection views. In this respect, the previously described iterative method was also very robust from this point of view.

Reconstruction of volumetric objects

Several investigators have published on the subject of 3-D reconstruction of a complete coronary vessel tree from the biplane angiogram. Indeed, 3-D reconstruction of the vessel skeleton from two views in which the coronary centerlines represent curves is rather straightforward applying the methods described above. However, reconstruction of a 3-D-vessel lumen using the projected luminal borders is not that obvious. In fact, such a reconstruction can be very approximate only and strict assumptions on the lumen cross sectional shape are required to make such a lumen reconstruction possible. Figure 6 illustrates how starting from the luminal borders of two perspective views only a cross sectional quadrangle can be reconstructed in 3-D that encompasses the searched lumen cross section. However, no additional information can be obtained on the exact luminal cross sectional shape. Generally an elliptical shape is assumed, but even then an additional specification as to the orientation of its major axes is needed.

Nevertheless, approximate lumen cross sections may be reconstructed in this way and using a series of neighboring cross sections a lumen approximation can be constructed. Figure 7 gives an example of such a lumen, reconstructed from

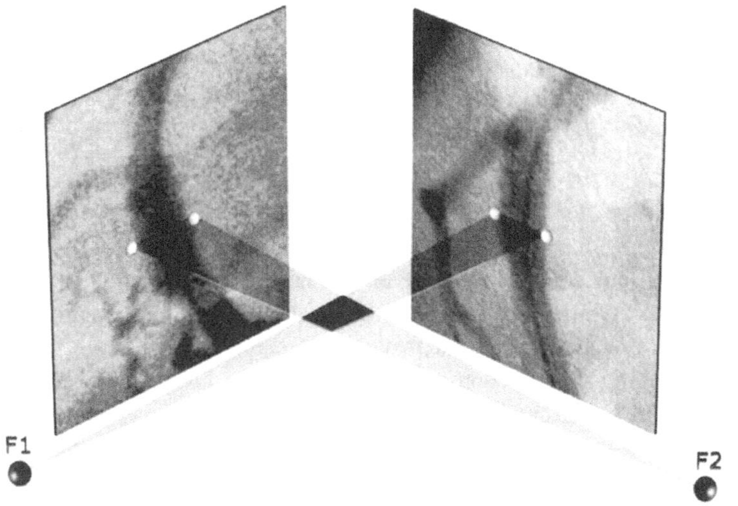

Figure 6. Starting from both foci F_1 and F_2 to corresponding luminal borders of two perspective views and assuming perpendicularity of the lumen centerline to the indicated X-ray planes, a cross sectional quadrangle can be reconstructed in 3-D that encompasses a searched lumen cross section. No information can be obtained of the true cross sectional lumen shape.

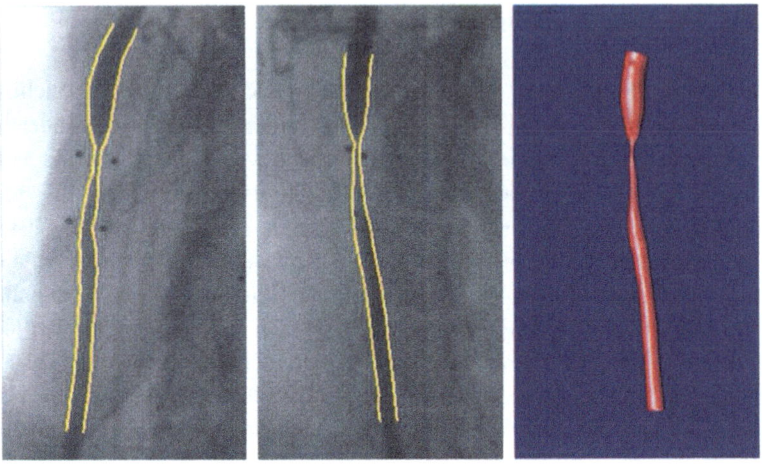

Figure 7. Biplane contrast angiograms of a rabbit carotid artery constricted by use of an external cast and corresponding 3-D lumen reconstruction assuming an elliptical cross section.

two angiographic perspective views of an experimental carotid narrowing in a rabbit, assuming an elliptical cross section and defining the longest axis to correspond with the greatest diameter in the projection. In patients 3D lumen reconstruction has been applied to estimate the degree of diffuse coronary artery disease from the 3D characteristics of coronary morphology[5].

Densitometry

It might be considered to improve on the lumen reconstruction by utilizing the densitometric information contained in the biplane contrast angiograms[7]. This, however, is a rather cumbersome method and requires several corrections to adjust for the imaging imperfections of the X-ray-image intensifier chain as well as subtraction of background structures, which unfortunately may have been affected by the contrast injection as well. Even taking those hurdles, no unique lumen cross sectional shape can be defined because two perspective densitographic cross sectional views do not contain sufficient information for this purpose. What can be reconstructed using the densitometric approach is the local lumen cross sectional area as this may be determined without knowing exact lumen shape. This knowledge has been applied in deriving flow information from the biplane angiogram[11].

ICUS

The advent of intracoronary ultrasound (ICUS) imaging[12,13] opened a new era in studying coronary arteries of patients in the cardiac catheterization laboratory. By this technique, the arterial wall and its various components were presented with unprecedented spatial resolution in thin local cross sections. However, this significant increase of cross sectional lumen and wall information had a very local character and occurred at the expense of loosing a true spatial overview of the coronary lumen and wall. To improve on this fragmented tomographic character of ICUS, soon after its invention, methods were designed to collect a series of neighboring cross sectional images by applying a manual pullback of the imaging catheter through the investigated arterial segment. These images were assembled in a single 3-D overview[14]. For this purpose brightness values of cross sectional pixels were transferred to a 3-D voxel brightness space assuming equidistant planes in the voxel space corresponding to an assumed linear pullback trajectory. Between the actually sampled cross sections, interpolation between neighboring cross sections was performed. To improve this 3-D overview, correct distance information between neighboring cross sections can be measured by determining pullback distance continuously and selecting only cross sectional images at equidistantly spaced locations. This also enables quantification of lumen volumes. Because of cardiac motion this solution is not accurate for ICUS application and its use has been restricted to the peripheral vessels[15]. Cardiac motion will move the imaging part of an ICUS catheter periodically back and forward in the axial lumen direction over the cardiac cycle. Consequently, in the coronaries the distance read out of any catheter pullback method, does not represent the actually varying axial transducer displacement relative to the coronary anatomy.

Automated catheter pullback methods

To improve the sampling of cross sectional images at equally spaced distances, pullback of the ultrasound transducer has been automated and motorized. The motorized drive pulls either continuously or in a stepped mode being triggered by the electrocardiogram. By triggering, the sampling is synchronized with the cardiac cycle. This highly improves the correlation between the external pullback distance with the actual intracoronary displacement relative to the coronary anatomy[16]. Moreover, a stepping system can be paused when irregular heartbeats occur, which further improves sampling accuracy. This pausing may be extended to include the breathing inspiration period to eliminate also this disturbing factor. The sampling moment needs not to be restricted to the R-wave; thus it is possible to collect ICUS images at any specified moment after the R-wave. This enables the study of the arterial geometry over the cardiac cycle.

Accuracy of motorized continuous pullback can be improved also by focusing on a certain cardiac phase i.e. by collecting images only at a specified moment related to the top of the R-wave. In this case, normal variations in heart rate will still cause axial variations in distance between successive samples, but the distances are easy to derive and interpolation techniques will still enable to compose a correct 3-D-voxel overview. Irregular heartbeats however, associated with significant variations in cardiac end diastolic and end systolic volumes, may be accompanied with unpredictable catheter displacements which still may induce errors in the sample distance.

Vasoactive state

The period of catheter pullback may last several minutes. For example, a motorized pullback, stepping with 0.5 mm increments, requires 150 steps to pass a quite common artery length of 75 mm. During this period, it is desired to maintain the artery in the same state of vasoactivity. For this purpose a short lasting nitrate injection can be applied as is also customarily done when performing contrast angiography. Otherwise, lumen size and shape may vary because of the catheter-induced disturbances of local blood velocity, which change local wall shear stress. Shear stress determines local nitric oxide production and low wall shear stress locations, induced by a catheter position close to the wall, may lead to a decrease in local nitric oxide production, which possibly results in local, asymmetric wall contraction. Similarly, locations with increased shear stress may dilate. For both acute and follow up studies of lumen size such variations are not desired. If shear stress variations could be induced locally in a controlled way, they might provide an interesting method to study local shear dependent endothelial function.

True 3-D reconstruction of the lumen and the vessel wall from ICUS information

For quantitative morphologic ICUS applications it is necessary to derive surface information from the luminal and external arterial wall. As a first step, contours of the luminal wall and external elastic lamina can be derived from the ICUS images in a less or more automated way[17]. For a true morphologic 3-D reconstruction of the lumen and wall the spatial orientation and location of those surfaces needs to be determined. Indeed, this implies the necessity of determining the exact spatial location and orientation of each sequentially collected ICUS cross section. The ICUS image orientation involves the spatial orientation of the imaging plane as well as the angular rotation of the image within this plane. Such information cannot be derived from ICUS itself and therefore other means, as discussed later, will be required.

Methods for ICUS transducer location determination

In parallel with the invention of ICUS, sophisticated techniques were proposed to determine the spatial location of cross sectional ICUS images[18]. For this purpose electromagnetic waves were generated from a few coils surrounding the patient and receiving coils mounted in the ICUS catheter. However, this technically interesting proposal has not come to age, probably because of its complexity, but also because in coronary applications tracking the transducer location in 3-D is of little help in getting a correct assembly of a series of ICUS images. This may seem paradoxical, but one should realize that catheter pullback will advance the transducer with relatively small axial distance increments compared to the much larger periodic displacements generated by the superimposed cardiac and respiratory motion. For this same reason, attempts utilizing biplane X-ray to record the ICUS transducer during pullback to reconstruct the 3-D transducer position using the marker reconstruction technique were also not very successful[19]. In addition, it must be realized that besides reconstruction of the transducer location, also the orientation of the imaging plane and the angular rotation of the images in this plane need to be determined and particularly the latter requires other means.

Fusion of 3-D angiography and 3-D ultrasound

From the above it will have become clear that neither 3-D angiography nor 3-D ultrasound itself will be able to present a true 3-D geometry of the coronary lumen and wall. Angiography has serious shortcomings in presenting cross sectional information, but has strong capabilities to present an overview of 3-D data in a 2-D projection format and therefore a biplane approach enables reconstruction of many 3-D features. Intravascular ultrasound presents high resolution local cross sectional information but fails in assembling this information in a spatially correct format. When we first presented[20,9] our method for true 3-D reconstruction using fusion of biplane angiographic and ICUS (ANGUS) contour information, the particular strength of each individual imaging technique was already fully utilized. This combination showed to contain sufficient information to solve all principal reconstruction problems mentioned before. We choose to predict the 3-D locations at which for a certain cardiac phase the subsequent ICUS images would be acquired, rather than trying to determine the 3-D transducer location during pullback. This goal could be achieved by relying on a sheath based ICUS catheter (2.9 F, MicroView, CVIS, USA). Such a catheter has an ultrasound translucent sheath, through which its rotating distal imaging part can be axially moved. The transducer rotates at 30 rev/s driven by an axially very flexible, but torsionally very stiff coiled axis. Because the sheath by nature of its tubing has a much higher axial stiffness than the imaging core, axial displacement of the core will not influence the lateral position of the sheath relative to the vessel. Furthermore, a sheath, once positioned in a curved vessel, strives to maintain a stationary position relative to the vessel, being determined by the condition of minimum energy stored in bending. This reduces the amount of lateral motion in an artery. Moreover, if position variations occur over the cardiac cycle these will be periodic and therefore predictable. For all of these reasons, at a certain phase of the cardiac cycle, a single recorded position of the sheath will accurately indicate the pullback path to be followed by the ICUS transducer relative to the lumen at that same cardiac phase. As one may realize, when using a non-sheath based ICUS catheter, which has more freedom for lateral motion, such a prediction will be less accurate[10]. Similarly, the angiographic lumen centerline, being only weakly related to the transducer path followed, cannot serve for an accurate ICUS path reconstruction. One of the main reasons why an accurate path reconstruction is essential, particularly its position relative to the vessel lumen, will be discussed later into detail and relates to reconstructing angular rotation of the ICUS image in the ICUS imaging plane.

Actual ANGUS imaging protocol and transducer path reconstruction

For the ICUS catheter pullback we apply a motorized, R wave triggered, stepped pullback (0.5 mm steps). Just after its start, when a possible backlash in the pullback system has been eliminated, we film in biplane (BICOR, Siemens A.G., Germany) the position of the sheath as indicated by the radio opaque distal imaging core. While filming (25 frames/s) contrast (Iopamiro 379, Bracco, Italy) diluted with saline (2:1 vol/vol) is injected, which allows simultaneous visualization of the lumen borders and the catheter path. During filming the patient is asked to stop breathing in order to grasp at least one cardiac cycle with as little as possible breathing motion. The biplane recordings of the catheter centerline, which we call the catheter coreline, are used to reconstruct the coreline in 3-D using the spatial curve reconstruction technique as described above. The coreline is reconstructed from its tip to its intersection with the entrance of the guiding catheter (7F). During pullback the timing of all events, i.e. all pullback steps and the X-ray filming pulses are recorded together with the electrocardiogram using a digital data recording system. This information is used to identify ICUS images that are recorded at the same cardiac phase as the coreline and to identify ICUS images that corresponds with the moment selected for 3-D coreline reconstruction. On the reconstructed 3-D coreline, equidistant locations are added at which the corresponding ICUS cross sections are repositioned[10]. Obviously, the ICUS image is always repositioned with its center on the coreline. The ICUS transducer location at filming is reconstructed on the 3-D coreline at a known distance proximal of the distal tip. The position of the ICUS image showing entrance of the guiding catheter coincides with the proximal coreline point. The length of the reconstructed coreline from transducer to guiding differs less than a few mm from the recorded cumulative stepping distance[10].

Reconstruction of the local ICUS imaging plane orientation

At the reconstructed ICUS locations, the imaging plane orientation is positioned perpendicular to the coreline direction. Because the direction of the imaging core in a sheath-based catheter is always aligned with the direction of the sheath, deriving this imaging plane orientation from the sheath orientation will be very accurate. However, when applying a similar approach to derive the imaging plane orientation from the transducer path of a non-sheath based catheter, the local catheter tip direction may deviate from the local path direction[10]. Otherwise, when using the angiographic centerline as an approximation for the transducer pullback path, the reconstruction of the ICUS imaging plane will be rather inaccurate.

Reconstruction of the angular rotation in the imaging plane

The last degree of freedom to be determined is the angular orientation of the ICUS image in the imaging plane. Technically, it might be possible to integrate a coil with the rotating ultrasound transducer, in order to sense an electromagnetic field[18] that determines continuously the transducer angular orientation. This also might serve to correct for errors resulting from non-uniform rotation in ICUS catheters applying a rotating transducer. Other means, for example a marked sheath, may also serve this goal and may improve speed and accuracy of 3-D ICUS reconstruction [21] . The ANGUS method, fusing angiographic and ICUS information, derives the angular rotation from the diluted contrast angiogram made at the start of pullback. These images show in biplane both the coreline and the lumen borders with the variation in their respective lateral distances over the pullback trajectory. Similar distance information is contained within the ICUS images. The correspondence between angiography and ICUS is used to determine the correct angular orientation. This will be explained in more detail. Figure 8 gives an example of a "catheter", which for clarity has a marked edge. The catheter stretches over a series of 3 successive planar bends. Interestingly, comparing both ends of the catheter a 90 degrees difference in its angular rotation is observed, while the catheter direction is identical. This phenomenon is geometry determined and does not introduce torsional stress in the catheter. Mathematically it is described by the Frenet-Serret formulas[22] and in our previous publications we called it twist. One should realize that the marked edge could indicate a reference direction for expressing the angular rotation of the ICUS images. When stacking successive ICUS images, one has to account for this twist to get an accurate 3-D reconstruction.

In our reconstruction algorithms, when repositioning a series of cross sectional contours onto a reconstructed catheter path, we always used the illustrated bending principle[20,9]. Each curved path, consisting of a series of reconstructed locations, can be subdivided in groups of three successive locations. For each trio it holds that related to the plane through this trio the angular orientation of the catheter is

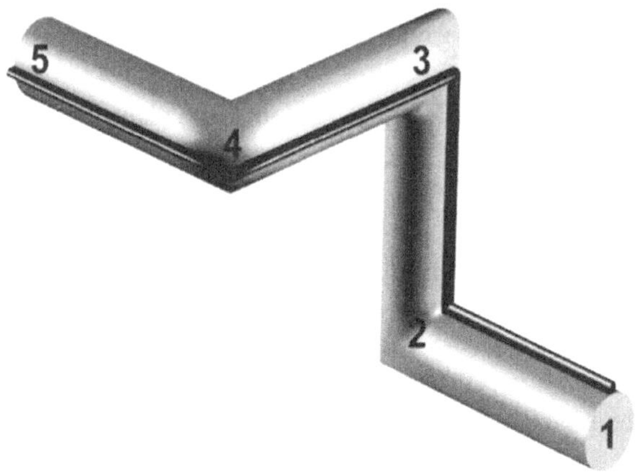

Figure 8. Following the course of a catheter having a marked edge through successive bends, rules are derived to determine the relative angular orientation of successive cross sectional ICUS images. Note that for each successive trio of points the angular orientation of the edge is parallel to the plane through this trio. This is in accordance with normal bending of the catheter without introducing torsion. Yet, at identical oriented parts 1-2 and 4-5 the catheter shows a 90° turn.

constant. Locations 1, 2 and 3 in Figure 8 might represent such a trio. Therefore, the angular orientation at location 1 relative to the plane through 1,2 and 3 determines the angular orientation up to point 3. Similarly the sequence of points 2, 3 and 4 determines a plane and the angular orientation with respect to this plane is similarly unchanged from 2 to 4. The amount of catheter twist in coronary phantoms and a more detailed description of this planar bending algorithm have been reported by others[23].

Knowing how to correct angular orientation between successive points leaves the question how to determine the absolute angular rotation of the complete set of contours. Earlier we resembled this problem with the realignment of a sock, put on in an unknown, generally wrongly rotated manner[9]. In Figure 9 this is illustrated.

Starting (Figure 9, top left) with the most distal lumen contour (in ICUS plane) at some arbitrarily chosen angular rotation, the subsequent contours are added according to the bending principle. Figure 9, top right shows how a closed lumen surface can be build up from a myriad of triangular surfaces connecting neighboring cross sections. When starting with some other angular rotation for the distal plane (Figure 9 bottom) another spatial lumen shape will be obtained. If this would be correct, we add to this set of contours the outer vessel wall data, to complete the reconstruction with the arterial wall.

To determine the correct "sock rotation" we build a series of 36 3-D lumen reconstructions, with 10 degrees increments in its angular rotation at the most

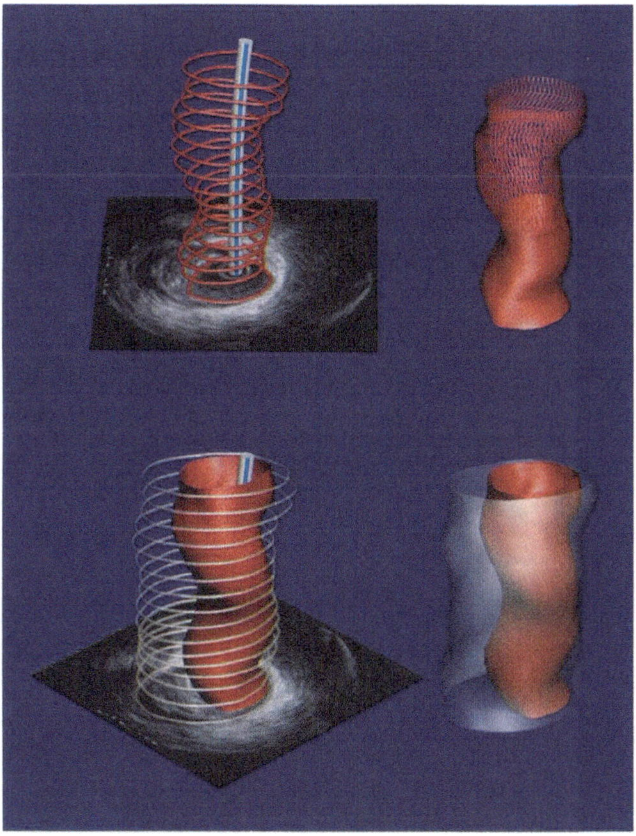

Figure 9. top left: Starting with the most distal contour (ICUS plane) at some arbitrary angular rotation, subsequent contours are stacked around the coreline according to a bending principle as described in the text. **top right**: Connecting neighboring contours with triangular surfaces allows generation of a closed lumen surface. **bottom left**: Starting with another angular rotation will result in a different lumen shape. The blue stripe on the coreline indicates that the orientation of the coreline does not change. This figure also illustrates how contours of the external elastic lamina may be added to complete the reconstruction of the arterial wall **(bottom right)**.

distal location. Subsequently, of all these 36 different sock rotations the biplane 2D projections are determined, which are quantitatively compared with the actual biplane angiogram[9,24]. Particularly the location of the coreline relative to the lumen borders is compared in both types of images. For this purpose, the lateral distance between the coreline and the lumen borders is measured at 64 equidistant locations in each view and compared using linear regression analysis. The highest positive correlation indicates the correct sock rotation. Figure 10 illustrates some distance measurements for various sock rotations of a 3-D stent reconstruction, with the accompanying angiographic measurements. Also the correlation results for the different sock rotations combined for both views are shown.

To determine the sock rotation, a local diameter comparison has limited value because varying the sock rotation will only affect the projected diameter in case of an elliptical lumen cross section, which occurs not that frequent. Obviously, the distance method is dependent on the validity of the path prediction. Clearly, the very high correlation factors obtained with the distance regression analysis confirm the validity of the assumptions made for the sheath based catheter. When applying this method to a free floating catheter, it is not expected that a reliable sock rotation can be determined. Some results of ANGUS reconstructions, together with the accompanying biplane angiograms, are shown in Figure 11.

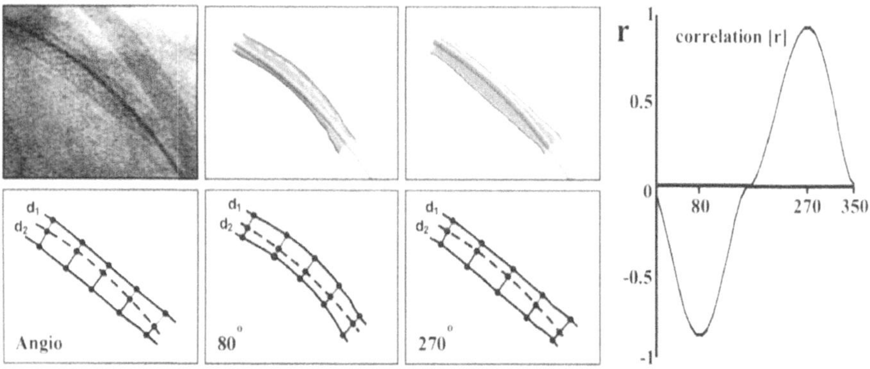

Figure 10. In the upper left panel an X-ray projection of a stent is shown as well as the coreline of an ICUS catheter. In the bottom left panel the stent contours and the course of the coreline are drawn. At 64 locations (only 5 indicated) distances d_1 and d_2 are determined which indicate the distance between coreline and stent contour in this projection. Two, out of 36 possible 3-D stent reconstructions at 10° angular rotation increments (details see text) are shown for respectively 80° and 270°. Note the marked differences in reconstructed stent curvature. For these two rotations, correlations in 2-D between distances of X-ray and the reconstruction projection are maximal. At maximal positive correlation (270°) the correct solution is obtained.

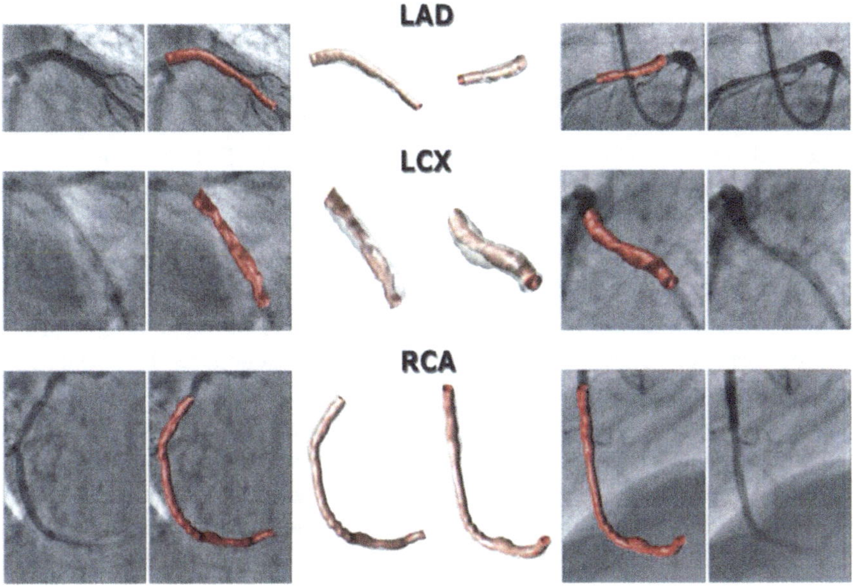

Figure 11. Compilation of representative Angus reconstructions of lumen and wall of a left anterior descending coronary artery (top), a left circumflex coronary artery (middle) and a right coronary artery (bottom). Viewing directions on the 3-D ANGUS reconstruction correspond as close as possible with the X-ray biplane projection views. Distance correlations for these three views were respectively: 0.96, 0.89 and 0.79.

Limitations

When applying the ANGUS method minor inaccuracies may still occur as a result of breathing. These may be due to small sheath displacements, to periodic variations in the ventricular end diastolic filling state or to lumen size variations during pullback. Severe stenoses or acute bends will pose problems if the driving coreline of the ICUS transducer perceives a local increase in friction, which may introduce an angular rotational shift in the ICUS image. Such locations may also introduce distortion of the ICUS image by temporary non-uniform rotation. In general, however, a coronary lumen with such a severe narrowing will not be investigated with ICUS. This however implies that ICUS is selectively applied and one must realize that this may affect the conclusions reached in patient studies.

Applications and future developments

We started the development of a method to achieve a morphologically true lumen and wall reconstruction in three dimensions with the final goal to study the relation between hemodynamic parameters and characteristics of the arterial wall in disease and healing in the patient. The lumen reconstruction then is used as input for computational hemodynamics. This implies that a multitude of 3-D brick like elements (each containing 27 nodes) needs to be generated which fill the lumen. After specifying boundary conditions, including flow and the type of fluid to be modeled, flow velocity is calculated at each node by a numerical computational approach that solves the Navier-Stokes equations. This yields the complete blood flow velocity field through the lumen. From the velocity field the important biomechanical parameter wall shear stress can be derived. This computed quantity is locally related to the measured wall thickness in the reconstruction. On this subject we have shown in patients that both the location of native atherosclerosis[25] as well as that of neointimal growth after stent placement[26] are related to wall shear stress. These studies confirmed in patients that wall shear stress, although being very weak (approximately 1/10000th of blood pressure), induces important vascular effects as now becomes commonly acknowledged[27].

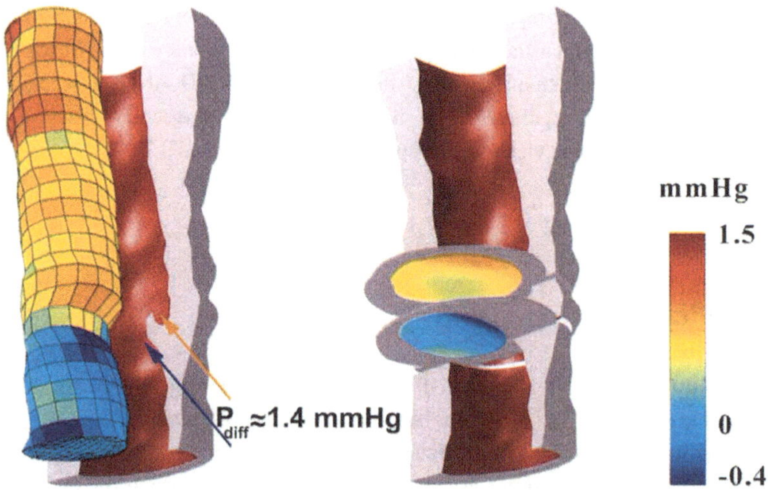

Figure 12. left part: Artery of a patient with a dissection at the edge of a stent (not visualized) observed just after stenting. Computational hemodynamics rendered the pressures at the luminal wall, here shown at the surface of the finite elements filling the lumen. Pressure is relative to the distal (bottom) cross section, flow is from top to bottom. **Right part:** pressure differences over the cross sections proximal and distal of the dissected cross section. The pressure difference (P_{diff}) acting on the dissected flap cannot explain propagation of the dissection as occurred later in this patient[28].

In another study we evaluated whether the flow induced pressure force acting on a dissected flap might be a stimulating factor for dissection propagation. It appeared that flow had only a minor role while changes in tensile stress must be the major determinants[28] of dissection propagation. For an artery presenting a dissection just distal of a stent, figure 12 (left part) shows calculated pressures at the surface of the luminal bricks relative to the distal (bottom) cross section. The right part shows the pressure variations over two cross sections proximal and distal of the dissected cross section. The indicated pressure difference of approximately 1.4 mmHg cannot explain the later observed distal propagation of the dissection in this patient[28]. Mostly, from a mechanical point of view vascular forces generated by flow induced pressure differences or wall shear stress are compared to blood pressure induced forces. Probably this reasoning also holds for new studies focusing on plaque rupture where the shear stress induced biological vascular effects and the blood pressure related mechanical forces will dominate the mechanical forces induced by flow. Obviously flow studies may be extended with other biomechanical factors like wall stress to further explore this subject. Also studies over time, which will show the effects of shear stress on healing after mechanical or pharmaceutical interventions will be performed.

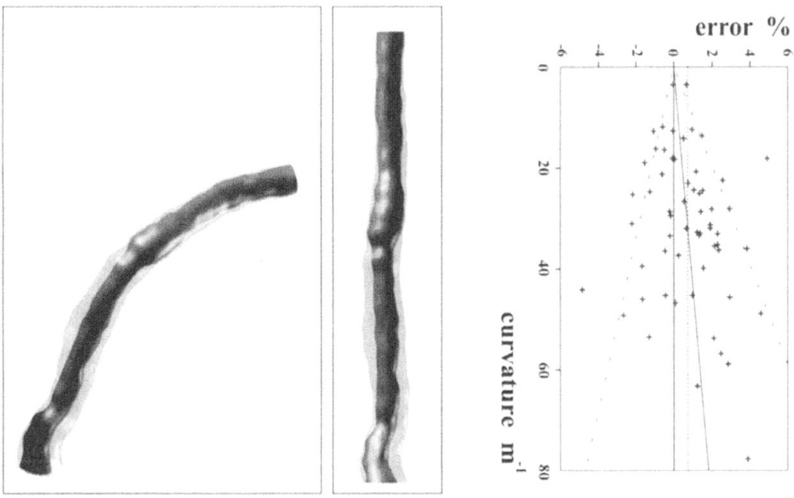

Figure 13. Errors in 3-D wall volume calculations from ICUS when neglecting curvature have been evaluated with ANGUS. The error increases with curvature of the ICUS catheter path, but remains below 6% for a representative sample of coronary arteries.

Meanwhile, several other applications have emerged from the 3-D reconstruction technique as well. In the field of quantitative ICUS, arterial wall volume is determined over a certain vessel length in order to have a measure for the degree of atherosclerosis. In these calculations the effects of vessel curvature are neglected because a linear pullback trajectory is assumed. Comparing the outcome of this approach with data obtained from 3-D ANGUS coronary reconstructions [29] showed that the errors introduced by this assumption are dependent on the curvature of the pullback trajectory rather than tortuosity of the artery. As would be expected the error indeed increases with increasing local coreline curvature (Figure 13).

In addition it will be possible to study the relationship between the 3-D location of ICUS derived tissue characteristics like lipid pools or calcium deposits and the biomechanical factors shear stress and wall stress. Also studies of coronary shape changes over the cardiac cycle may be envisioned in order to study the effect of periodic bending of a coronary segment and its hypothesized relation with local atherosclerosis. In one of our studies performed in pigs, we determined coronary shape changes induced by placement of Wallstents and their effects on shear stress[30] (Figure 14).

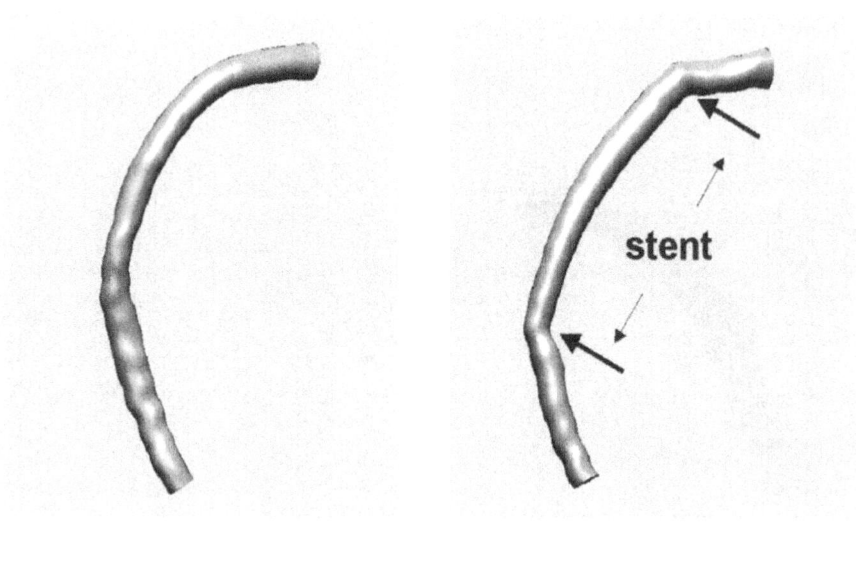

Figure 14. Coronary 3-D shape changes induced by placement of Wallstents in pig coronary arteries have been evaluated with ANGUS. Such changes will induce local shear stress deviations, which play a role in the distribution of intimal hyperplasia.

Besides the future extended use of ANGUS in patient studies as described above, efforts will also be directed to extend the current reconstruction possibilities. Biplane installations are not that widely available and if present, logistic problems pose some limitation to frequent application of ANGUS. Therefore by applying respiration registration it might be possible to use serial monoplane acquisition of the catheter coreline to predict the spatial 3-D pullback trajectory with sufficient accuracy.

Currently, only rather short coronary segments, not containing major side branches, are used in the shear stress calculations. Major side branches will have a significant effect on the outcome of the flow velocity calculations near branches. Therefore we will investigate whether the already locally available information in ICUS near branches can be used to extend the reconstruction to include thebifurcation geometry. In addition, a sequential pullback may be applied in related arterial branches and after fusing this information it will yield a complete geometry of a main arterial segment with a major branch.

Further it may be desirable to extend studying arterial geometry over the heart cycle rather than at a single moment as is currently performed. This is not a principal problem, but considering the required analysis effort, further automation of the reconstruction procedures will be necessary for such applications. New, improved automated ICUS segmentation methods will give an important contribution to this subject.

References

1. Thomson E. Stereoscopic roentgen pictures. Electrical Engineer 1896, 21, 256
2. Selvik G, Roentgen stereophotogrammetric analysis. Acta Radiol 1990; 31(2):113-26
3. McKay S, Potel MJ, Rubin JM. Graphics methods for tracking three-dimensional heart wall motion. Comp Biomed Res. 1982;15:455-473
4. Wollschläger H, Lee P, Zeiher A, Solzbach U, Bonzel T, Just H. Mathematical tools for spatial computations with biplane isocentric X-ray equipment. Biomed Tech (Berl). 1986; 31(5):101
5. Wahle A, Wellnhofer E, Mugaragu I, Sauer HU, Oswald H, Fleck E. Assessment of diffuse coronary artery disease by quantitative analysis of coronary morphology based upon 3-D reconstruction from biplane angiograms. IEEE Transactions on medical imaging, 1995;14:230-241
6. Reiber JHC, Gerbrands JJ, Boomań F, Troost GJ, Boer A den, Slager CJ, Schuurbiers JCH. "Objective characterization of coronary obstructions from monoplane cineangiograms and three-dimensional reconstruction of an arterial segment from two orthogonal views" In: Appl Comp in Med. MD Schwartz (ed.) 1982; IEEE cat. TH0095-0, 93-100.
7. Parker DL, Pope DL, Van Bree R, Marshall HW. Three-dimensional reconstruction of moving arterial beds from digital subtraction angiography. Comput Biomed Res 1987;20:166-85

8. Metz CE, Fencil LE. Determination of three-dimensional structure in biplane radiography without prior knowledge of the relationship between the two views: theory. Med Phys. 1989;16(1):45-51
9. Laban M, Oomen JA, Slager CJ, Wentzel JJ, Krams R, Schuurbiers JCH, den Boer A, von Birgelen C, Serruys PW, de Feyter PJ. ANGUS: A new approach to three dimensional reconstruction of coronary vessels by combined use of angiography and intravascular ultrasound. Computers in Cardiology, IEEE Comp. Soc, Piscataway, NJ. 1995:325-328
10. Slager CJ, Wentzel JJ, Schuurbiers JC, Oomen JA, Kloet J, Krams R, von Birgelen C, van der Giessen WJ, Serruys PW, de Feyter PJ. True 3-Dimensional reconstruction of coronary arteries in patients by fusion of angiography and IVUS (ANGUS) and its quantitative validation. Circulation, 2000, 102(5):511-6
11. Guggenheim N, Chappuis F, Suilen C, Doriot PA, Dorsaz PA, Descouts P, Rutishauser W. 3-D reconstruction of coronary arteries in view of flow measurement. Int J Card Imaging, 1992, 8(4): 265-272
12. Slager CJ. "Echo-vonkerosie rekanalisatie inrichting" Dutch patent application 1987, 87.00632
13. Bom N, Slager CJ, Phaff AC. "Sensing methods for selective recanalization by spark erosion" In: Proc. Ninth Annual Conf of the IEEE Engineering in Medicine and Biology Soc. 1987, 205-206
14. Rosenfield K, Losordo W, RAmaswamy K, Pastore JO, Langevin E, Razvi S, Kosowsky BD, Isner JM. Three-Dimensional reconstruction of human coronary and peripheral arteries from images recorded during two-dimensional intravascular ultrasound examination. Circulation 1991;84:1938-1956
15. Li W, Bom N, van Egmond FC. Three-dimensional quantification of intravascular ultrasound images. J Vasc Invest 1995, 1(2): 57-61,
16. Bruining N, von Birgelen C, di Mario C, Prati F, Li W, den Hoed W, Patijn M, De Feijter PJ, Serruys PW, Roelandt JRTC. Dynamic three-dimensional reconstruction of ICUS images based on an ECG-gated pull back device. Computers in Cardiology, IEEE Comp. Soc, Piscataway, NJ, 1995:633-636
17. Li W, Bosch JG, Zhong Y, van Urk H, Gussenhoven EJ, Mastik F, van Egmond F, Rijsterborgh H, Reiber JHC, Bom N. Image segmentation and 3-D reconstruction of intravascular ultrasound images. In: Acoustical Imaging, ed. Y. Wei and B. Gu. Plenum Press, NY,1993:489-496
18. Martinelli MA, von Thuna P, Acoustic image system and method. USA patent 1987, 4821731
19. Evans JL, Ng KH, Wiet SG, Vonesh MJ, Burns WB, Radvany MG, Kane BJ, Davidson CJ, Roth SI, Kramer BL, Meyers SN, McPherson DD. Accurate three-dimensional reconstruction of intravascular ultrasound data. Spatially correct three-dimensional reconstructions. Circulation. 1996;93(3):567-76.
20. Slager CJ, Laban M, von Birgelen C, Krams R, Oomen JAF, den Boer A, Li W, de Feijter PJ, Serruys PW, Roelandt JRTC. ANGUS: A new approach to three-dimensional reconstruction of geometry and orientation of coronary lumen and plaque by combined use of coronary angiography and IVUS. J Am Coll Cardiol, 1995, 25:144A.
21. Slager CJ. Catheter for obtaining three-dimensional reconstruction of a vascular lumen and wall. United States Patent 5,771,895, 1996.

22.In: Davis HF, Snider AD. Introduction to Vector analysis. Wm. C. Brown Publishers, 1988, 86: 88

23.Prause GP, DeJong SC, McKay CR, Sonka M. Towards a geometrically correct 3-D reconstruction of tortuous coronary arteries based on biplane angiography and intravascular ultrasound. Int J Card Imaging 1997;13:451-62

24.Slager CJ, Wentzel JJ, Oomen JA, Schuurbiers JC, Krams R, von Birgelen C, Tjon A, Serruys PW, de Feyter PJ. "True reconstruction of vessel geometry from combined X-ray angiographic and intracoronary ultrasound data." Semin Interv Cardiol 2(1):43-47, 1997

25.Krams R, Wentzel JJ, Oomen JA, Vinke R, Schuurbiers JC, de Feyter PJ, Serruys PW, Slager CJ. Evaluation of endothelial shear stress and 3-D geometry as factors determining the development of atherosclerosis and remodeling in human coronary arteries in vivo. Combining 3-D reconstruction from angiography and IVUS (ANGUS) with computational fluid dynamics. Arterioscler Thromb Vasc Biol 1997;17(10):2061-2065

26.Wentzel JJ, Krams R, Schuurbiers JC, Oomen JA, Kloet J, van der Giessen WJ, Serruys PW, Slager CJ. Relationship between neointimal thickness and shear stress after Wallstent implantation in human coronary arteries. Circulation. 2001,103(13):1740-5.

27.Malek AM, Alper SL, Izumo S. Hemodynamic shear stress and its role in atherosclerosis. JAMA. 1999;282(21):2035-42.

28.Thury A, Wentzel JJ, Schuurbiers JC, Ligthart JM, Krams R, de Feyter PJ, Serruys PW, Slager CJ. Prominent role of tensile stress in propagation of a dissection after coronary stenting: computational fluid dynamic analysis on true 3-D reconstructed segment. Circulation. 2001 Sep 11;104(11):E53-4.

29.Schuurbiers JC, von Birgelen C, Wentzel JJ, Bom N, Serruys PW, de Feyter PJ, Slager CJ. On the IVUS plaque volume error in coronary arteries when neglecting curvature. Ultrasound Med Biol. 2000; 26(9):1403-11.

30.Wentzel JJ, Whelan DM, van der Giessen WJ, van Beusekom HM, Andhyiswara I, Serruys PW, Slager CJ, Krams R. Coronary stent implantation changes 3-D vessel geometry and 3-D shear stress distribution. J Biomech. 2000, 33(10):1287-95.

Shear stress and the IVUS derived vessel wall thickness

J.J. Wentzel[1,2], C. Cheng[2], R. de Crom[3,4], N. Stergiopulos[5], P.W. Serruys[1], C.J. Slager[1], and R. Krams[1]

[1]Thoraxcenter, [2]ICIN, [3]Cell Biology and Genetics, [4]Vascular Surgery, Erasmus-MC, The Netherlands and [5]Ecole Polytechnique, Lausanne, Switzerland

Summary: The role of shear stress in atherosclerosis has been well documented in vitro. In this chapter a novel technique and several applications will be described. The technique consists of a combination of 3D imaging of blood vessels and Computational Fluid Dynamics. The 3D imaging techniques consist either of 3D-IVUS or a combination of 3D IVUS and Angiography ('ANGUS'). In the applications of these techniques it will be demonstrated that shear stress plays a role in the prediction of plaque location in in-stent restenosis, in vascular remodeling after balloon angioplasty, and in plaque location in primary atherosclerosis. Attempts to locally increase shear stress by a newly developed flow divider indicate that shear stress accounts for 50% of restenosis.

In conclusion, this chapter describes several new techniques to estimate shear stress and describes the role of this parameter in several conditions, both in humans and in animals in vivo.

Key words: biomechanics, atherosclerosis, shear stress, vascular remodeling, in-stent restenosis, PTA

Introduction

Atherosclerosis is still the leading cause of mortality in the western world. Several risk factors have been identified that contribute to the progression of the disease. The most important risk factors include smoking, hypertension, diabetes and increased levels of plasma cholesterol. However, since these risk factors are systemic of nature, they cannot explain the observation that atherosclerosis occurs predominantly at certain locations like bifurcations and side branches. As these predilection sites are locations with deviations of the normal velocity field, it has been postulated that the coupling of the velocity field to the endothelial cells (i.e. the shear stress) plays an important role in plaque localization and plaque growth [1-3].

Shear stress is the frictional force, exerted by the flowing blood on the endothelial cell surface. Under normal conditions a value close to 1.5 Pa (1 Pa =1 N/m^2) is measured in the arterial tree. This value is thought to result from a tight regulation intending at keeping this value constant during a wide range of physiological conditions.

To study the influence of shear stress on the location of athero sclerotic plaques and the development of neointima and vascular remodeling after interventions, the local shear stress, and the local wall (neointimal) thickness are required. A technique that enables to do so is developed by Brands *et.al.*, who estimates the local shear rate and vessel diameter by high resolution ultrasound [4]. This technique is applicable to all superficially located arteries and Kornet *et al.* showed in the carotid arteries of healthy patients that the intimal – medial thickness was inversely related to shear stress[5].

For deeper located blood vessels no technique was available to determine local shear stress values accurately. Hence we developed a method, to obtain shear stress in the combination of an imaging technique and computational fluid dynamics (CFD). As imaging technique, IVUS was used. However, for curved vessels the bending of the IVUS catheter introduced straightening effects, which are not allowed for CFD techniques. Hence, a new technique (ANGUS) based on biplane angiography and Intravascular Ultrasound (IVUS) was developed. ANGUS provides high resolution of both the lumen geometry and the local wall thickness for curved blood vessels. The combination of ANGUS with computational fluid dynamics offers the unique possibility to acquire a detailed measurement of the local wall geometry and to combine this with local shear stress values in humans in vivo. In animal studies similar techniques are applied to study shear stress in atherosclerosis and restenosis. Due to the extra possibilities offered by these studies, the underlying mechanism are studied

Methods

3-D reconstruction techniques

Straight arteries

3-D reconstruction of straight arteries, like the external iliac artery may be based on intravascular ultrasound (IVUS) or other techniques. In this chapter we will discuss IVUS based techniques only. Intravascular ultrasound is a technique that provides cross sectional images in vivo of the vessel wall and lumen. Intravascular ultrasound (IVUS) catheters (2.9 or 4.3 F; 30 MHz; Du-MED, Rotterdam, the Netherlands) are used to record the target area by a continuous pull back, starting from a distal reference point, like a side branch, to the entrance of the guiding catheter. The IVUS images are recorded on VHS video-tape and digitized at intervals of 0.5 mm. Lumen and media are traced semi-automatically by a well-validated software package. The lumen and media contours are positioned perpendicular to the IVUS catheter axis at intervals of 0.5 mm and a 3-D reconstruction of the lumen and wall of a straight artery, like the iliac arteries, is obtained.

Curved arteries

3D-reconstruction of the coronary arteries is performed applying a combination of ANGiography and intravascular UltraSound (ANGUS). A detailed description of the 3D-reconstruction is presented elsewhere in this volume [6]. Briefly, a sheath based intravascular ultrasound (IVUS) catheter (CVIS 2.9F, Sunnyvale CA, USA) is positioned distally from the vessel segment and is filmed with a biplane angiographic system (Siemens, Bicor, Erlangen, Germany) just after the start of pull back. To eliminate respiratory and cardiac motion artifacts, a single biplane view at end-diastole of the catheter position is selected and digitized. From the biplane views, the transducer path is reconstructed in 3D space[7]. In addition, (IVUS) images are collected at end-diastole using an ECG triggered, motorized pull back operating with a stepsize of 0.5 mm (TomTec, Munich, Germany). Subsequently, the frames are digitized and analyzed with a semi-automatic contour detection program[8]. Output of the program consists of lumen contours, signifying the blood-vessel interface and media contours signifying the media – adventitia interface.

Subsequently, the lumen contours are filtered and positioned perpendicular onto the reconstructed 3-D catheter path, which serves as a backbone for the reconstruction. The angular position of the ultrasound transducer, and thus of the ultrasound images, is determined from a comparison between simulated silhouette images derived from the 3-D reconstruction with the actual coronary angiogram[6, 9]. Finally, a 3-D reconstruction is obtained containing vessel lumen and vessel wall.

Computational fluid dynamics

The numerical modeling of incompressible and isothermal flow of a (generalized) Newtonian fluid, requires the solution of the Navier-Stokes equations and the continuity equation:

$$\rho \dot{v} + \rho(v \cdot \nabla)v - \nabla \cdot \sigma = f$$

$$\nabla \cdot v = 0$$

where v is the velocity vector, σ the Cauchy stress tensor, f the body force per unit mass, ∇ the gradient operator, ρ the density and $\dot{v}$ is the partial time derivative. Discretization using Galerkin's finite element method leads to a set of non-linear differential equations. Linearization of the non-linear convective term is performed applying the Newton-Raphson method[10]:

$$(v \cdot \nabla v)^{n+1} \approx v^n \cdot \nabla v^{n+1} + v^{n+1} \cdot \nabla v^n - v^n \cdot \nabla v^n$$

Mesh definition

In order to use a finite element method to obtain an approximate solution of the incompressible 3-D Navier Stokes equations, the lumen of the reconstructed arteries is filled with a 3-D mesh consisting of 32 elements in each cross-section. The Crouzeix-Raviart type elements used in this mesh contain 27 nodes, with a quadratic approximation for the three velocity components. Furthermore, the pressure is defined in the centroid. The resolution of the applied mesh is dependent on the amount of cross sections. The axial mesh resolution of the utilized meshes ranged from 0.43 mm to 0.74 mm.

Boundary conditions and Assumptions

In order to solve the Navier Stokes equations, we need to determine the properties of the applied fluid, the inflow and outflow velocity conditions and the velocity conditions at the vessel wall. We assume that blood behaves as a Newtonian fluid with a viscosity of $3*10^{-3}$ Pa.s and with a density of 1050 kg/m^3. Since the shear stress is linearly related to the flow, the flow should be determined from measurements obtained from the flow through the original arteries. However, flow measurements are not always applicable in an in vivo situation, so we need to determine a flow close to the original situation. However, one should realize that the relative shear stress distribution in a geometry is hardly dependent on the flow [27]. The basal flow through the coronaries was derived such to obtain an average shear stress of 1.5 N/m^2 [11, 12] and for the iliac arteries a flow of 80 ml/min [13] was assumed. A parabolic inflow profile was implemented. As the entrance diameter of each artery was different, mean entrance velocity varied.

The other boundary conditions are: zero stress in the normal and tangential direction at the exit of the blood vessel and no slip at the vessel wall. The resulting equations are solved with a validated finite element package (Sepran, Sepra, Leiden, the Netherlands)[14] applying a direct solver. The iteration process is tested for convergence using the following criteria:

$$\left\| v^{l+1} - v^l \right\| \leq \varepsilon ; \varepsilon = 10^{-4}$$

With v is velocity, l the iteration step, and || || the maximum norm.

Post Processing

The lumen and media contours allow to derive the local lumen area (LA), the wall area (WA), the media bounded area (MBA) and the local wall thickness or neointimal thickness. At the same locations the shear stress is obtained by multiplying the shear rate determined by computational fluid dynamics with the viscosity. Furthermore regional wall stress (WS) for each cross section is derived from the Laplace formula. In order to compare the parameters before, after intervention and at follow up, we aligned the arteries according to a distal reference side branch.

Results and Discussion

Shear stress in atherosclerosis

Shear stress and its localizing role in atherosclerosis has been postulated by Caro *et. al.* in 1971. They proposed a shear stress dependent transfer of lipids into the vessel wall[15]. According to that theory, *low shear stress* promotes the interaction between blood borne elements and the vessel wall. Fry, on the other hand, provided experimental evidence for a *high shear stress* related desquamation of the endothelium [16]. Due to the lost barrier function of the endothelium, diffusion of blood borne elements into the vessel wall might be facilitated [16]. Later studies clearly indicated that low shear stress, but not high shear stress promotes atherosclerotic development [17-19]. According to numerical modeling studies, low shear stress regions are accompanied by local high LDL concentrations in the blood near the arterial wall [20]. In contrast, the permeability of an artery is positively related to the shear stress [21, 22]. Consequently, the accumulation of LDL in the vessel wall, which is measured in low shear stress regions [23], is due to the increase in local LDL concentration partly antagonized by the low endothelial permeability. Despite the consensus that low shear stress regions favor atherosclerosis, evidence for the validity of this theory originates mainly from animal experiments and from post mortem studies in humans [24-26]. The post mortem specimens were used to construct a 3-D-model of the artery. With the help of laser Doppler techniques or computational fluid dynamics [25], the shear stress at the wall was derived by measuring the velocity profile near the wall and a relation was found between shear stress and the location of atherosclerotic plaques. All these techniques do not allow estimating the shear stress in deeper located arteries like the aorta or the coronary arteries in vivo. We studied the relationship between the localization of atherosclerotic plaques and shear stress in human coronary arteries in vivo by applying the above described techniques [27]. Figure 1 shows the 3-D reconstruction of an atherosclerotic right coronary artery of a patient. This 3-D

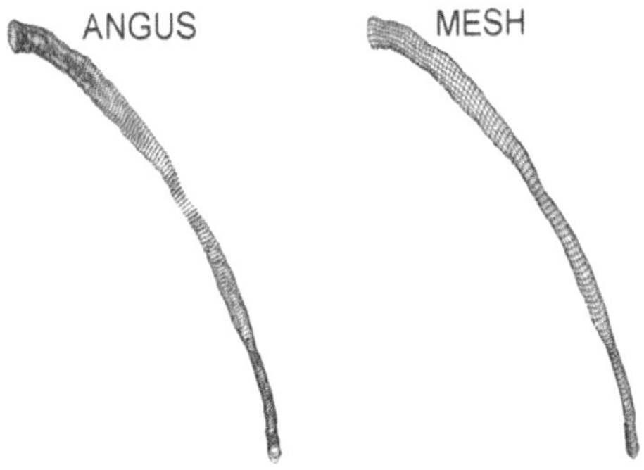

Fig. 1. 3D reconstruction of a right coronary artery from a combination of ANGiography and IVUS (ANGUS). The left panel displays the applied luminal mesh for application of computational fluid dynamics

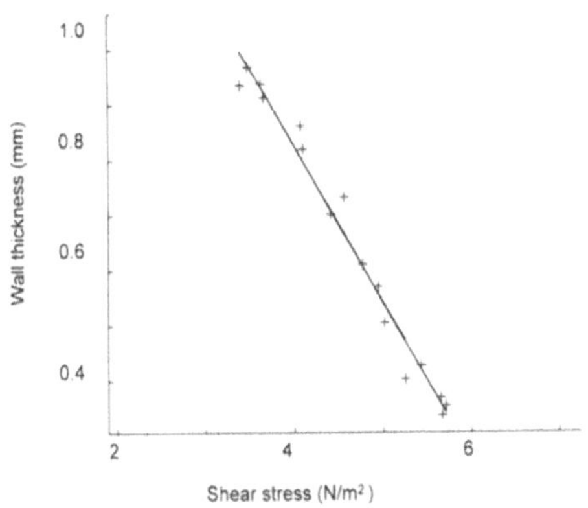

Fig. 2. Relationship between shear stress and echographic derived vessel wall thickness after averaging in the axial direction of a human right coronary artery. Inflow velocity of 20 cm/sec.

reconstruction was applied to perform the shear stress calculation by computational fluid dynamics. This study confirmed earlier observations, that the amount of atherosclerotic plaque is inversely related to shear stress, implying that at low shear stress regions, often observed in the inner curve of a coronary artery, atherosclerotic plaque is most prominent (Figure 2). This observation was regardless the assumed inflow conditions as can be appreciated from the three regression lines referring to conditions where average coronary flow velocity equaled 10, 20 and 30 cm/sec (wall thickness=-1.28* shear stress +3.18 [mm]; wall thickness=-0.30 * shear stress+2.00 [mm]; wall thickness=-0.10 * shear stress+2.00 [mm]).

Shear stress and vascular remodeling

Sustained increments in flow increase the vessel diameter mediated by the activation of several genes [11, 28]. Such a structural adaptation of the vessel wall is entitled 'vascular remodeling' and has a much larger effect than the acute response. For instance, 2 to 3 weeks of steady flow increment the diameter of the carotid artery in atherosclerotic rabbits increases by 83% of the original diameter. This contrasts with the range of adaptation to acute increments in flow, which increased the vessel diameter only in the order of 10% [29]. The above-mentioned 83% increase of vessel diameter was accompanied by a reduction of shear stress to its baseline values. Hence, it has been postulated that a negative feedback loop exists between shear stress and the vessel lumen with shear stress as the controlling variable [28]. Glagov's group showed that this negative feedback loop indeed exists and also remains intact in experimental atherosclerosis [11]. Therefore, due to the high compensatory potential of the feedback loop, this mechanism is able to adjust for most of the atherosclerotic plaques found in patients. As shown by the same group of researchers, human coronary vessels show expansion of their outer vessel circumference until the plaque area occupies approximately 40% of total wall area. Thereafter, compensation is lost and lumen narrowing occurs [30]. Later studies confirmed this post mortem study and extended the findings in vivo by intra vascular ultrasound (IVUS) measurements in humans [31].

Vascular remodeling was not only observed during physiological adaptation of the arteries or during atherosclerotic plaque development but was also recognized as the main factor responsible for restenosis after balloon angioplasty[32]. Post showed in the normal rabbit and the hypercholesterolemic Yucatan micropig that 75% of the lumen narrowing was caused by constrictive vascular remodeling compared to only 25% neointimal formation [32]. However, the question remained whether this constrictive remodeling after balloon angioplasty is similar to the physiological adaptation, regulated by shear stress and or wall stress. We argued that if shear stress and/or wall stress are important regulators in vascular remodeling after balloon angioplasty than they should comply to the following constraints. Firstly, shear or wall stress should change after balloon angioplasty, secondly the change of wall stress or shear stress induced by balloon angioplasty should be predictive of vascular remodeling, and thirdly, vascular remodeling should stop when shear stress or wall stress have been returned to 'normal values'.

In order to study whether a control system is operable during constrictive remodeling after balloon angioplasty, we applied the above described technique in the iliac arteries of atherosclerotic Yucatan pigs and tested the above described requirements[33]. Firstly, our results showed that after balloon angioplasty in these arteries indeed the shear stress decreased up to 87% and wall stress increased up to 20% for high acute gains as shown in Figure 3B, 3E respectively. (Note normalized shear stress is the shear stress normalized to the shear stress at the reference segments). Secondly, a multivariate model including shear stress and wall stress significantly predicted vascular remodeling (Figure 4 Vascular remodeling= (0.40±0.95) + (0.11±0.04)*normalized shear $stress_{post}$ - (0.58±0.08) *normalized wall $stress_{post}$; r^2=0.73, p<0.05). Finally, we could show that the arteries stop to remodel at the moment the shear stress and wall stress equalizes the shear stress at the reference segments. Figure 3C and 3F show respectively the shear stress and wall stress at 6 weeks follow up normalized for the reference segments, for which the number 1 (one) means that the values in the balloon dilated segment are equal to the reference segment. *In conclusion* constrictive vascular remodeling after PTA is controlled by a shear- and a wall stress negative feedback mechanism, aiming at keeping shear stress and wall stress constant.

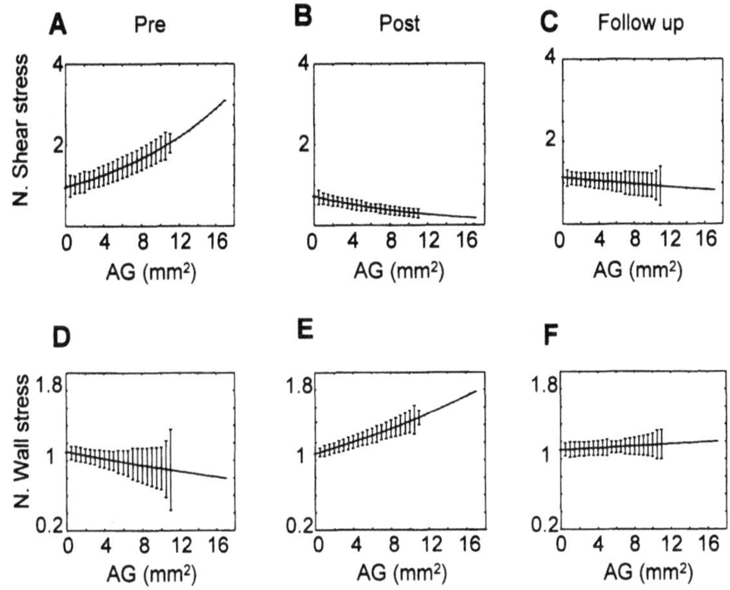

Fig. 3. Normalized shear stress and wall stress before (A,D), after PTA (B,E) and at follow up (C,F) versus acute lumen gain (AG)

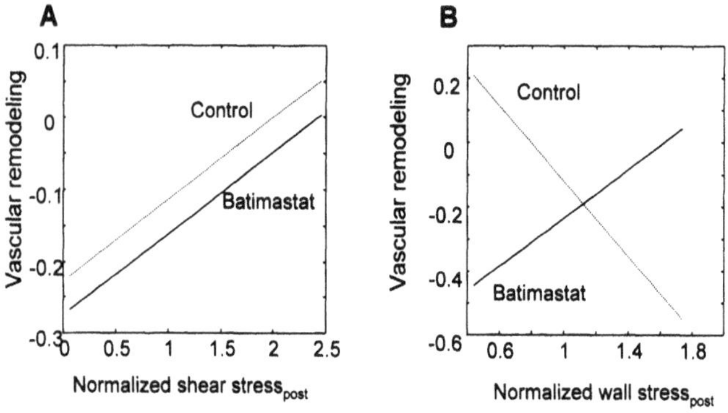

Fig. 4. Vascular remodeling versus normalized shear stress$_{post}$(A) and versus normalized wall stress$_{post}$(B) after PTA for the Control group (--) and for the Batimastat group (-) derived from the multivariate prediction model

Shear stress, wall stress and Matrix Metalloproteinases

Matrix metalloproteinases (MMP) are a group of enzymes, which are responsible for the degeneration of the extracellular matrices in both normal physiological and patho-physiological processes. Nine types of MMPs are identified and are divided into three groups based on the substrate preferences. The *collagenases* degrade type I and III collagens, *gelatinases* act mainly on the basement membrane components and *stromelysins* cover a broad range of substrates including proteoglycans, laminin, fibronectin, and basement membrane collagens [34]. Several studies showed that the MMPs are involved in the flow mediated remodeling of arteries by applying specific MMP-inhibitors [35, 36]. Others showed that even in a period of 7 days of changed flow conditions the MMP-2 mRNA expression was upregulated for both low and high flow conditions [36], which suggests that changes in shear stress mediate the production of MMPs.

Not only during remodeling, but also after vessel injury, for instance after balloon angioplasty, the MMP expression was increased [37]. It was proposed that vessel stretch itself caused the expression of MMP-2 and MMP-9, as studied by Meng [38]. The pharmaceutical inhibition of MMPs after vessel injury by, for instance, balloon angioplasty, showed to prevent the smooth muscle cells from migration [39], but did not reduce neointimal formation [40]. Since MMPs are

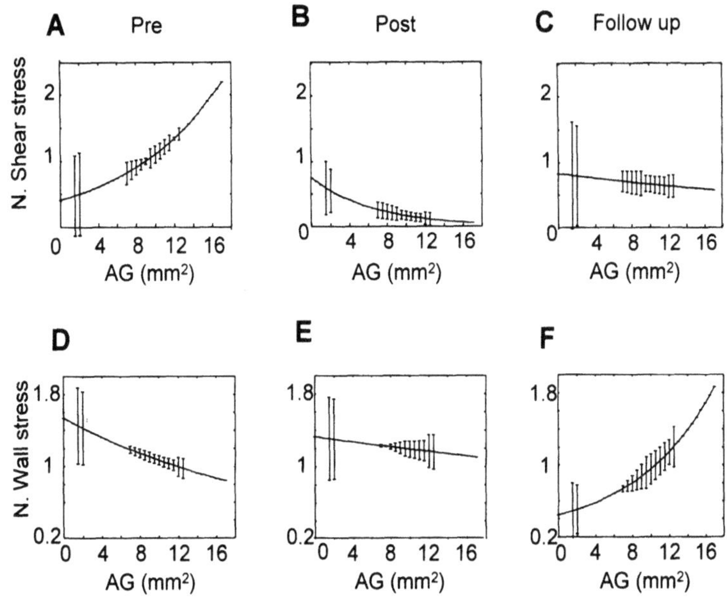

Fig. 5. Normalized shear stress and wall stress before (A,D), after PTA (B,E) and at follow up (C,F) versus acute lumen gain (AG) for the Batimastat group. Note that the range of the figures of the Control group (Fig. 3) and the Batimasbt group (Fig. 5) are different.

involved in the processes of extra cellular matrix remodeling and smooth muscle migration [40], recent experiments showed that MMP-inhibition could prevent the artery from vascular remodeling after balloon angioplasty in the atherosclerotic pig [41]. The question what effect MMP-inhibition would exert on the shear stress and wall stress regulation after balloon angioplasty became of interest. To answer that question we applied the above described techniques and investigated the influence of MMP-inhibiton of shear stress and wall stress regulation after balloon angioplasty in the external iliac arteries of atherosclerotic Yucatan pigs[33]. Before, and after the intervention and at 6 weeks follow up (fup), intravascular ultrasound pullback was performed allowing 3-D reconstruction of the treated segment and computational fluid dynamics to calculate media bounded area and shear stress. Wall stress was derived from the Laplace fomula. Batimastat, a MMP-inhibitor, was administered 3 times with 2 weeks intervals starting from the moment of balloon angioplasty. We observed for the Batiamstat group, that shear stress regulation after balloon angioplasty after administration of MMP-inhibition was still operable. Figure 5 shows that, similar to the above described measurements, the shear stress at 6 weeks follow up, was equal to the shear stress at the reference segments. This means that despite Batimastat reduced the contribution of constrictive vascular remodeling to lumen decrement in the follow up period, the shear stress was regulated. This regulation was the result of an increase in wall area. In geometrical terms, Batimastat induced iliac arteries with a 'normal' lumen and a thick vessel wall. Consequently, the wall stress regulation was hampered (Figure 5). Furthermore, a multivariate model including wall stress and shear stress, significantly predicted vascular remodeling (Vascular remodeling = (0.68±0.15) + (0.11±0.04)*normalized shear $stress_{post}$ + (0.38±0.13)*normalized wall $stress_{post}$ $r^2=0.73$, $p<0.05$). After administration of Batimastat, the contribution of wall stress to vascular remodeling was changed such that now a positive relationship between wall stress and vascular remodeling was observed (Figure 4).

Shear stress and neointimal formation

Kohler reported one of the first experiments investigating the influence of shear stress on neointimal hyperplasia (NIH) after intervention [42]. He showed in endothelialized grafts placed in baboons, that increasing shear stress in grafts reduced the neointimal formation. In contrast, at locations of reduced flow, neointimal thickness was increased. Later on, Salam in PTFE bypass grafts implanted in dogs reported similar observations. [43]. In addition, they demonstrated that increasing shear stress 2 months after graft implantation induced a regression of the neointimal thickness [44]. Shear stress related neointimal formation has also been reported at the edges of anastomosis [45, 46]. At these edges, not only low shear stress regions occur, but also vortices were observed[46]. These latter findings suggest that deviations of normal shear stress values are involved in the generation of neointimal formation under a variety of conditions. However, it is presently unclear whether vortices or low shear stress sec is the predictor of neointimal growth.

Studies into the mechanism of shear stress related NIH indicated that a decreased NO production or growth factor production, or an increased cytokines (TNF-α and IL-1) production in low shear stress regions are related to neointimal hyperplasia [47-49]. Nitric oxide production is also related to vascular remodeling explaining the close relationship between both processes. Until now the only experiments suggesting shear stress to be related to in-stent neointimal formation refer to in vitro experiments. As an example the reendothelialization of stent material was studied in vitro in relation to the local shear stress and appeared to be shear stress dependent, such that the low shear stress regions are delayed in endothelialization compared to high shear stress regions [50]. Since reendothelialization is assumed to play a key role in controlling neointimal formation after stent implantation [51, 52], this observation suggest that also in stent neointimal formation may be related to shear stress. Furthermore, in accordance with the neointimal formation observed in vascular bypass grafts or anastomosis, it is not unlikely that shear stress has similar influence on the

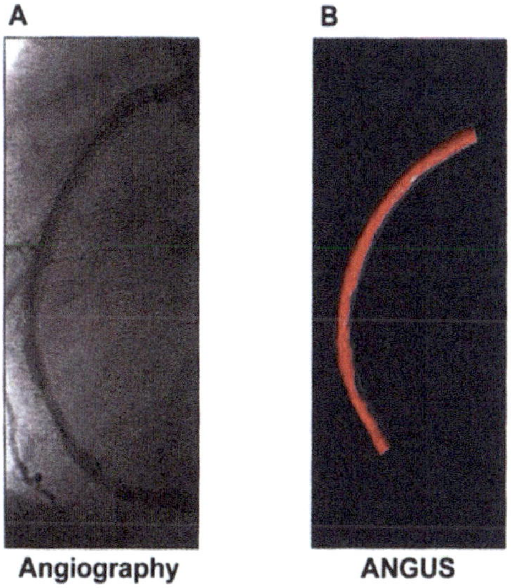

Fig. 6. A) Angiogram of a stented right coronary artery B) 3-D ANGUS reconstruction of the right coronary artery showing vessel lumen and wall.

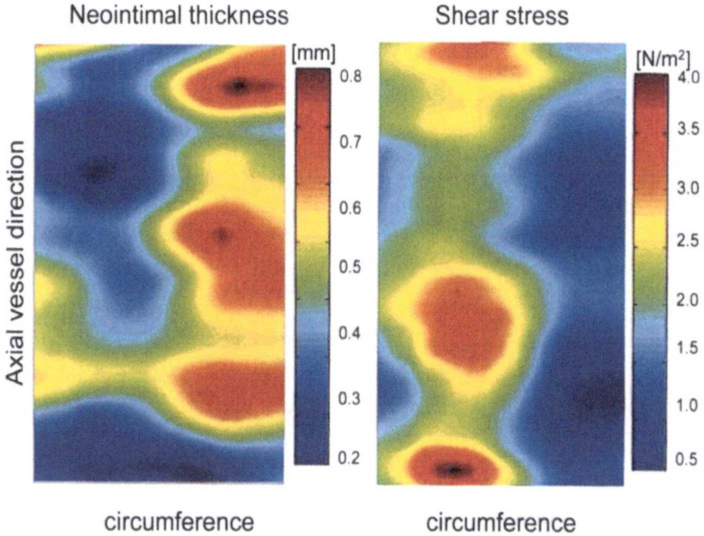

Fig. 7. 2-D map of neointimal thickness and shear stress of human coronary artery shown in Figure 6. Left to right: vessel circumference. Bottom to top: proximal to distal axial vessel location

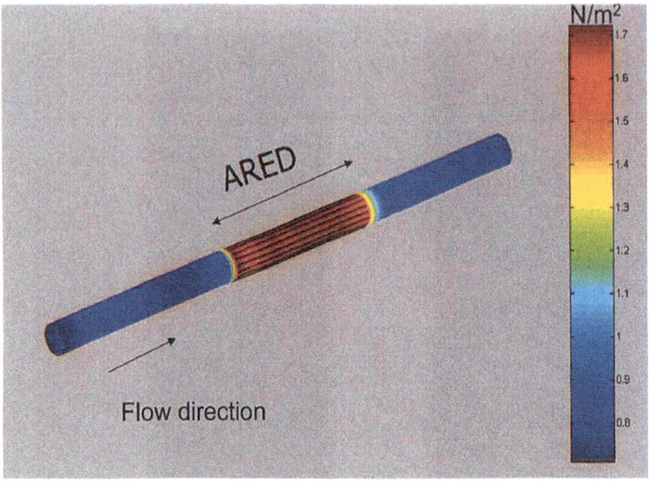

Fig. 8. Numerical analysis of a vessel with an ARED flow divider placed inside the vessel. Boundary conditions were: no slip conditions on the wall, zero stress at the outlet and parabolic velocity at the entrance with a maximal velocity of0.25 cm/sec.

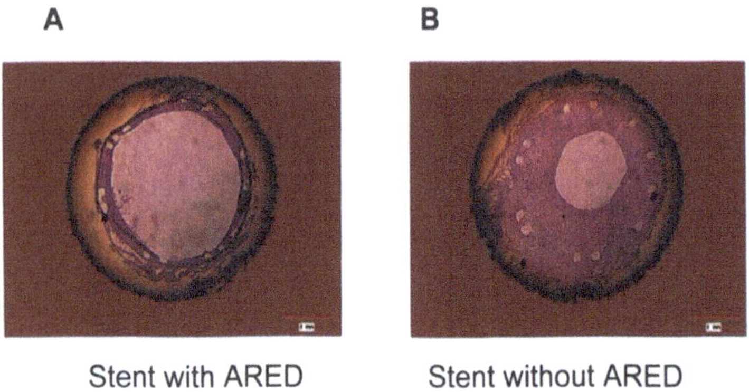

Fig. 9. A representative example of a histological cross section obtained from a stented arterial segment with (Panel A) and without the ARED flow divider (Panel B)

vascular repair process after stent implantation, thereby influencing the neointimal formation.

Hence, we studied for the first time the relationship between shear stress and neointimal hyperplasia 6 months after Wallstent implantation in 14 coronary arteries of patients [53]. Figure 6 shows an example of a 3-D reconstruction of a stented right coronary artery of a patient. The neointimal thickness and shear stress of the same patient are shown in Figure 7. From this figure can be observed that the neointimal growth was inversely related to the shear stress. Intimal thickness at the low shear stress locations was higher than at the high shear stress locations ($p<0.05$). The average relationship between neointimal thickness and shear stress for all the 14 patients was: $Th=(0.59\pm0.24)-(0.08\pm0.10)*SS$ mm, ($p<0.05$, 95% confidence interval (CI) slope: -0.14 and -0.02; intercept (IC): 0.45 and 0.72). Our results show similarity with the results obtained from animal studies, where neointima formation in by pass grafts[42, 43] was found to be related to the shear stress distribution.

In addition, we studied several parameters obtained from blood samples. Of the investigated parameters and risk factors, only a history of hypercholesteremia was found to abolish the slope of the relationship between neointimal thickness and shear stress, independent of neointimal thickness. For the patients not showing this history (n=9) the relationship was: $Th=(0.69\pm0.19)-(0.12\pm0.10)*SS$ mm, $p<0.05$, (95% CI slope: -0.18 and -0.05; IC:0.57 and 0.81) and for the patients with a history of hypercholesteremia (n=5): $Th=(0.40\pm0.23)-(0.007\pm0.06)*SS$ mm, (p=NS, 95% CI slope:-0.05 and 0.06; IC: 0.2 and 0.6).

In order to locally increase low shear stress regions in stented regions, we performed an animal study applying a flow divider developed by Endoart (Endoart, Lausanne, Switzerland). The principle is that a small cylinder placed in the middle of a (stented) arterial segment acts as a velocity divider, thereby increasing the shear stress at the vessel wall while pressure drop is negligible. The experiment indicated that shear stress of the stented segments was reduced to 70% of pre-stent conditions. Placement of the ARED flow divider returned shear stress to baseline levels as confirmed by CFD based upon the average diameters obtained in 9 rabbits (Figure 8). Consequently, the remaining NIH in the stented segment is due to factors not related to shear stress, but to tissue damage and reaction to the foreign stent material.

The preliminary analysis on the NIH in the animals indicated that NIH was reduced by 50%. (Figure 9). The large effect clearly identifies shear stress as a major modulating factor in restenosis.

References

1. Shaaban AM and Duerinckx AJ (2000) Wall shear stress and early atherosclerosis: a review. AJR Am J Roentgenol 174: 1657-1665.
2. Zarins CK, Giddens DP, Bharadvaj BK, et al. (1983) Carotid bifurcation atherosclerosis. Quantitative correlation of plaque localization with flow velocity profiles and wall shear stress. Circ Res 53: 502-514.
3. Gnasso A, Irace C, Carallo C, et al. (1997) In vivo association between low wall shear stress and plaque in subjects with asymmetrical carotid atherosclerosis. Stroke 28: 993-998.
4. Brands P, Hoeks A, Hofstra L, et al. (1995) A noninvasive method to estimate wall shear rate using ultrasound. Ultrasound in medicine and biology 21: 171-185.
5. Kornet L, Lambregts J, Hoeks AP, et al. (1998) Differences in near-wall shear rate in the carotid artery within subjects are associated with different intima-media thicknesses. Arterioscler Thromb Vasc Biol 18: 1877-1884.
6. Slager CJ, Wentzel JJ, Schuurbiers JC, et al. (2000) True 3-dimensional reconstruction of coronary arteries in patients by fusion of angiography and IVUS (ANGUS) and its quantitative validation. Circulation 102: 511-516.
7. Laban M, Oomen J, Slager C, et al. (1995) ANGUS: A new Approach to Three-Dimensional Reconstruction of Coronary Vessels by Combined Use of Angiography and Intravascular Ultrasound. IEEE, Computers in Cardiology 95CH35874: 325-328.
8. Li W, Bosch J, Zhong Y, et al. (1993) Image segmentation and 3-D reconstruction of intravascular ultrasound images. In: Acoustical Imaging, ed. Y. Wei and B. Gu. Plenum Press, New York : 489-496.
9. Wentzel JJ, Whelan MD, van der Giessen WJ, et al. (2000) Coronary stent implantation changes 3-D vessel geometry and 3-D shear stress distribution. J Biomech 33: 1287-1295.
10. Cuvelier C, Segal A, and van Steenhoven A, *Finite Element Methods and Navier-Stokes equations*. 1986, Dordrecht: Reidel, 1986.
11. Zarins CK, Zatina MA, Giddens DP, et al. (1987) Shear stress regulation of artery lumen diameter in experimental atherogenesis J Vasc Surg 5: 413-420.
12. Malek AM, Alper SL, and Izumo S (1999) Hemodynamic shear stress and its role in atherosclerosis. Jama 282: 2035-2042.
13. Bagshaw RJ, Cox RH, Karreman G, et al. (1986) Baroreceptor control of pressure-flow relationships during hypoxemia. J. Appl. Physiol. 60: 166-175.
14. van de Vosse FN, van Steenhoven AA, segal A, et al. (1989) A finite element analysis of the steady laminar entrance flow in a 90 curved tube. Int. J. Num. Method in Fluids 9: 275-287.
15. Caro CG, Fitz-Gerald JM, and Schroter RC (1971) Atheroma and arterial wall shear. Observation, correlation and proposal of a shear dependent mass transfer mechanismfor atherogenesis. Proc R Soc Lond B Biol Sci 177: 109-159.
16. Fry DL (1968) Acute vascular endothelial changes associated with increased blood velocity gradients. Circ Res 22: 165-197.
17. Giddens DP, Zarins CK, and Glagov S (1993) The role of fluid mechanics in the localization and detection of atherosclerosis J Biomech Eng 115: 588-594.
18. Sabbah HN, Khaja F, Hawkins ET, et al. (1986) Relation of atherosclerosis to arterial wall shear in the left anterior descending coronary artery of man Am Heart J 112: 453-458.

19.Friedman MH, Deters OJ, Bargeron CB, et al. (1986) Shear-dependent thickening of the human arterial intima Atherosclerosis 60: 161-171.
20.Deng X, Marois Y, How T, et al. (1995) Luminal surface concentration of lipoprotein (LDL) and its effect on the wall uptake of cholesterol by canine carotid arteries [published erratum appears in J Vasc Surg 1995 Dec;22(6):648] J Vasc Surg 21: 135-145.
21.Waters CM (1996) Flow-induced modulation of the permeability of endothelial cells cultured on microcarrier beads J Cell Physiol 168: 403-411.
22.Jo H, Dull RO, Hollis TM, et al. (1991) Endothelial albumin permeability is shear dependent, time dependent, and reversible Am J Physiol 260: H1992-1996.
23.Gibson CM, Diaz L, Kandarpa K, et al. (1993) Relation of vessel wall shear stress to atherosclerosis progression in human coronary arteries Arterioscler Thromb 13: 310-315.
24.Friedman MH, Bargeron CB, Deters OJ, et al. (1987) Correlation between wall shear and intimal thickness at a coronary artery branch. Atherosclerosis 68: 27-33.
25.Friedman MH, Bargeron CB, Duncan DD, et al. (1992) Effects of arterial compliance and non-Newtonian rheology on correlations between intimal thickness and wall shear J Biomech Eng 114: 317-320.
26.Moore J, Jr., Xu C, Glagov S, et al. (1994) Fluid wall shear stress measurements in a model of the human abdominal aorta: oscillatory behavior and relationship to atherosclerosis. Atherosclerosis 110: 225-240.
27.Krams R, Wentzel J, Oomen J, et al. (1997) Evaluation of endothelial shear stress and 3D geometry as factors determining the development of atherosclerosis and remodeling in human coronary arteries in vivo. Combining 3D reconstruction from angiography and IVUS (ANGUS) with computational fluid dynamics. Arterioscler Thromb Vasc Biol 17: 2061-2065.
28.Kamiya A and Togawa T (1980) Adaptive regulation of wall shear stress to flow change in the canine carotid artery Am J Physiol 239: H14-21.
29.Tronc F, Wassef M, Esposito B, et al. (1996) Role of NO in flow-induced remodeling of the rabbit common carotid artery Arterioscler Thromb Vasc Biol 16: 1256-1262.
30.Glagov S, Weisenberg E, Zarins CK, et al. (1987) Compensatory enlargement of human atherosclerotic coronary arteries N Engl J Med 316: 1371-1375.
31.Mintz GS, Kent KM, Pichard AD, et al. (1997) Intravascular ultrasound insights into mechanisms of stenosis formation and restenosis Cardiol Clin 15: 17-29.
32.Post MJ, Borst C, and Kuntz RE (1994) The relative importance of arterial remodeling compared with intimal hyperplasia in lumen renarrowing after balloon angioplasty. A study in the normal rabbit and the hypercholesterolemic Yucatan micropig [see comments]. Circulation 89: 2816-2821.
33.Wentzel JJ, Kloet J, Andhyiswara I, et al. (2001) Shear stress and wall stress regulation of vascular remodeling after balloon angioplasty: effect of matrix metalloproteinase inhibition. Circulation,2001,3:104(1):91-96.
34.Dollery CM, McEwan JR, and Henney AM (1995) Matrix metalloproteinases and cardiovascular disease. Circ Res 77: 863-868.
35.Abbruzzese TA, Guzman RJ, Martin RL, et al. (1998) Matrix metalloproteinase inhibition limits arterial enlargements in a rodent arteriovenous fistula model Surgery 124: 328-334; discussion 334-325.
36.Bassiouny HS, Song RH, Hong XF, et al. (1998) Flow regulation of 72-kD collagenase IV (MMP-2) after experimental arterial injury Circulation 98: 157-163.

37.Zempo N, Kenagy RD, Au YP, et al. (1994) Matrix metalloproteinases of vascular wall cells are increased in balloon-injured rat carotid artery. J Vasc Surg 20: 209-217.
38.Meng X, Mavromatis K, and Galis ZS (1999) Mechanical stretching of human saphenous vein grafts induces expression and activation of matrix-degrading enzymes associated with vascular tissue injury and repair [In Process Citation]. Exp Mol Pathol 66: 227-237.
39.Zempo N, Koyama N, Kenagy RD, et al. (1996) Regulation of vascular smooth muscle cell migration and proliferation in vitro and in injured rat arteries by a synthetic matrix metalloproteinase inhibitor. Arterioscler Thromb Vasc Biol 16: 28-33.
40.Bendeck MP, Irvin C, and Reidy MA (1996) Inhibition of matrix metalloproteinase activity inhibits smooth muscle cell migration but not neointimal thickeningafter arterial injury. Circ Res 78: 38-43.
41.de Smet BJGL (1998) A new Paradigm in Restenosis: Morphometric and Molecular Characteristics. dissertation .
42.Kohler TR, Kirkman TR, Kraiss LW, et al. (1991) Increased blood flow inhibits neointimal hyperplasia in endothelialized vascular grafts Circ Res 69: 1557-1565.
43.Salam T, Lumsden A, Suggs W, et al. (1996) low shear stress promotes intimal hyperplasia thickening. Journal of vascular investigation 2: 12-22.
44.Mattsson EJ, Kohler TR, Vergel SM, et al. (1997) Increased blood flow induces regression of intimal hyperplasia Arterioscler Thromb Vasc Biol 17: 2245-2249.
45.Ethier CR, Steinman DA, Zhang X, et al. (1998) Flow waveform effects on end-to-side anastomotic flow patterns. J Biomech 31: 609-617.
46.Liu SQ (1998) Prevention of focal intimal hyperplasia in rat vein grafts by using a tissue engineering approach. Atherosclerosis 140: 365-377.
47.Marano G, Palazzesi S, Vergari A, et al. (1999) Protection by shear stress from collar-induced intimal thickening: role of nitric oxide. Arterioscler Thromb Vasc Biol 19: 2609-2614.
48.Rectenwald JE, Moldawer LL, Huber TS, et al. (2000) Direct evidence for cytokine involvement in neointimal hyperplasia. Circulation 102: 1697-1702.
49.Song RH, Kocharyan HK, Fortunato JE, et al. (2000) Increased flow and shear stress enhance in vivo transforming growth factor-beta1 after experimental arterial injury. Arterioscler Thromb Vasc Biol 20: 923-930.
50.Sprague EA, Luo J, and Palmaz JC (1997) Human aortic endothelial cell migration onto stent surfaces under static and flow conditions J Vasc Interv Radiol 8: 83-92.
51.Asahara T, Bauters C, Pastore C, et al. (1995) Local delivery of vascular endothelial growth factor accelerates reendothelialization and attenuates intimal hyperplasia in balloon-injured rat carotid artery [see comments]. Circulation 91: 2793-2801.
52.Chen FE, Huang DB, Chen YQ, et al. (1998) Crystal structure of p50/p65 heterodimer of transcription factor NF- kappaB bound to DNA. Nature 391: 410-413.
53.Wentzel JJ, Krams R, Schuurbiers JC, et al. (2001) Relationship betwee Circulati neointimal thickness and shear stress after Wallstent implantation in human coronary arteries on 103: 1740-1745.

Part 3

Clinical Applications of IVUS

WHAT HAVE WE LEARNED FROM 10 YEARS PERIPHERAL INTRAVASCULAR ULTRASOUND?

Elma J. Gussenhoven[1,2], Tjebbe Hagenaars[1,2], Jeroen A. van Essen[1,2], Trude C. Leertouwer[1], Jan Honkoop[1], Nicolaas Bom[1,2]

[1]Erasmus Medical Center Rotterdam and the [2]Interuniversity Cardiology Institute the Netherlands.

Summary: Although angiography has been considered as the 'gold standard' in the assessment of the human vasculature, the technique only displays the contrast-filled lumen and gives no information on the arterial wall. The necessity for more detailed information on the arterial anatomy accelerated development of high frequency intravascular ultrasound (IVUS) systems in the 1980s. It soon became evident that, compared with other imaging techniques, the capability to provide information on the arterial wall is a unique advantage of IVUS. This chapter will answer questions related to 1) the accuracy of IVUS as assessed in *in vitro* studies and 2) the value of the clinical use of IVUS in peripheral arteries, as experienced in the last 10 years at the authors' center. Results from studies in the femoropopliteal artery, the renal artery and the abdominal aorta will be presented. In addition, the role of three-dimensional (3-D) reconstruction of IVUS will be discussed.

Key words: intravascular ultrasound, balloon angioplasty, vascular interventions, peripheral arteries, three-dimensional reconstruction.

What did we learn from *in vitro* studies?

From the *in vitro* studies we learned that IVUS is a unique imaging modality to distinguish two main artery types: muscular arteries like coronaries, renals and femorals versus elastic arteries like carotids and the aorta. Additionally, the composition of atherosclerotic plaque (calcified or not calcified) can be distinguished (Fig. 1)[1].

Initially we claimed the likelihood to identify lipid inside the plaque, recognized as a hypoechoic area, with IVUS. Nowadays, we are reluctant to use this terminology

---

This work was supported by the Interuniversity Cardiology Institute the Netherlands.

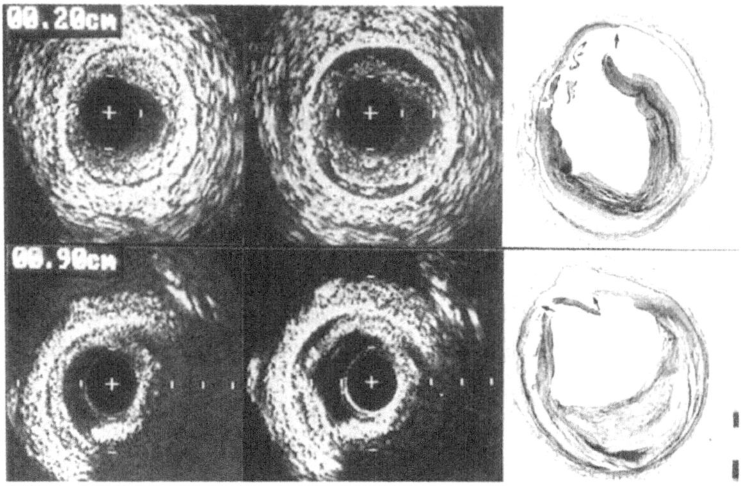

Fig. 1. Matched intravascular ultrasound (IVUS) cross-sections obtained *in vitro* before and after balloon angioplasty of the coronary artery and their histologic counterpart. The displacement sensing device indicated the distance between the 2 cross-sections (0.7 cm). A soft non-calcified lesion was involved in the upper panel and a calcified lesion was present in the lower panel characterized by a bright echostructure with distal shadowing. As result of the balloon angioplasty lumen area increased due to an increase in vessel area. Note that the dissection occurred at the thinnest region of the plaque. The arrow indicates the media rupture Calibration = 1 mm; + = catheter.

for two reasons. First, *in vitro* studies showed that the echogenicity of a lesion was dependent both on the position of the IVUS catheter and the angle of the insonification of the ultrasound beam (Fig. 2). Second, although the presence of a hypoechoic 'vulnerable plaque' region would imply a typical rupture site at the vessel wall during an intervention, *in vitro* studies and clinical studies showed this not to be the case (Fig. 3).

In addition, from the *in vitro* studies with IVUS we learned that the basic mechanism related to balloon angioplasty [2,3] was unequivocally different than initially proposed by Dotter and Judkins in 1964 [4]. Although these IVUS studies showed that lumen area increase was solely caused by vessel stretch rather than plaque compression, in coronary arteries we encountered that both vessel stretch (81%) and plaque compression (19%) contributed to lumen area enlargement [2]. By correlating the IVUS findings obtained at the most stenotic site with histology, we learned that the sensitivity of IVUS in detecting vascular damage was good for dissection (74%) and low for media rupture (59%) [2]. A higher incidence of vascular damage was found when the whole segment was analyzed, rather than one single cross-section at the most stenotic site. Dissections were found exclusively at the thinnest site of the plaque and no relation between plaque calcification and dissection was found (Fig. 1). Similar features were encountered in clinical studies performed by our group in iliac and femoropopliteal arteries (Fig. 3) [5-7].

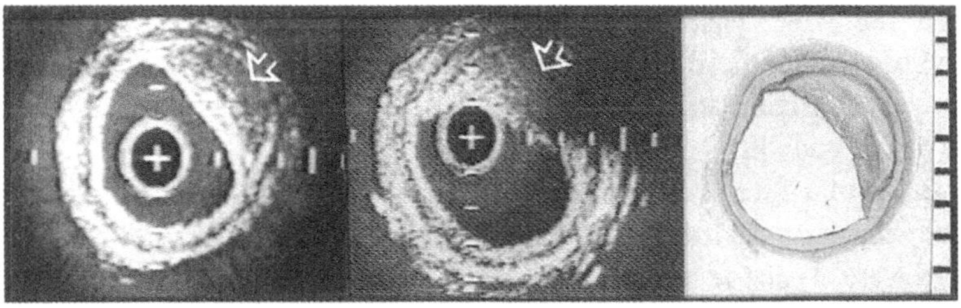

Fig. 2. Intravascular ultrasound cross-sections and corresponding histologic section showing that the ultrasound characteristics of a lesion may depend on the position of the ultrasound catheter relative to the lesion. The hypoechoic area in the lesion (open arrow) did not correspond to lipid. Calibration = 1 mm; + = catheter.

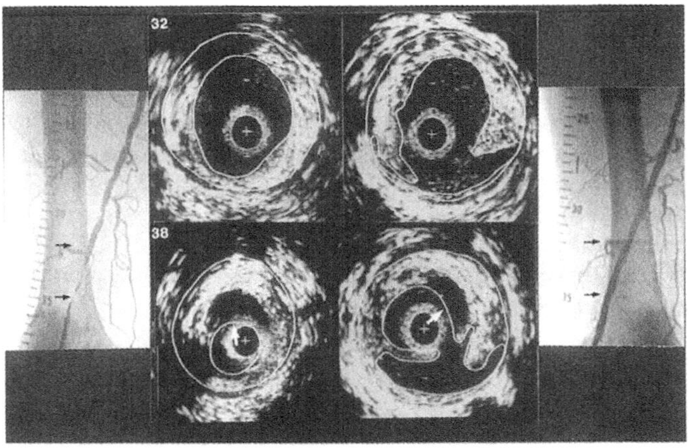

Fig. 3. Intravascular ultrasound cross-sections and angiographic records obtained from the superficial femoral artery at the corresponding levels before (left) and after balloon angioplasty (left). As result of the balloon angioplasty, lumen area increased due to an increase in vessel area while the plaque area remained virtually unchanged. Note that the dissection occurred at the thinnest region of the plaque. Despite the thin fibrotic cap covering the hypoechoic region the cap did not rupture following the intervention (level 38: arrow). Contour detection was used to determine the change in lumen area (inner contour) and vessel area (outer contour). Calibration = 1 mm; + = catheter.

PERIPHERAL ARTERIES

Can IVUS be used to predict peripheral artery restenosis?

To answer the question whether IVUS data on morphologic and quantitative parameters are predictors of clinical outcome, patients were studied with IVUS prior to and following angiographic successful percutaneous transluminal angioplasty (PTA) of the superficial femoral artery. Analysis of the IVUS data combined with angiographic, clinical and hemodynamic data up to 6 months follow-up, formed the basis of this study [8].

Intravascular ultrasound predictors of 1 month **anatomic outcome** evidenced with duplex scanning included; 1) percentage (%) lumen area stenosis after PTA, 2) lumen area increase, 3) plaque area decrease, and 4) % area stenosis decrease. This suggests that early anatomic restenosis encountered at 1-month follow-up is in fact the result of an inadequate PTA procedure. The only predictor of 6 months anatomic outcome was % area stenosis after PTA. Multivariate analysis revealed that % area stenosis after PTA was the only independent predictor of both 1 and 6 months anatomic outcome.

Intravascular ultrasound predictors of 1 month **clinical outcome** after PTA based on Rutherford criteria were the presence of hard lesion (i.e. calcified lesion) and the mean arc of hard lesion. Multivariate analysis revealed that the mean arc of hard lesion was the only independent predictor of 1 month clinical outcome. However, no predictors for 6 months clinical outcome were found. In summary, the strongest IVUS parameter predictive of anatomic restenosis was a larger % area stenosis after PTA [8].

Whether interventional management based on IVUS parameters found in these studies will result in a better clinical outcome should be further investigated: these future studies will define the true utility of IVUS.

Can IVUS disclose the mechanism of restenosis following peripheral vascular intervention?

To determine whether IVUS may reveal the parameters responsible for lumen area reduction following vascular intervention, including PTA and stent placement and medical intervention using statin therapy, we conducted serial IVUS studies.

Restenosis after balloon angioplasty

To assess the contribution of plaque growth (i.e. intimal hyperplasia) and vascular remodeling to the change in lumen area 1 year after angiographic successful PTA of the femoropopliteal artery, patients were studied with IVUS at baseline immediately after PTA and at 1 year follow-up [9]. Matched IVUS cross-sections derived from the target site (i.e. most stenotic site) at follow-up, the adjacent "treated" sites and the not dilated "reference" sites were analyzed for change in lumen area, vessel area and plaque area (Figs. 4 and 5). It was found that lumen narrowing at the target site seen at follow-up was caused by plaque growth and vessel shrinkage. At both the treated and reference sites a significant increase in plaque area was evidenced, whereas the type of vascular remodeling determined lumen area change (Fig. 5). Surprisingly, the presence of a dissection did not provoke plaque growth nor did it influence vascular remodeling.

It is worth mentioning that restenosis may be due to either a not dilated lesion or an insufficiently dilated lesion (Fig. 6). We observed that in procedures using a balloon diameter of 5 mm, lumen area was smaller, than in procedures using a balloon diameter exceeding 5 mm [9]. Given the fact that the majority (92%) of PTAs performed with a 5 mm balloon presented a restenotic lesion at follow-up, one should seriously question whether this latter balloon size should remain an option. Similarly, in 47% of the PTAs performed with a 6 mm balloon size a restenotic lesion was involved (unpublished observations). Altogether, these findings support the idea that a balloon with a larger diameter than currently used in routine practice should be recommended for PTA in peripheral arteries.

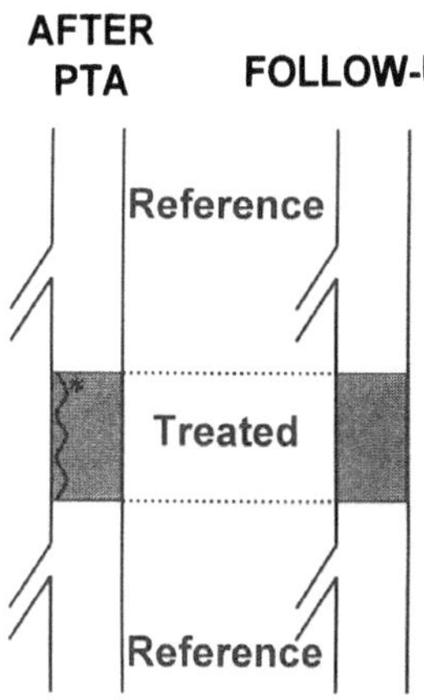

	Lumen	Vessel	Plaque
AREA			
Target site	↓ 39 %*	↓ 7 %**	↑ 18 %*
Treated site	↓ 13 %*	↑ 2 %**	↑ 19 %*
Reference site	↓ 1 %	↑ 5 %*	↑ 12 %*
VOLUME			
Reference site	↑ 2 %	↑ 2 %	↑ 3 %

(↑ = increase; ↓ = decrease; * = $p<0.001$; ** = $p<0.05$)

Fig. 4. Changes in lumen, plaque and vessel area expressed in percentages derived from matched intravascular ultrasound cross-sections obtained after percutaneous transluminal angioplasty (PTA) and at 1-year follow-up. The data show that lumen area at the target site is the net result of both plaque area increase and vessel area decrease. At the treated site and reference site the plaque area increase is significant and not different from the area increase seen at the target site. Calibration = 1 mm; + = catheter.

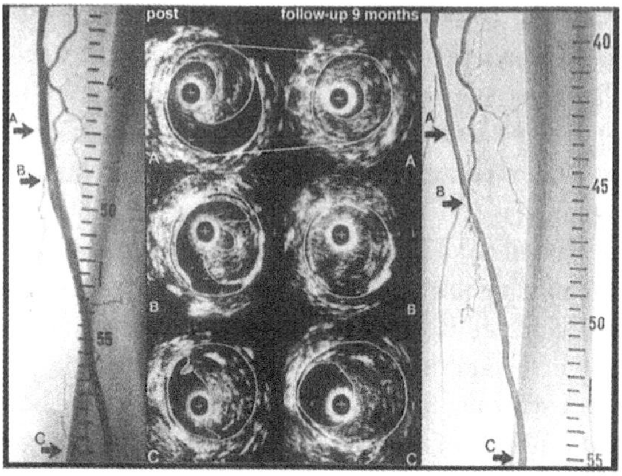

Fig. 5. Intravascular ultrasound cross-sections and angiographic records obtained from the superficial femoral artery at the corresponding levels after balloon angioplasty (left) and at 9 months follow-up (right). Dissection seen as result of the intervention (A, B, C) was absent at follow-up. At follow-up lumen area decrease was associated with vessel area decrease (A) and plaque area increase (A, B). Vessel area enlargement was seen at level C. Contour detection was used to determine the change in lumen area (inner contour) and vessel area (outer contour). Calibration = 1 mm; + = catheter.

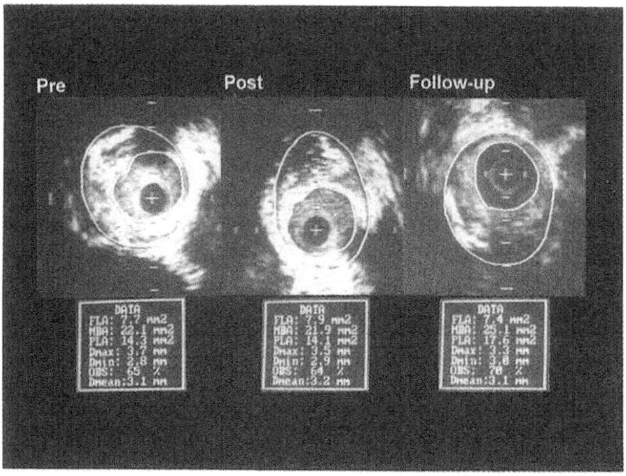

Fig. 6. Corresponding intravascular ultrasound cross-sections of the superficial femoral artery obtained before and immediately after percutaneous transluminal angioplasty and at 6 month follow-up for restenosis. Note that the relative small lumen area after intervention did not change at follow-up; in contrast, vessel area increased slightly. Calibration = 1 mm; + = catheter.

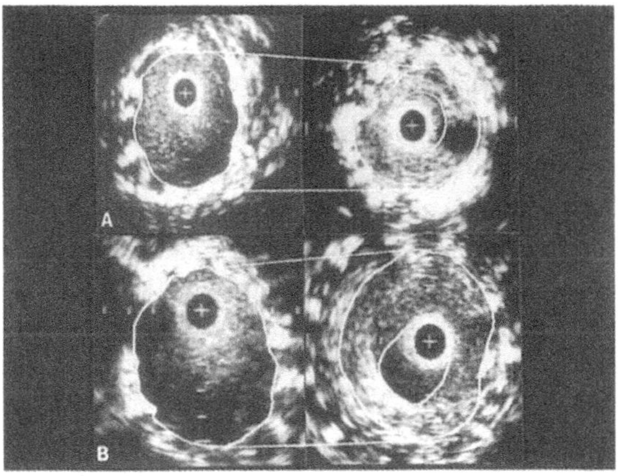

Fig. 7. Matched intravascular ultrasound cross-sections of the femoropopliteal artery obtained after Palmaz stent deployment (left panel) and at follow-up (right panel). Lumen area (inner contour) and stent area (outer contour) display quantitative results. A, Intimal hyperplasia (arrow) and stent area reduction were the predominant mechanisms responsible for restenosis seen at the junction of two stents in a Palmaz stent. B, Restenosis in a Memotherm stent was solely due to intimal hyperplasia. Calibration = 1 mm; + = catheter.

Restenosis after stent placement

The contribution of intimal hyperplasia and stent area change to the change in lumen area seen at follow-up was studied in a patient who received multiple Palmaz stents placed in the femoropopliteal artery. It was found that in the presence of a Palmaz stent both intimal hyperplasia and stent area reduction resulted in lumen area reduction (Fig. 7) [10]. The extent of the changes seen at the stent junctions of two stents was greater than those seen within the stents: lumen area reduction 67% vs. 23%; stent area reduction 26% vs. 11%; and intimal hyperplasia 10.8 vs. 3.3 mm^2, respectively. Conversely, in Memotherm stents lumen area reduction seen at the junction of two stents was solely caused by intimal hyperplasia [11]. These IVUS studies demonstrated a higher risk of restenosis at stent junctions both in Palmaz and Memotherm stents [10,11]; this suggests that, as far as possible, use of multiple stents should be avoided. In addition to intimal hyperplasia, stent remodeling contributes to restenosis in Palmaz stents but not in Memotherm stents.

Restenosis after PTA followed by 1 year statin therapy

From clinical studies using angiography we have learned that HMG-CoA reductase inhibitors (statins) may reduce restenosis after coronary angioplasty [12-

14]. However, the precise effect of statin therapy on the mechanism of restenosis after vascular intervention remains to be elucidated. Therefore, in a retrospective observational IVUS study we evaluated whether simvastatin therapy limits lumen area reduction 1 year after PTA by reducing reactive plaque growth, reducing reactive vasoconstriction or both [15]. The results of this study revealed a significant difference between patients with and without simvastatin in lumen area reduction at 1 year follow-up: -4% in patients with simvastatin and -25% in patients without simvastatin. Overall, both groups of patients showed an identical increase in plaque area. Surprisingly, plaque area increase was associated with an increase in vessel area in patients with simvastatin and a decrease in vessel area in patients without simvastatin. In other words, the use of simvastatin appeared to have a beneficial effect on lumen size due to its effect on the mode of vascular remodeling rather than on the degree of plaque growth. Angiographic records confirmed the difference in lumen area reduction between both groups of patients; patients with simvastatin showed no angiographic restenosis, while in 8 patients without simvastatin angiographic restenosis was present [13]. The importance of IVUS in this respect is that vessel area measurement is unique to IVUS and has no angiographic equivalent.

RENAL ARTERIES

Can IVUS define the nature of renal artery stenosis?

After having established that in autopsy specimens renal artery narrowing could be due not only to atherosclerotic plaque compensated for by adaptive remodeling, but also to hitherto undescribed focal narrowing of an otherwise normal renal arterial wall (i.e. coarctation) [16], we designed a clinical study to establish whether vessel coarctation (i.e. vessel constriction) could be identified in patients with symptomatic renal artery stenosis [17]. Consecutive symptomatic patients with angiographically proven atherosclerotic renal artery stenosis and referred for stent placement were studied with IVUS. Analysis included assessment of lumen, plaque and vessel area at the most stenotic (target) site and at a distal reference site (most distal cross-section in the main renal artery with normal appearance). Renal artery coarctation was considered present whenever the target/reference vessel area was ≤85%. The incidence of coarctation lesions and adaptive remodeling lesions was equal (50%). Coarctation lesions had a significantly smaller plaque area than adaptive remodeling lesions (Fig. 8) [17]. Thus, low-plaque coarctation may cause a considerable proportion of symptomatic renal artery stenosis, which is angiographically and clinically indistinguishable from plaque-rich renal artery stenosis.

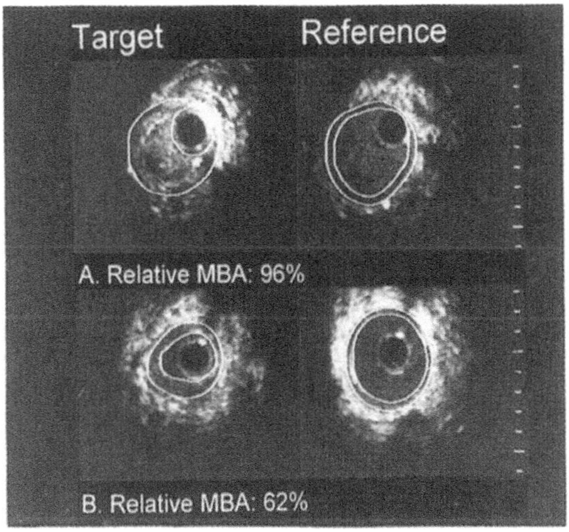

Fig. 8. Intravascular ultrasound cross-sections obtained from renal arteries showing the adpadtive remodeling (A) and the coarctation type of renal artery stenosis (B). The left panel represents the target site and the right panel the reference site. Contour detection was used to determine the lumen area (inner contour) and vessel area (outer contour). Note the small relative media-bounded area (MBA) and little plaque in the coarctation stenosis (B). Calibration = 1mm; black circle = catheter.

What is the impact of IVUS during of renal artery stenting?

To investigate the impact of IVUS during renal artery stenting, patients with atherosclerotic renal artery stenosis (n=22) were studied with IVUS after predilatation and after angiographic successful stent deployment (diameter stenosis <10%)[18]. After predilatation IVUS was used to assess whether the balloon size selected angiographically was correct (discrepancy balloon-reference lumen diameter <20%). After stent placement IVUS images were reviewed for 1) complete stent-vessel wall apposition; 2) complete stent expansion (discrepancy stent-reference lumen diameter <20%), and 3) complete lesion covering by the stent. Modifications based on IVUS included selection of a larger balloon, additional dilatation and placement of a second stent.

After predilatation IVUS warranted the use of a larger balloon in 5 patients. After stent placement IVUS had the ability to show either incomplete stent apposition (n=1) (Fig. 9a), discrepancy between stent and reference lumen diameter (n=3), or a persistent lesion distal to the stent (n=2). These latter patients were treated with additional dilatation and/or with placement of a second stent; in 3 patients a larger balloon was used (Fig. 9b).

To determine the ultimate quantitative effect of the intervention the IVUS images were analyzed off-line. It was found that stent placement resulted in a

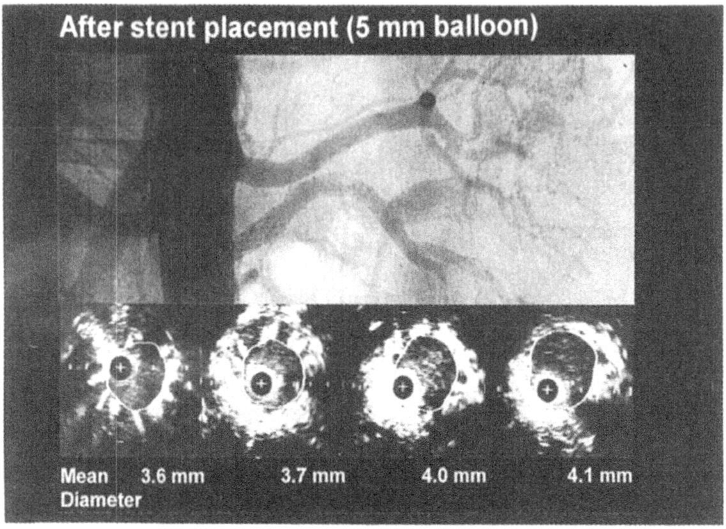

A

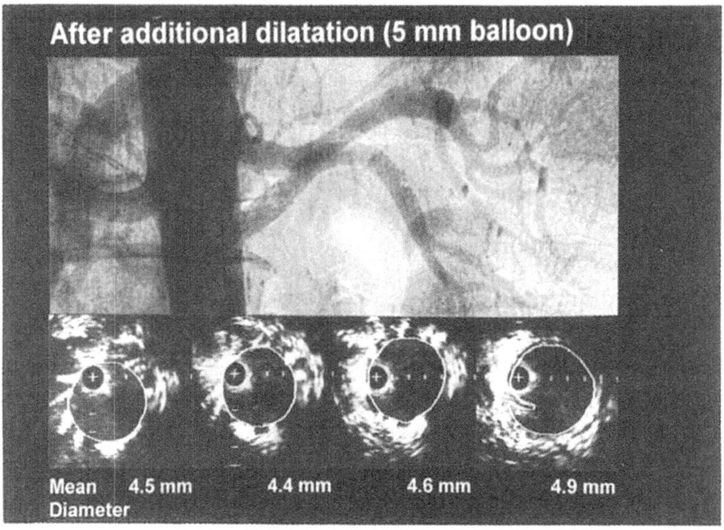

B

Fig. 9. Angiograms and corresponding intravascular ultrasound (IVUS) cross-sections obtained after stent placement (A) and after additional dilatation (B). The contour in the IVUS cross-section represents the lumen or stent area. IVUS images display (from right to left) the reference site, the distal stent edge, in-stent site, and the proximal stent edge in the aorta. After stent placement IVUS showed incomplete stent-vessel wall apposition (arrow), which was not seen on the angiogram. Note the discrepancy between the balloon size used (5 mm) and the smallest in-stent diameter (3.6 mm), and that the target site is located at the proximal stent edge. Additional dilatation performed because of incomplete stent-vessel wall apposition, resulted in an increase in lumen diamter and complete stent apposition. + = catheter; calibration = 1 mm..

significant increase in lumen area (46%): the mean residual area stenosis decreased from 48% before stent placement to 32% after stent placement. In patients with a satisfactory semiquantitative IVUS result after stent placement, quantitative analysis revealed a residual area stenosis of 25%. In patients who underwent additional intervention based on IVUS after stent placement, lumen area increased significantly (24%) and residual area stenosis decreased from 33% to 28%. Final results after stent placement showed that the smallest intra-stent lumen area was 29% smaller than the reference lumen area. Overall the intra-stent lumen diameter was smaller than the balloon diameter used (mean difference 22%) [18].

In conclusion, IVUS monitoring during renal artery stent placement warranted additional intervention in order to achieve additional lumen gain, that was not considered necessary angiographically.

Can IVUS define the nature of restenosis 1 year after renal artery stent placement?

Although stent placement is in common use for revascularization of renal artery stenosis, restenosis at follow-up remains a limiting factor. To characterize the restenotic process, we used IVUS as an adjunct to standard angiography after stent placement and at 1 year follow-up. Consecutive patients (n=41) presenting with symptomatic renal artery stenosis were studied. At follow-up, IVUS showed a significant lumen area decrease in the stent from 24 mm^2 to 17 mm^2 solely due to plaque accumulation. Similarly, the lumen area at the distal main renal artery was significantly decreased (29 mm^2 vs. 20 mm^2) due to unexplained vessel area decrease; this was confirmed angiographically. The nature of this shrinkage remains to be determined.

ABDOMINAL AORTA

Is IVUS as accurate as computed tomographic angiography to assess the nature of an abdominal aortic aneurysm?

To assess the potential diagnostic role of intraoperative IVUS to guide the endovascular stent-graft procedure for AAA, we designed a validation study with CTA [19]. Patients with AAA were studied with IVUS (12.5 MHz) and CTA. The length of the aneurysm and the length and diameter of the proximal and distal neck, obtained with IVUS, were compared with the data obtained with CTA. Tomographic IVUS images of the proximal and distal neck were reconstructed into a 3-D format [20].

The results showed that IVUS identified 97% of the renal arteries and 80% of the accessory renal arteries. Comparison of the length measurements of the aneurysm and the proximal and the distal neck, obtained with IVUS and CTA,

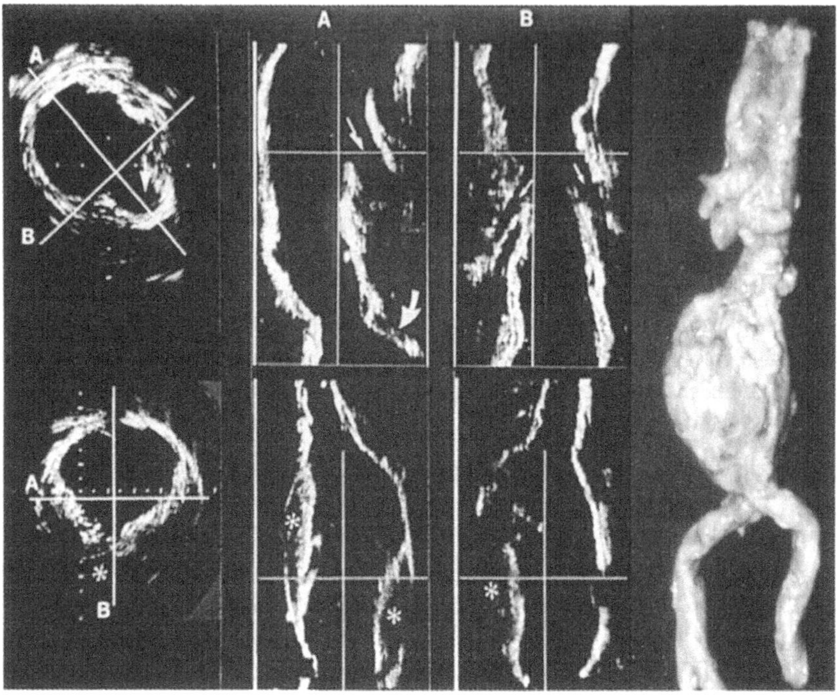

Fig. 10. Three-dimensional (3-D) reconstruction of an abdominal aortic aneurysm obtained with intravascular ultrasound. Lines A and B represent longitudinal sections that stand perpendicular to each other as demonstrated in the axial images shown on the left side. The 3D reconstructions of the proximal neck show the superior mesenteric artery (arrow), the renal arteries (small arrows) and the proximal part of the aneurysm (curved arrow). The aneurysm presents with a thrombus (asterisk). Open arrow: aortic bifurcation. On the right the anatomic specimen studied is shown.

revealed a correlation of 0.99 ($p<0.001$), with a coefficient of variation of 9%; IVUS tended to underestimate the length compared to CTA (0.48 ± 0.52 cm; $p<0.001$). This difference may be attributed to the non-coaxial position and the stiffness of the IVUS catheter, leading to a different path between the IVUS catheter and the central lumen line. Comparison of lumen diameter measurements of the proximal and distal neck, derived from IVUS and CTA, showed a correlation of 0.93 ($p<0.001$), with a coefficient of variation of 9%; IVUS tended to underestimate aneurysm neck diameter compared to CTA (0.68 ± 1.76 mm; $p=0.006$). This latter difference may be due to the fact that on IVUS the minimum lumen diameter was taken to represent the actual lumen diameter in order to avoid overestimation due to non-coaxial positioning of the IVUS catheter. Finally, tomographic IVUS images of the proximal and distal neck reconstructed into a 3-D format facilitated anatomic understanding of neck anatomy and improved lumen diameter and length measurements of multiple IVUS cross-sections (Fig. 10).

What is the role of IVUS during endovascular repair of abdominal aortic aneurysm?

Several studies have proven the feasibility and efficacy of endovascular stent-grafts for the treatment of abdominal aortic aneurysm (AAA) [21,22]. We assessed the role of IVUS during endovascular repair of AAA using an AneuRx, an EndoVascular Technologies (EVT), an Excluder, a Talent or a Vanguard endograft [23]. Before placement, IVUS was used to determine the location of the renal arteries and aortic bifurcation, the lumen diameter of proximal and distal fixation sites, and the length of the aneurysm (Fig. 11). Based on these data the definitive size of the modular components of the AneuRx endovascular graft was selected. After placement, IVUS was used to determine the success of the procedure.

The results showed that, based on the data obtained before endograft placement, the definitive dimensions of the AneuRx endograft components could be selected. In 1 patient IVUS revealed that a proximal neck was too wide for endovascular treatment. In addition, we experienced that IVUS influenced the definitive length of the EVT endograft, could detect accessory renal arteries, warranted predilatation of the common iliac artery, and may influence the planned side of introduction. After endovascular grafting, IVUS may reveal incomplete.

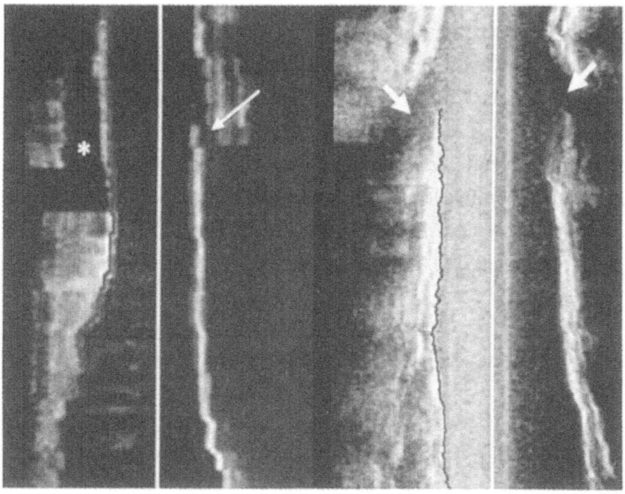

Fig. 11. Three-dimensional (3D) reconstructions of the proximal neck of an abdominal aortic aneurysm obtained with intravascular ultrasound before and after placement of an endovascular stent-graft. The left panel shows the proximal neck before intervention. Note the superior mesenteric artery (long arrow) and the renal vein (asterisk) crossing over the aorta. The right panel shows the proximal neck after intervention. Note that the renal arteries are open (short arrows); there is good apposition between the endovascular stent-graft and the native aortic wall..

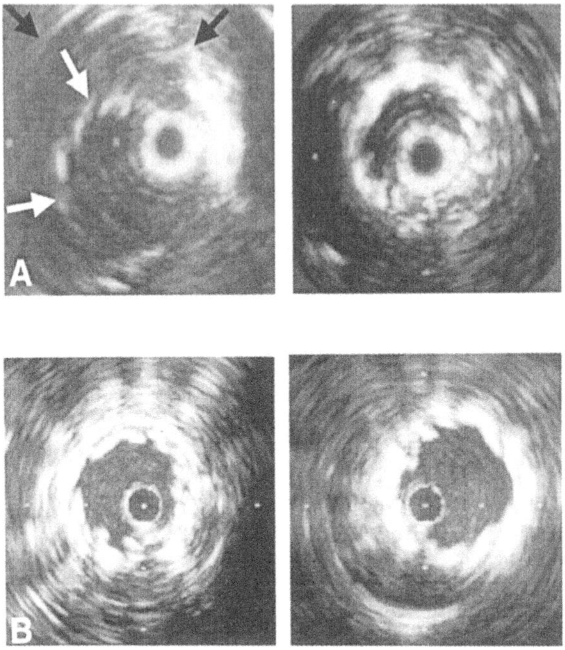

Fig. 12. Intravascular ultrasound cross-sections showing inadequate stent-graft placement for an abdominal aortic aneurysm (left panel) and the outcome after additional intervention (right panel). Panel A, shows the iliac leg of an endovascular stent-graft (white arrows) to end inside the aneurysm (black arrows). After placement of an extension cuff a satisfactory apposition is reached. Panel B, shows stenosis of the iliac leg of an endovascular stent-graft and folding of abundant graft material. After additional balloon angioplasty the diameter increased from 6.5 to 8.5 mm adequately determined with IVUS. After placement, IVUS detects parameters that may determine the outcome of the procedure and allow additional interventions to be performed during the same session.

apposition at the proximal or distal part of the endograft, overstenting or occlusion of the renal arteries, stenosis and folding of the iliac legs or damage to the access vessels. Detection of these features may necessitate additional interventions including placement of extension cuffs or additional PTA (Fig. 12). We conclude that definitive dimensions of modular endovascular prosthesis components can be adequitely determined with IVUS. After placement, IVUS detects parameters that may determine the outcome of the procedure and allow additional interventions to be performed during the same session.

FUTURE PERSPECTIVES

Ten years experience has confirmed that the value of IVUS imaging for basic and clinical research purposes is beyond dispute; however, although the ultrasonic information is clearly superior to angiographic data, as yet, IVUS remains an adjunct to angiography and not a substitute.

Future application of IVUS should be directed at steering the introduction of new interventional devices such as stent placement, endovascular grafts and endovascular radiotherapy. There is no doubt that IVUS is perfectly suited for localizing the exact region of stent or stent-graft placement, as well as for evaluating ambiguous angiographic results. In addition, there is already evidence that the use of IVUS in longitudinal studies may disclose the key factors responsible for restenosis - information which cannot be derived angiographically.

Finally, in order to gain insight into the progression of atherosclerosis, we developed a dedicated computer system that offers the potential for automated quantitative assessment of lumen, vessel and plaque volume [24]. This system, validated by two observers using separate pull-back maneuvers, revealed that volume measurements derived from IVUS images were highly reproducible. Therefore, for future longitudinal studies the combination of IVUS together with the dedicated computer analysis system will enable to monitor the effect of medical intervention, both pharmaceutical and endovascular, on the progression/regression of the atherosclerotic plaque and vessel volume.

REFERENCES

1. Gussenhoven EJ, Essed CE, Lancée CT, et al (1989) Arterial wall characteristics determined by intravascular ultrasound imaging: an *in vitro* study. J Am Coll Cardiol 14:947-952.
2. Lugt van der A, Gussenhoven EJ, Stijnen T, et al (1995) Comparison of intravascular ultrasonic findings after coronary balloon angioplasty evaluated *in vitro* with histology. Am J Cardiol 76:661-666.
3. Lankeren van W, Gussenhoven EJ, Qureshi A, et al (1999) Effect of balloon angioplasty of the iliac artery assessed with intravascular ultrasound and histology. Cardiovasc Int Radiol 22:50-55.
4. Dotter CT, Judkins MP (1964) Transluminal treatment of arteriosclerotic obstruction. Description of a new technic and a preliminary report of its application. Circulation 30:654-670.
5. Lankeren van W, Gussenhoven EJ, Lugt van der A, et al (1996) Intravascular sonographic evaluation of iliac artery angioplasty: what is the mechanism of angioplasty and can intravascular sonography predict clinical outcome? Am J Roentgenol 166:1355-1360.
6. The SHK, Gussenhoven EJ, Zhong Y, et al (1992) The effect of balloon angioplasty on the femoral artery evaluated with intravascular ultrasound imaging. Circulation 86:483-493.

7. Lugt van der A, Gussenhoven EJ, Reekers JA, et al (1997) Effect of balloon angioplasty on iliofemoral arteries evaluated with intravascular ultrasound. Eur J Vasc Endovasc Surg 13:549-556.
8. Lugt van der A, Gussenhoven EJ, Pasterkamp G, et al (1998) Intravascular ultrasound predictors of restenosis after balloon angioplasty of the femoropopliteal artery. Eur J Vasc Endovasc Surg 16:110-119.
9. Lankeren van W, Gussenhoven EJ, Honkoop J, et al (1999) Plaque area increase and vascular remodeling contribute to lumen area change after percutaneous transluminal angioplasty of the femoropopliteal artery: an intravascular ultrasound study. J Vasc Surg 29:430-441.
10. Lankeren van W, Gussenhoven EJ, Kints van MJ, et al (1997) Stent remodeling contributes to femoropopliteal artery restenosis: an intravascular ultrasound study. J Vasc Surg 25:753-756.
11. Leertouwer TC, Gussenhoven EJ, Lankeren van W, et al (1998) Memotherm in-stent restenosis due to intimal hyperplasia: an intravascular ultrasound study. J Ang Vasc Surg 4:54-58.
12. Sahni R, Maniet AR, Voci G, et al (1991) Prevention of restenosis by lovastatin after successful coronary angioplasty. Am Heart J 121:1600-1608.
13. Iwasaki K, Kusachi S, Hina K, et al (1995) Preventive effect of pravastatin on restenosis following coronary angioplasty: prospective randomized trial. J Cardiol 25:15-21.
14. Nakamura Y, Yamaoka O, Uchida K, et al (1996) Pravastatin reduces restenosis after coronary angioplasty of high grade stenotic lesions: results of SHIPS (SHIga Pravastatin Study). Cardiovasc Drugs Ther 10:475-483.
15. Hagenaars T, Gussenhoven EJ, van Sambeek MRHM, et al (2000) Effect of simvastatin on restenosis after percutaneous transluminal angioplasty of femoropopliteal arterial obstruction Am J Cardiol 86:774-776.
16. Leertouwer TC, Gussenhoven EJ, Jaarsveld van BC, et al (1999) In-vitro validation, with histology, of intravascular ultrasound in renal arteries. J Hypertens 17:271-277.
17. Leertouwer TC, Gussenhoven EJ, Dijk van LC, et al (1999) Intravascular ultrasound evidence for coarctation causing symptomatic renal artery stenosis. Circulation 99:2976-2978.
18. Leertouwer TC, Gussenhoven EJ, Overhagen van H, et al (1998) Stent placement for renal artery stenosis guided by intravascular ultrasound. J Vasc Interv Radiol 9:945-952.
19. Essen van JA, Gussenhoven EJ, Lugt van der A, et al (1999) Accurate assessment of abdominal aortic aneurysm by intravascular ultrasound scanning: validation with computer tomographic angiography. J Vasc Surg 29:631-638.
20. Essen van JA, Gussenhoven EJ, Blankensteijn JD, et al (in press) Three-dimensional intravascular ultrasound assessment of the proximal and distal neck of abdominal aortic aneurysms. J Endovasc Surg.
21. Parodi JC (1995) Endovascular repair of abdominal aortic aneurysms and other arterial lesions. J Vasc Surg 21:549-557.
22. Balm R, Eikelboom BC, May J, et al (1996) Early experience with transfemoral endovascular aneurysm management (TEAM) in the treatment of abdominal aortic aneurysms. Eur J Vasc Endovasc Surg 11:214-220.
23. Essen van JA, Smet de AAEA, Blankensteijn JD, et al (1999) Intravascular ultrasound (IVUS) to guide endovascular stent-graft placement in abdominal aortic aneurysm (AAA). CVIR 22(suppl 2):S112-155.

24.Hagenaars T, Gussenhoven EJ, Essen JA, et al (2000) Reproducibility of volumetric quantification in intravascular ultrasound images Ultrasound Med Biol 26:367-374.

Intra-coronary Ultrasound to Guide Percutaneous Coronary Intervention

J. Ligthart[1], P.J. de Feyter[1]

[1] University Hospital Rotterdam, Thoraxcenter,
Dr. Molenwaterplein 40, 3015 GD, Rotterdam, the Netherlands

Summary: Intra-coronary ultrasound has a unique value to guide percutaneous coronary intervention procedures. Pre-intervention ultrasound imaging may help determine device selection. Intra-coronary ultrasound during interventions often results in improved immediate outcome during balloon angioplastyor atherectomy and in particular during stent implantation. Intra-coronary ultrasound may offer a unique opportunity for precise radiation dosage during brachytherapy.

Keywords: ultrasound, coronary intervention, stents, balloon angioplasty, atherectomy

Introduction

Contrast coronary angiography is the most important imaging technique to detect coronary obstructive atherosclerosis and to guide percutaneous coronary interventions.

However, it is now widely acknowledged that coronary angiography has its limitations, due to the fact that it is a silhouette technique, visualizing the coronary lumen in a 2-dimensional plane. Intracoronary ultrasound is a unique diagnostic tool that provides extended insights beyond the coronary lumen, because it provides a tomographic perspective of the coronary anatomy and it is able to directly image the coronary wall.

Coronary angiography and intracoronary ultrasound each have their own unique value to guide percutaneous coronary intervention and in this review we will highlight the additional value which ultrasound does offer during percutaneous treatment of obstructive coronary atherosclerosis.

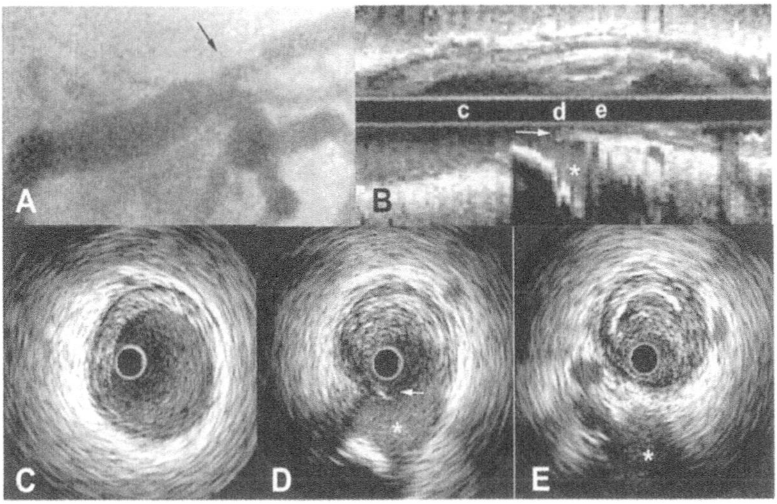

Fig.1. Top left panel A shows an angiogram of a lesion in the proximal LAD. The bottom right panel E shows the IVUS cross-section at the place of the minimal luminal area. From 19.00 hrs – 5.00 hrs an eccentric mixed plaque is seen, consisting of soft material and a 90 degrees fibrous ring. Bottom panel D is the IVUS cross-section at the carina of the LAD and the circumflex artery. The LCX is marked with *. Opposite of the LCX, the plaque is extending from 19.00hrs-4.00 hrs. An arrow marks an intracoronary thrombus, partly overlapping the carina. The bottom left panel C is an IVUS cross-section in the main stem, showing a concentric intimal thickening from 08.00 hrs- 15.00 hrs. The top right panel B is a longitudinal reconstruction, indicating the location of the described cross-section by their corresponding letters. The LCX is marked by *. Above d and e the eccentric plaque is visible, while the arrow indicates the thrombus, covering partly the LCX. Based on these IVUS images atherectomy was chosen as PCI.

Pre-interventional imaging

Pre-interventional imaging may offer valuable information to the investigator to select the most optimal interventional device or approach for a specific lesion [1]. This is based on ultrasound obtained information about plague distribution, depth and extent of calcification, presence of thrombi or precise assessment of vessel size.

More precise knowledge of plaque distribution, not available by coronary angiography, such as involvement of ostial lesions or side branches, may lead to an initial approach by an atherectomy device to prevent plaque shift (figure 1).

The existence of a large, extensive calcified plaque may trigger the need for an initial approach with rotational atherectomy (figure 2).

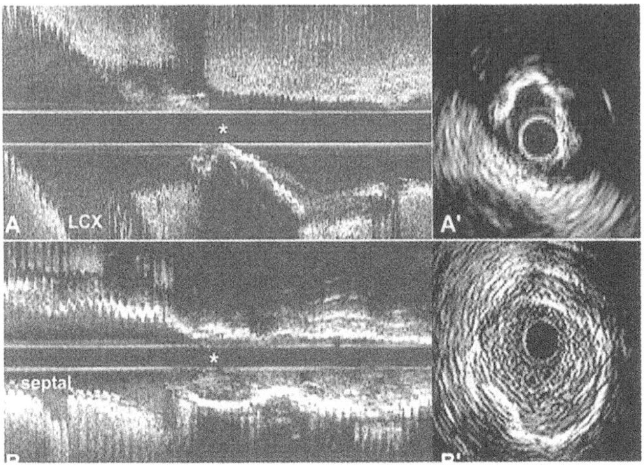

Fig. 2. Different morphology of calcific plaques. Panel A: Longitudinal reconstruction of the proximal LAD with at * a 180 degrees superficial calcific plaque (A') from 9-4 o'clock. Panel B: Longitudinal reconstruction showing that the stenosis is caused by a 180 degrees deep calcific plaque (B'). Note the soft plaque upon the calcium at *. The extensive superficial calcified plaque as shown in panels A and A' may indicate an approach with rotational atherectomy

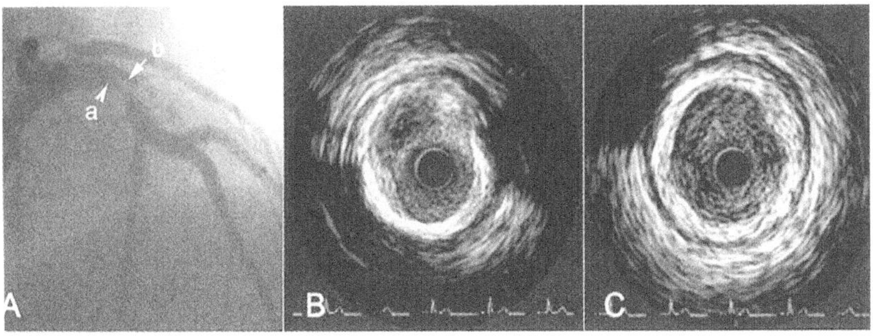

Fig.3. Panel A shows an angiogram of a lesion in the proximal LAD. Arrow a indicates the position of the IVUS cross-section in panel B an arrow b indicates the cross-section in panel C. Cross section B shows a concetric mixed fibro-calcific plaque with superficial calcium from 6 o'clock till 9 o'clock and a spot of calcium at 3 o'clock. At 11 o'clock an eccentric mass of subacute and chronic organized thrombus is visible. At the minimal luminal diameter (Panel C) , an almost 360 degrees fibrous plaque is visible with calcification at 10 o' clock. The lumen is filled with a concentric mass of organized thrombus. The IVUS catheter is occluding the remaining lumen..

Angiographically unrecognized intracoronary thrombi may call for adjunctive treatment with a platelet GP 2b/3a receptor inhibitor or use of a distal embolization protection device (figure 3). The determination of the true vessel size with IVUS will secure the choice of a properly sized balloon or stent and may distinguish between a truly small vessel and a small lumen of a diffusely diseased vessel, as can be seen in diabetic patients.

However, although pre-interventional imaging provides more anatomical details, it still remains to be determined whether this truly helps to improve the immediate and long-term outcome of percutaneous coronary interventions.

IVUS-Guidance for balloon angioplasty

IVUS determined morphological features such as presence and extent of dissections, plaque burden or vessel and lumen size, showed no statistically significant impact on the long term outcome of balloon angioplasty [2], although the residual percent plaque burden was the most powerful predictor of clinical outcome [3].

IVUS is superior to angiography to detect the extent and severity of balloon angioplasty induced dissections (figures 4, 5). In particular, 3-D reconstructions of extensive dissections are helpful to guide in the size and length of a stent to 'tack' the dissection (figure 6).

The Clout-Pilot Trial demonstrated that ultrasound guided balloon angioplasty resulted to result in immediate larger angiographic minimum lumen diameter, because IVUS aided to select a larger balloon size, without the occurrence of dissections [4]. Unfortunately, no data are available to demonstrate that this initial larger lumen indeed is associated with a lesser late restenosis rate.

A strategy of provisional stenting using IVUS to select aggressive balloon sizing (based on media-to-media dimension of the vessel) demonstrated that the occurrence of major adverse events was similar to stenting [5, 6]. In these two non-randomized studies, stenting was used if the aggressive balloon angioplasty resulted in a sub-optimal result, in the majority due to the occurrence of extensive dissection.

IVUS guided atherectomy

The presence and depth of calcification, as can be established with IVUS, plays an important role for the selection of the use of a directional atherectomy device or rotational atherectomy. Superficial calcium (figure 7) is often associated with poor retrieval of tissue when using DCA, although the latest upgraded version of DCA, the flexcut, is able to cut through calcium. And, the preferred option for superficial calcium is high-speed rotational atherectomy (figure 8). Most striking is the fact that following either DCA or rotablation, there is an abundant remaining plaque

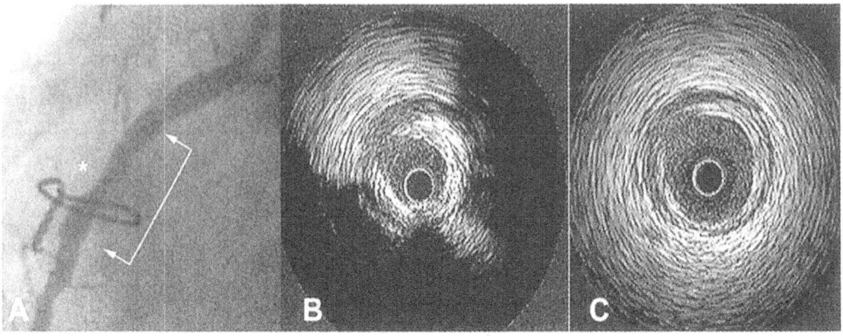

Fig. 4. Panel A shows an acceptable angiographic result post balloon angioplasty between the arrows in the proximal RCA. IVUS cross-sections B and C show a submedial dissection , not visible in the angiographic image. The * shows the position of cross-section B with a dissection from 9-12 o'clock. Cross section C is more proximal with a dissection from 8-2 o'clock.

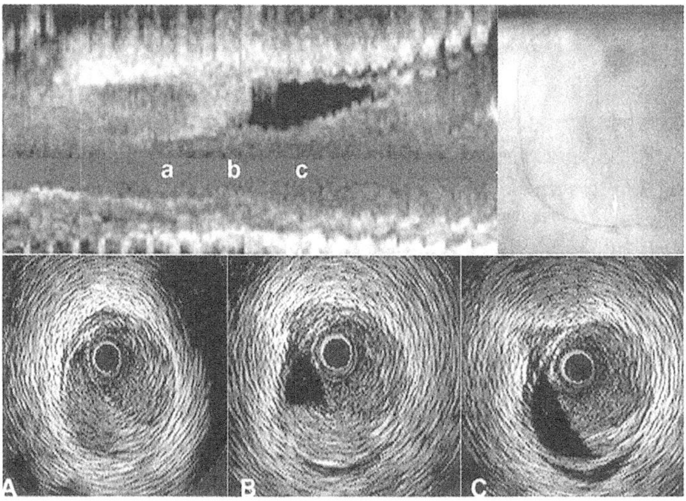

Fig.5. An IVUS impression of an antegrade subintimal dissection in the distal right coronary artery after balloon dilatation. The angiography in the top right panel reveals some contrast staining (arrow) . Cross section A shows the proximal part of the dissection. The flap is visible from 5 till 11 '0 clock. Behind this flap a false lumen is present. At 5 'o' clock non-moving blood is present in this false lumen. Cross section B is located more distally, showing the false lumen from 5 till 11 'o'clock. Non-moving blood is observed at 6 'o'clock. The black spot at 10 'o'clock is staining X-ray contrast in the false lumen. Cross-section C shows the distal end of the dissection , completely filled with staining X-ray contrast. At the longitudinal reconstruction (top left) the bifurcation of the posterior descending artery and the posterolateral is just visible at the right edge. A, b and c correspond with the position of the cross-sections A,B, and C. The dissected area displays from distal to proximal : staining contrast (c) , staining blood (b) and moving blood in the false lumen (a).

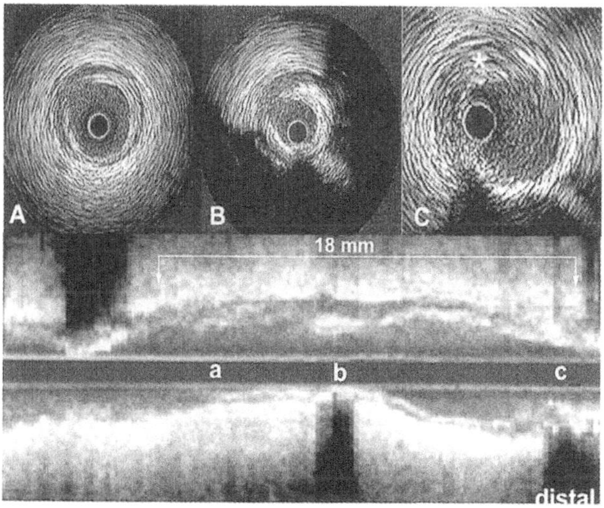

Fig.6. An IVUS impression of an 18 mm long combined retro- and antegrade submedial dissection. Cross sections A and B show the ruptured plaque at 10 'o'clock with the submedial dissection. Cross-section C shows the distal end of the dissection with at * staining blood in a false lumen. The lower section shows a longitudinal reconstruction . The entrance of the dissection begins at a. Cross-sections A, B and C correspond with a, b and c in the longitudinal reconstruction.

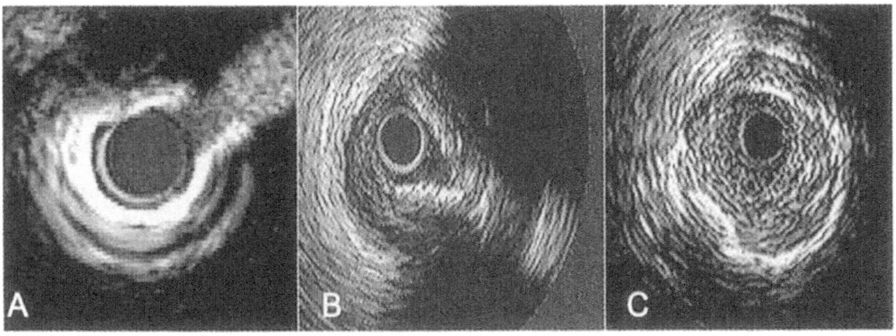

Fig.7. IVUS cross sections with three examples of calcified plaque distribution. Panel A shows an ostial lesion with 360 degrees of superficial calcium. Panel B shows a concentric mixed plaque with almost 180 degrees of superficial calcium from 1 o'clock till 7 o'clock. Panel C shows a 200 degrees deep calcific plaque from 1 o' clock till 9 o'clock. The lesion is formed by a concentric mass of soft material upon the calcium. For treatment of the lesions in panel A and B rotational atherectomy may be the preferred approach, while directional atherectomy may be sufficient for the lesion in panel C.

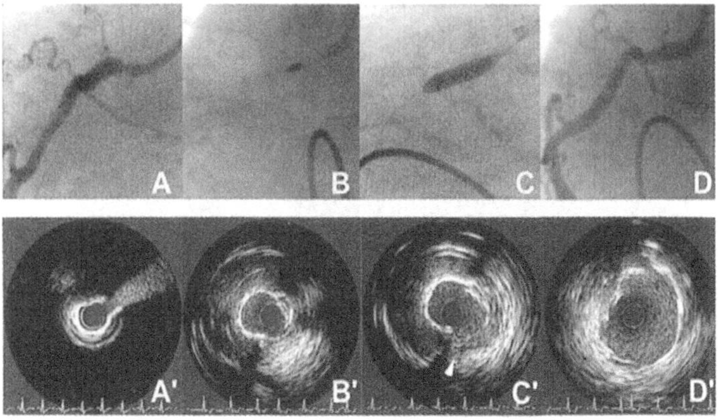

Fig. 8. Panel A: Angiogram of an ostial lesion in the right coronary artery. IVUS (A') shows a concentric 360° superficial calcific ring at the ostium. The * indicates the aorta. Post rotablator, burr-size 1.5 – 2.15mm burr (B), the lumen significantly has increased as shown on IVUS (B'). A 360° calcific ring still remains. Comparing both IVUS figures (A'and B') gives an indication of the thickness of this calcific ring. Balloon angioplasty (4.0x10mm World pass®, (B)), rupture of the calcific ring indicated by the arrowhead (B'). Angiography (D) and IVUS (D') show the final result post stenting (Cordis BX Velocity® 4.0x8mm)

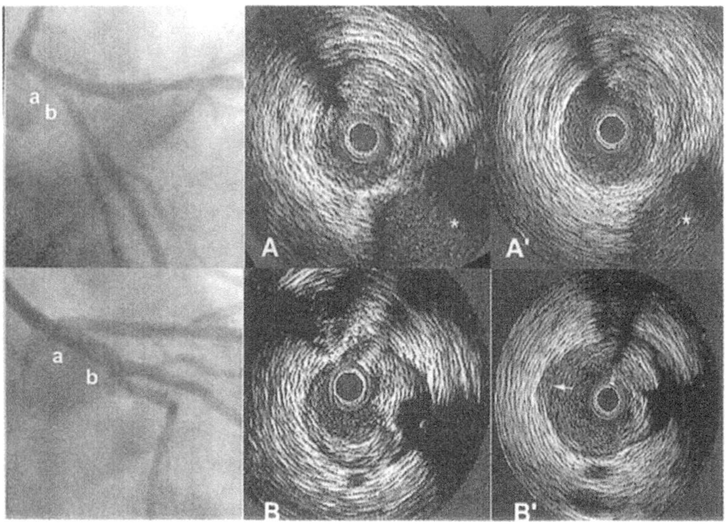

Fig. 9. The angiogram at the top left panel shows a stenosis at the proximal circumflex artery. Cross-sections A and B show an eccentric soft plaque from 10-5 o'clock. (A) and an eccentric mixed plaque with 10 degrees ARC of superficial calcium at 3 o'clock (B). In both cross-sections a wire is visible (11 o' clock at A, 1 o' clock at B) The left angiogram shows the result after directional atherectomy (BANTAM 7F). Cross-section A' shows that atherectomy removed plaque from 10-12 o'clock. Cross-section B' shows plaque removal at 6 o'clock and a misdirected cut at 9 o'clock. Note the residual plaque at 1 o' clock in both sections A' and B'.

burden, even when angiography demonstrates a wide lumen (figure 9). Ultrasound guided aggressive DCA does achieve more aggressive tissue removal, but this has not resulted in statistically significant reduction of late restenosis in the OARS and ABACAS studies [7, 8]. However, the START study, using an aggressive IVUS guided atherectomy strategy, demonstrated a significant reduction of the restenosis rate in the atherectomy group (16%) compared with stenting (33%) [9].

IVUS guided high-speed rotational atherectomy has demonstrated the differential cutting ability of the device (i.e. removing calcium rather than soft tissue) (figure 8) and IVUS measurement of the true lumen does help to select the correct burr size to optimize the result.

IVUS guided stenting

Intra-coronary ultrasound imaging has played a significant role in understanding and optimizing stent implantation.

Coronary angiography largely underestimates, or does not recognize, malapposition, underexpansion or asymmetric stent implantation.

IVUS providing a tomographic perspective, and in particular 3-D reconstruction-imaging obtained from stacking the individual IVUS slices, clearly and often spectacularly delineates the morphologic details of stent implantation.

Malapposition occurs regularly at focal areas within the stent, whereas large areas of stent malapposition occur less frequently (figure 10).

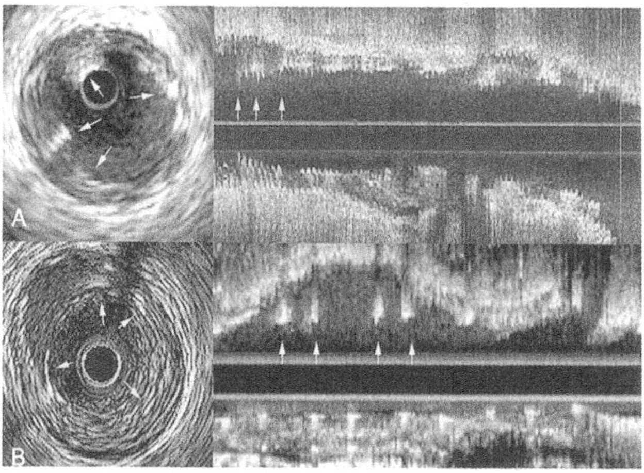

Fig. 10. Panel A shows an IVUS cross section with a stent malapposition from 7 o'clock till 12 o'clock. The stent struts are indicated by the arrows. The longitudinal reconstruction on the top right panel shows this focal malapposition, indicated by the arrows. Panel B and the longitudinal reconstruction in the bottom right panel show a stent malapposition extending over a larger area, which is caused by an aneurysm after a spontaneous dissection.

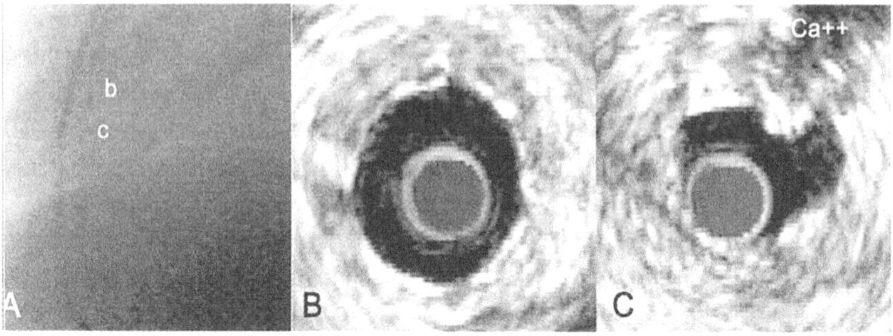

Fig. 11. At the X-ray image in panel A, after stentimplantation , the position of the IVUS cross sections in panel B and C are indicated by b and c. The IVUS probe in this image is located at c. Panel C shows an asymmetric stent deployment in a calcific lesion, despite of high pressure balloon inflation (24 bar). The calcium ring is visible from 11.00-14.00 hrs. Panel B shows the same stent more proximal with an almost circular deployment. The asymmetry of the stent is hardly visible in the X-ray image.

Asymmetric stent expansion occurs quite often at calcific sites, in particular when the opposite vascular segment is normal or when the opposite plaque consists of soft tissue, which is much more compliant than the calcific plaque. This asymmetric stent expansion is hard to remedy because high-pressure stenting does not prevent this problem, in fact it may worsen the situation (figure 11).

Underexpansion of the stent can occur at local areas within the stent, or the stent over its entire length is underexpanded. This phenomenon is easily detected by IVUS and the severity and extent of underexpansion is easily recognized by 3-D reconstructed imaging (figures 12, 13). IVUS guided optimized stent implantation should lead to a better acute outcome (less sub-acute thrombosis), although this is difficult to demonstrate in convincing studies due to the use of more refined stents, use of high pressure and the adjunctive treatment with thienopyridines (tidopidine or clopidogrel) all of which acknowlegded to reduce (sub) acute thrombosis.

Intuitively, one may argue that optimized IVUS guided stenting should result in a larger stent lumen, which logically should translate into a lower clinically significant restenosis rate, because in stent restenosis is largely due to the degree of neo-intimal hyperplasia which is better accomodated in a larger stent.

Several multicenter trials have addressed the issue whether an IVUS guided stent implantation (assuming an optimal stent implantation with large lumen) indeed results in a late lower restenosis rate compared to non IVUS guided stent implantation.

The Cruise study demonstrated a lower target vessel revascularization rate in IVUS guided stenting from 14.9% to 8.9% [10]. The OPTICUS Trial showed no significant difference in ultrasound guided versus angiographic guided stent implantation [11]. The French Resist Trial, a rather small randomized trial of 155 patients, showed a non-significant trend towards a reduction in the restenosis rate [12].

The largest trial – the AVID - 759 randomized patients demonstrated a trend toward lower revascularization rates with IVUS guidance (8%) versus angiographic guidance (12%). In a few subsets, vessels <3.25 mm, severe lesions and lesions in saphenous venous bygrafts, there was a statistically significant lower restenosis rate after IVUS guidance [13]. IVUS obtained parameters may predict late in-stent restenosis.

A reference chart has been constructed showing that IVUS measurements post-stent implantation: length and minimal in-stent lumen area, accurately predict the 6 month restenosis rate [14] (reference chart figure 14).

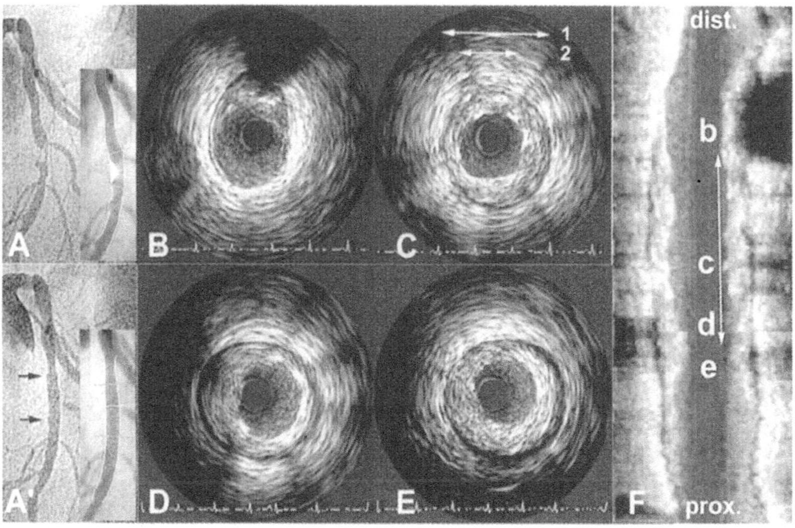

Fig. 12. Panel A shows an acceptable angiographic result after stentimplantation of the distal right coronary artery. The IVUS cross-section however at panel A' reveals a smaller instent lumen , with a measured minimal luminal area of 4.62 mm^2 . Additional dilatation with high pressure (18 bar) made the minimal luminal area increase to 8.3 mm^2. (Panel B') The arrows in both panels A and B indicate the position of the IVUS cross-sections.

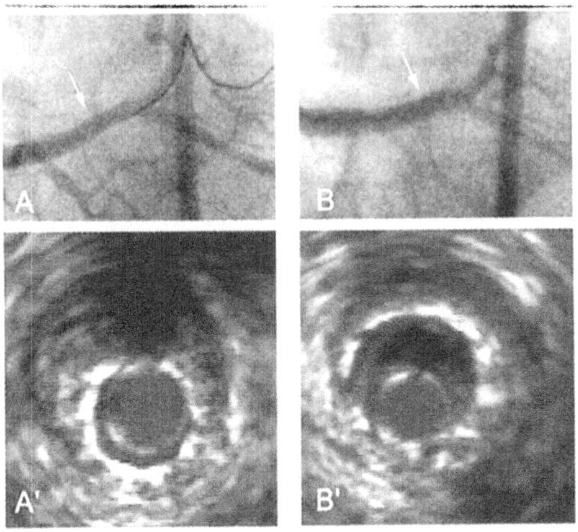

Fig. 13. Panels A and A' show a left superior oblique angiography of a severe lesion in the LAD before and after stenting. The stent size was guided by QCA (inserts). QCA shows a reference diameter of 2,45mm, resulting in a choice of a 3.0x20mm stent. The result on A' shows indeed a "step up-step down" effect indicated by the arrowheads. QCA (insert) indicates a stent MLD of 2,75 mm. IVUS however, shows a different view. Panels B, C, D and E show respectively cross-sections of the distal reference, the minimal stent area, the proximal side of the stent and the proximal reference. Panel F shows a longitudinal reconstruction with the position of the shown cross-sections indicated by their corresponding letters. The stent length (20mm) is indicated by the arrow. IVUS shows a well-deployed but undersized stent in a diffusely diseased vessel. The minimal luminal diameter and the total vessel diameter measured in section C are respectively 2,45mm and 4,1 mm. This may justify a stent size of 3,5 or even 4.0mm, according to the CLOUT-study, to minimize the chance for restenosis within 6 months. (See also table in figure 14)

IVUS Predictors of 6-month In-stent Restenosis

stent length	minimum in-stent area, mm² 3.0-3.9	3.9-4.8	4.8-5.7	5.7-6.6	6.6-7.5	7.5-8.4	8.4-9.3	9.3-10.2	10.2-11.1	11.1-12.0
10–15	0.30	0.25	0.21	0.17	0.13	0.11	0.08	0.07	0.05	0.04
15–20	0.30	0.28	0.23	0.19	0.15	0.12	0.10	0.08	0.06	0.05
20–25	0.36	0.31	0.25	0.21	0.17	0.14	0.11	0.09	0.07	0.05
25–30	0.40	0.34	0.28	0.23	0.19	0.15	0.12	0.10	0.08	0.06
30–35	0.43	0.30	0.31	0.26	0.21	0.17	0.14	0.11	0.09	0.07
35–40	0.46	0.40	0.34	0.29	0.24	0.19	0.16	0.13	0.10	0.08
40–45	0.50	0.43	0.37	0.31	0.26	0.22	0.18	0.14	0.11	0.09
45–50	0.53*	0.48	0.40	0.34	0.29	0.24	0.20	0.16	0.13	0.10
50–55	0.57*	0.50	0.44	0.38	0.32*	0.27*	0.22*	0.18*	0.14*	0.11*
55–60	0.60*	0.54*	0.47*	0.41*	0.35	0.29*	0.24*	0.20*	0.16*	0.13*

P.J.de Feyter et all, ***Circulation. 1999; 100: 1777-1783***

Fig. 14. Reference chart

IVUS guided in-stent restenosis treatment

In-stent restenosis is often difficult to treat because, and irrespective of the use of any device (balloon, re-stenting, laser, directional atherectomy, high speed rotational atherectomy or cutting balloon), there is a high (up to 60%) re-restenosis rate. However, ultrasound may be of great value because IVUS may reveal that the initial stent implantation was inadequatc with underexpansion (local or general). In a study of 1090 consecutive patients with in-stent restenosis stent underexpansion was present in 38 % of the treated lesions with an in-stent cross-sectimal area < 6.0 mm2 [15]. Proper balloon sizing may now lead to an improved stent lumen area [14].

IVUS guided brachytherapy

Intra-coronary brachytherapy for in-stent restenosis does significantly reduce the high repeat restenosis rate of approximately 60% with balloon angioplasty to 30% with brachytherapy. Dose delivery to media and/or adventitia may determine the efficacy of brachytherapy, which in turn is dependent on the thickness and composition of the plaque. IVUS based algorithms to deliver a specific radiation dose to a specific lesion may optimize the efficacy of the therapy (figure 15).

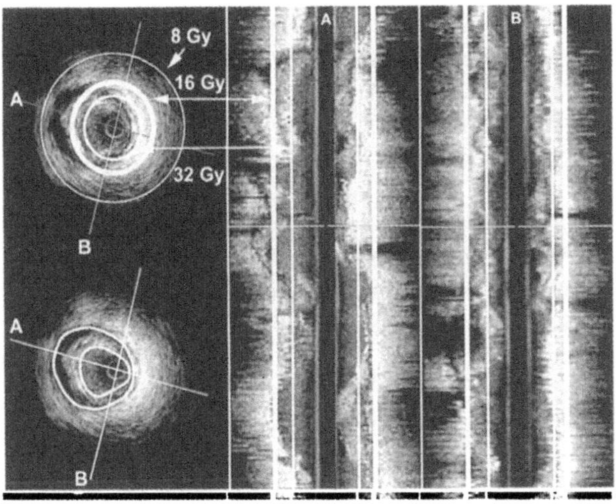

Fig.15. Considering that a non-centred radioactive source for intracoronary brachytherapy behaves like an intracoronary ultrasound catheter in a tortuous artery, precise dosimetry can be derived. The top left panel shows an IVUS cross-section surrounded by three isodose lines, indicating the areas that received 32, 16 and 8 Gy. The longitudinal reconstruction (midpanel) shows the often-eccentric position of the catheter in the irradiated segment. The isodose lines now indicate the inappropriate dose delivery at the vessel wall .

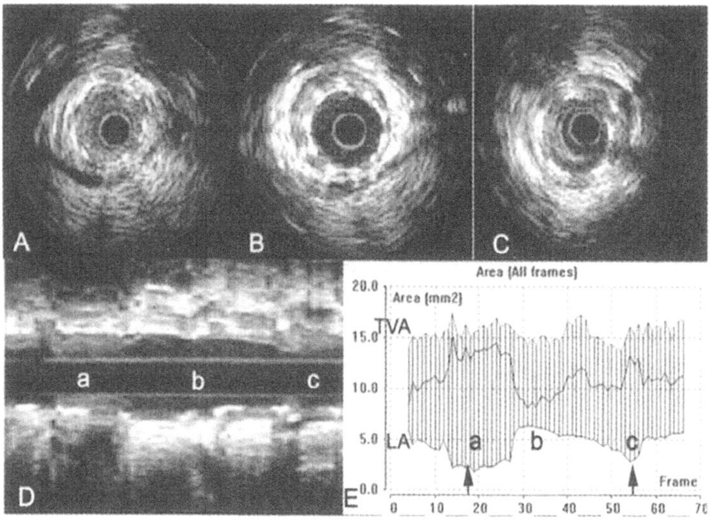

Fig. 16. "Candy wrapper effect" in a radioactive stent, 6 month post implantation. Panel A shows the IVUS cross-section at the proximal edge of the stent, with an severe eccentric instent hyperplasia. The IVUS catheter occludes the remaining lumen. Panel B shows an IVUS cross-section of the mid-part of this stent without hyperplasia, while panel C shows severe eccentric instent hyperplasia (from 5 till 12 o'clock) at the distal edge of the stent. At the longitudinal reconstruction (panel D) the position of the cross-sections are indicated by a,b and c. Panel D shows a graphic representation derived from measurements of the Total Vessel Area (TVA) and the Lumen Area (LA) of the stented segment. Both arrows indicate the distal and proximal edges of the stent. The increment between the cross-sections (Frames) is 0.5mm. Note the small lumen areas at a and c, compared to the lumen area at b,

Brachytherapy for de novo coronary lesions (either stented or non-stented) may be effective, but has been shown to be severely hampered by the accelerated restenosis at the edges: "the candy wrapper effect" (figure 16). IVUS is often an invaluable tool to assess the exact localization of in-stent restenosis after brachytherapy, i.e. within the stent or the edges, which is almost impossible to discern with angiography and use of non-radiopaque stents.

Conclusion

Intracoronary ultrasound has significantly contributed to our understanding of the mechanisms of balloon angioplasty, tissue removal devices and stenting. IVUS guided stenting is helpful in cases of angiographic uncertainty. On the basis of the available data it cannot be strongly recommended that IVUS should be routinely used to optimize stent implantation and hence to reduce late in-stent restenosis.

Intravascular ultrasound is a significant research tool which has greatly helped to understand lesion progression regression, vessel remodeling and restenosis mechanisms after various percutaneous treatment modalities. New mechanical and pharmacological interventions require IVUS imaging to assess mechanisms and efficacy.

References

1. Mintz GS, Pichard AD, Kovach JA, et al (1994) Impact of pre-intervention intra-vascular ultrasound imaging on transcatheter treatment strategies in coronary artery disease. Am J Cardiol 73: 423-30
2. Peters RJ, Kok WEM, Di Mario, et al (1997) Prediction of restenosis after coronary balloon angioplasty. Circulation 95: 2254-2261
3. Fitzgerald PJ, Yock PG (1993) Mechanisms and outcomes of angioplasty and atherectomy assessed by intravascular ultrasound imaging. J Clin Ultrasound 21: 579-588
4. Stone GW, Hodgon JM, St Goar FG, et al (1997) Improved prodedural results of coronary angioplasty with intra-vascular ultra-sound guided balloon sizing – The Clout-Pilot Trial. Circulation 95: 2044-2052
5. Abizaid A, Pichard AD, Mintz GS, et al (1999) Acute and long-term results of an intravascular ultrasound-guided percutaneous transluminal coronary angioplasty provisional stent implantation strategy. Am J Cardiol 84: 1298-1303
6. Schroeder S, Baumbach, A. Haase KK, et al (1999) Reduction of restenosis by vessel size adapted percutaneous transluminal coronary angioplasty using intravascular ultrasound. Am J Cardiol 83: 875-79
7. Simonton CA, Leon MB, Baim DS, et al (1998) Optimal directional coronary artherectomy. Final results of the Optimal Atherectomy Restenosis Study (OARS). Circulation 97: 332-339
8. Suzuki T, Hosokawa, Katoh O, et al (1999) Effects of adjunctive balloon angioplasty after intra-vascular ultrasound guided optimal directional coronary atherectomy. JACC 34: 1028-1035
9. Tsuchikane E, Sumitsuji s, Awanta N, et al (1999) Final results of the stent versus directional coronary atherectomy randomized trial (START). JACC 34: 1050-1057
10. Fitzgerald PJ, Oshima A, Hayase M, et al (2000) Final results of the Can Routine Ultrasound Influence Stent Expansion (CRUISE) study. Circulation 102: 523-530
11. Mudra H, Macaya C, Zahn R et al (1998) Interim analysis of the "Optimization with IVUS to reduce in-stent restenosis" (OPTICUS) trial. Circulation 98: I-363 abstract
12. Schiele F, Meneveau N, Vuillemenot A, et al (1998) Impact of intravascular ultrasound guidance in stent deployment on 6 month restenosis, Rate: a multicenter, randomized study comparing two strategies, with and without intravascular ultrasound guidance. JACC 32: 320-328
13. Russo RJ, Nicosia A, Teirstein PS (1997) Angiography versus intravascular ultrasound directed stent placement. JACC 29: 369A abstract

14.de Feijter PJ, Kay P, Disco C, Serruys PW (1999) Reference Chart derived from post-stent implantation intravascular ultrasound predictors of 6 month expected on quantitative coronary angiography. Circulation 100 : 1777 - 1783
15.Castagora M, Mintz G.S. Weissman N.J. et al (2001) The contribution of mechanical problems to in-stent restenosis: an intravascular ultrasound analysis of 1090 consecutive in-stent restenosis lesions. JACC 37: Suppl A 42A (abstract)

Detection of Vulnerable Coronary Plaque; The Emerging Role of Intravascular Ultrasound

Paul Schoenhagen, MD; E. Murat Tuzcu, MD; Steven E. Nissen, MD

The Cleveland Clinic Foundation, Department of Cardiology, Desk-F 15
9500 Euclid Ave., Cleveland OH 44195, USA

Summary: Sudden rupture of vulnerable coronary plaques is the initiating event of most acute coronary syndromes. However, *asymptomatic* episodes of plaque vulnerability, rupture and fibrosis may be far more common and related to atherosclerotic disease progression. The identification and eventual treatment of these vulnerable coronary plaques is a major challenge in cardiology [1]. Current diagnostic techniques can not yet reliably identify vulnerable plaques. Intravascular ultrasound, an established interventional technique, is one of the most promising imaging approaches to identify these plaques *in vivo*. This article will describe the current role of IVUS in the detection of vulnerable plaques based on a comparison to the underlying vascular biology.

Key Words: Coronary Artery Disease, Vulnerable Plaque, Plaque Rupture, Acute Coronary Syndrome, Intravascular Ultrasound, Imaging

Introduction: Rational for the Use of an In Vivo Imaging Method

Traditional models of coronary artery disease (CAD) focus on the diagnosis and treatment of focal, highly stenotic lesions, assuming that these lesions will progress, causing acute clinical events. However, it is now well known that most acute coronary syndromes (ACS) are caused by "vulnerable" lesions, which are usually not highly stenotic before the event [2-5] (figure 1).

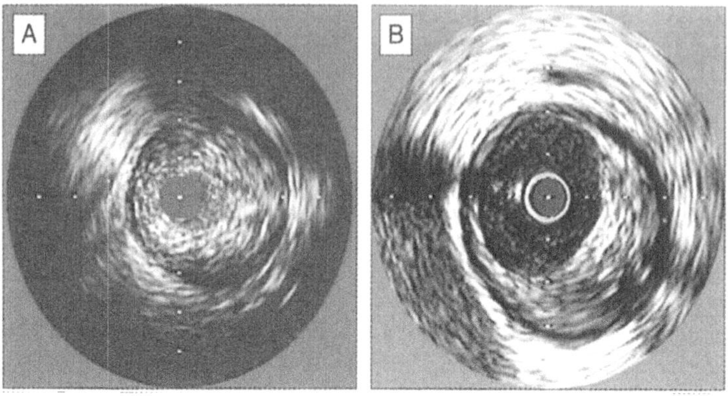

Fig. 1: Example of a severely stenotic (A) and a mildly stenotic (B) focal lesion. Most acute coronary syndromes are initiated from mildly stenotic lesions. However, because of outward expansion of the vessel wall, these lesions are often relatively large.

Plaque vulnerability describes biochemical and physical properties predisposing these lesions to an increased risk of rupture and thrombosis. Most current knowledge about plaque vulnerability has been derived from histologic studies of plaques that had already ruptured. The morphology of ruptured plaques has been described and the potential cellular and molecular triggers of rupture, erosion and thrombosis are partially known [6 -8].

However, it is not clear if these features reflect the characteristics of vulnerable plaques before rupture. We believe that "vulnerable" lesions are likely highly dynamic and numerous throughout the coronary tree. It is also probable that one of the mechanisms of disease progression is symptomatic and asymptomatic rupture with subsequent fibrosis. Therefore, in order to identify vulnerable plaques, prospective, morphological observation of large numbers of lesions in large parts of the coronary tree in comparison to clinical presentation will be necessary.

In vivo imaging techniques are best suited to allow this prospective observation. Optimally, such a technique should be able to observe morphologic changes related to plaque progression/ regression in individual lesions and simultaneous assess changes of disease burden in large parts of the coronary tree. Intravascular ultrasound is currently the most promising technique (figure 2).

In this article we will summarize current knowledge about the biology of vulnerable plaques, and analyze how these characteristics are reflected by

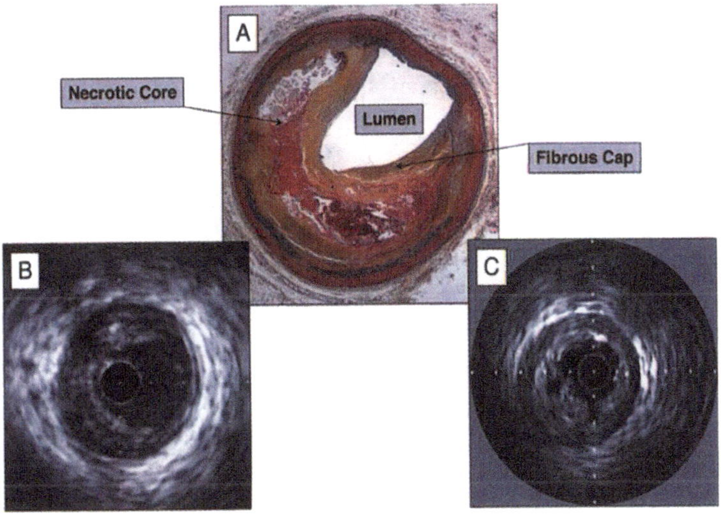

Fig.2: IVUS images reflects the anatomy of the vessel wall. Histologic (A) and intravascular ultrasound (B) images of lesions with features of vulnerability are shown. Panel C shows a ruptured lesion in a patient with an acute myocardial infarction.

intravascular ultrasound (IVUS). This will be the basis to describe observations in patients with acute coronary syndromes and the emergent use of IVUS in serial, prospective studies of lesions at risk.

Biology of Plaque Progression and Vulnerability

The formation of atherosclerotic lesions is initiated by the recruitment of mononuclear leukocytes to the intima. Recruitment is mediated by specific adhesion molecules, expressed on the surface of vascular endothelial cells [9,10]. Chemoattractant chemokines facilitate the entry into the intima, where these cells accumulate lipid and become foam cells. These cells are characteristic of the early atheromatous precursor, the fatty streak [11].

The accumulation of macrophage foam cells may either be temporary or initiate atheroma progression and evolution into more advanced plaques that may eventually cause clinical disease [12]. Atheroma progression is associated with the accumulation of a large, often eccentric, lipid pool (necrotic core) separated from the lumen by a fibrous cap [13,14] (figure 3). Positive arterial remodeling, the outward expansion of the vessel at the lesion site, allows plaque accumulation without lumen obstruction [15]. The accumulation of inflammatory cells and

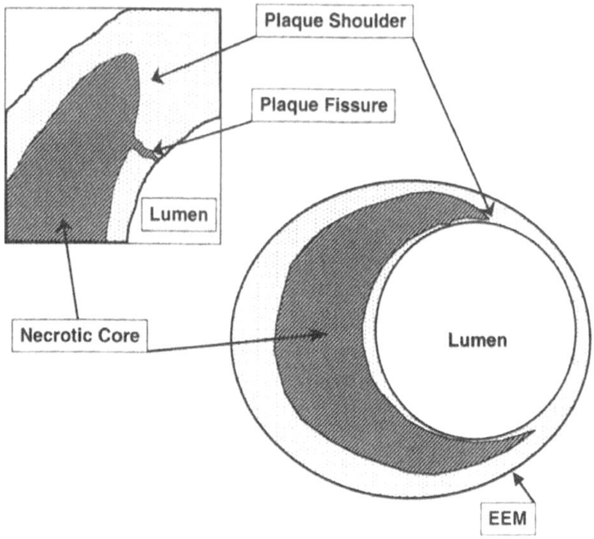

Fig.3: Cartoon of a vulnerable plaque. Features of vulnerability are the large lipid pool (necrotic core) separated from the lumen by a thin fibrous cap.

changes in the metabolism of extracellular matrix macromolecules contributes to these complex changes in vascular morphology. [16,17,18]. Progressing atheroma often accumulate calcium. This lesion "mineralization" appears to be an actively controlled process [19,20,21].

Importantly, this sequence of events begins early in life as has been documented in post-mortem and in-vivo IVUS studies [22,23], and eventually leads to the formation of advanced lesions. However, advanced atheromata may remain silent for decades. Sudden changes initiated by as yet incompletely understood triggers lead to plaque rupture and may result in an acute coronary syndrome. Recent research has furnished new insight into the molecular mechanisms that cause transition from the chronic to the acute phase of atherosclerosis [7]. The size of the lipid core and the thickness of the fibrous cap influence stress distribution across the lesion. The border between the fibrous cap and normal vessel wall at the edge of a large lipid core (the so-called shoulder area) is subjected to particular high stress and therefore is prone to rupture [24,25]. The metabolism of the macromolecules of the extracellular matrix is an important mechanism contributing to atherosclerotic plaque stability. The tensile strength of the fibrous cap depends upon a balance between collagen biosynthesis by smooth muscle cells [26-28] and collagen degradation by members of several proteinase families and their endogenous inhibitors [29,30]. The secretion of these proteinases is induced by the accumulation and activation of macrophages and other inflammatory cells at the lesion site [31]. In particular, the group of matrix-metalloproteinases (MMPs) plays a central role in the degradation of the extracellular matrix weakening the connective tissue framework (extracellular matrix) of the fibrous cap [32-35].

In contrast to the accumulation of inflammatory cells smooth muscle cells are found in reduced density in plaques that have ruptured. Apoptosis, or programmed cell death, appears to contribute to the paucity of smooth muscle cells in vulnerable plaques [36-38].

he transition from an early to an advanced, vulnerable intimal lesioTn is associated with changes of the media and adventitia. Medial thinning appears to be an early finding in atherosclerotic lesions [39-43]. At the same time changes in vessel size (arterial remodeling) accompany lesion development and may be related to changes of the adventitia [15, 44-46]. It is an attractive hypothesis that large, positive remodeled plaques are more susceptible to mechanical forces leading to plaque rupture. The recent finding that increased MMP expression is found more frequently in positively remodeled lesions [47,48] suggests that the same pathophysiologic mechanisms may be responsible for expansion of the vessel wall and plaque vulnerability [49].

Initiated by rupture or erosion of vulnerable lesions [50, 51], thrombosis is the terminal event causing most of the acute manifestation of CAD [7,8,52,53]. Platelets accumulate on the exposed plaque. Activated platelets increase the expression of the glycoprotein IIb/IIIa membrane receptor complex on the cell surface, initiating further platelet aggregation and fibrin deposition, eventually forming a luminal thrombus. The size of the thrombus is determined by the balance between this prothrombotic milieu and local, intrinsic thrombolytic activity.

Some of these episodes of vulnerability, rupture and thrombosis are clinically symptomatic (acute coronary syndromes) but most episodes are silent. Further plaque development is characterized by fibrosis [27,28], leading to a thick fibrous cap and a small lipid core. It has been suggested, that the fibrotic changes and vessel shrinkage, while causing more severe luminal stenosis, render lesions more resistant to rupture and can therefore be regarded as a " healing" process.

In summary, stable and unstable plaques are likely temporary phases in the development of coronary lesions. Lesions progress through repeated cycles of vulnerability and rupture. Silent episodes of plaque rupture with subsequent lesion stabilization are likely the most common form of disease progression.

Assessment of Corresponding Morphology with Intravascular Ultrasound

Intravascular ultrasound (IVUS) is one of the most promising techniques for the detection of vulnerable plaques, because it allows in vivo observation of large portions of the coronary tree and individual lesions. Intravascular ultrasound is an established imaging technique, which, in the hand of experienced operators has a favorable safety record. Technical details and new approaches will be described in other chapters of this monograph. Briefly, IVUS uses high ultrasound frequencies, typically centered at 20 to 50 MHz. At 30 MHz, corresponding to a wavelength of 50 μm, axial resolution is usually <150 μm. Lateral resolution is influenced by imaging depth, and averages 250 μm at typical coronary diameters. Several morphologic features of advanced coronary plaques can be recognized with IVUS.

Normal Arterial Appearance

The acoustic properties of the arterial wall determine how its anatomical tissue layers (endothelium, intima, media, adventitia) are reflected in the IVUS image. This has been studied in excised pressure-distended vessels and in vivo [54-57]. Tissue interfaces with an abrupt change in acoustic impedance are usually well visualized. In the normal artery, two such tissue interfaces are found, one at the border between blood and the leading edge of the intima and a second at the external elastic membrane (EEM), which is located at the media -adventitia border. In normal arteries the intima is thin with an acoustic impedance similar to blood. Therefore the intima can not be differentiated from the media in 30% to 50% of normal coronary arteries. However, in good-quality images the tunica media can sometimes be visualized as a distinct, relatively sonolucent layer, separated from the lumen by a more echodense intima. The EEM area at the border of media and adventitia is well visualized and allows the assessment of remodeling. The outer border of the adventitia is indistinct, merging imperceptibly into the surrounding tissues.

Appearance of atherosclerotic lesions

Initial lesion growth leads to the expansion of the intimal thickness. As described for the normal vessel wall, the differentiation between intima and media is not consistently possible. Therefore most investigators define an *intima-media* thickness of greater than 0.25 to 0.50 mm as pathological. In some lesions the media can be observed as a separate layer, which may be a characteristic of early atherosclerosis [58].

The echodensity of atherosclerotic lesions reflects its histologic composition (figure 4). In comparative histologic studies, echolucent appearance by intravascular ultrasound was correlated with the lipid content of plaques [57]. Lipid-laden lesions appear echolucent while fibrous and calcified tissues are relatively echogenic. Calcified lesions have a characteristic "shadow" caused by signal dropout.

Structural, morphologic details of advanced lesions are reflected in the IVUS image. Lipid-laden lesions often exhibit a prominent echolucent core and an echogenic fibrous cap. Fibrous cap thickness ranges at 50-150μM and may in

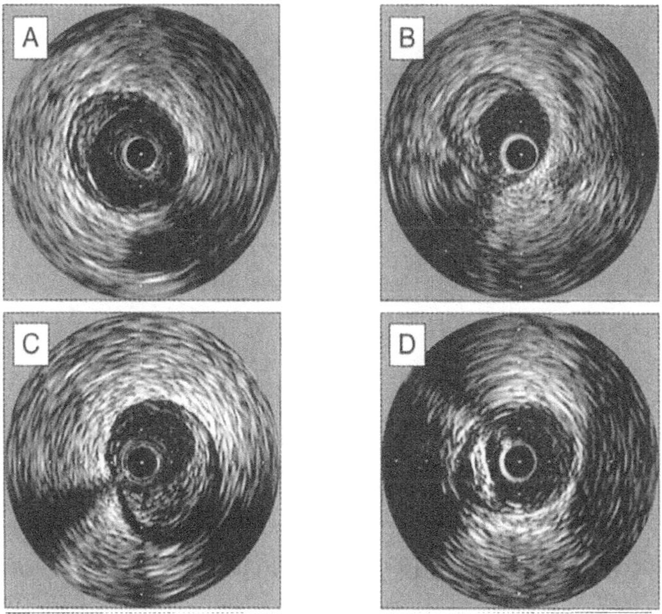

Fig.4: Examples of plaques with different acoustic properties. The comparison of the plaque with the surrounding adventitial tissue allows a coarse classification. The "soft" plaque in panel A is echolucent and the calcified plaque in panel D is echodense with acoustic shadowing.

many cases be below the resolution of IVUS. The presence and distribution of lesion calcification can be described with IVUS [59, 60].

Arterial Remodeling

The term arterial remodeling refers to changes in vascular dimensions during the development of atherosclerosis. Direction and extent of remodeling can be assessed in individual lesions by comparing lesion and reference site (figure 5,6,7). Arterial remodeling was initially described in necropsy specimens by Glagov et al [15], and later confirmed in IVUS studies [61-66]. At diseased sites, the EEM expands or shrinks in size, preventing or contributing to luminal stenosis. Initially considered as a purely compensatory mechanism to preserve lumen size, the direction of remodeling may be related to plaque development and stability [67,68,69].

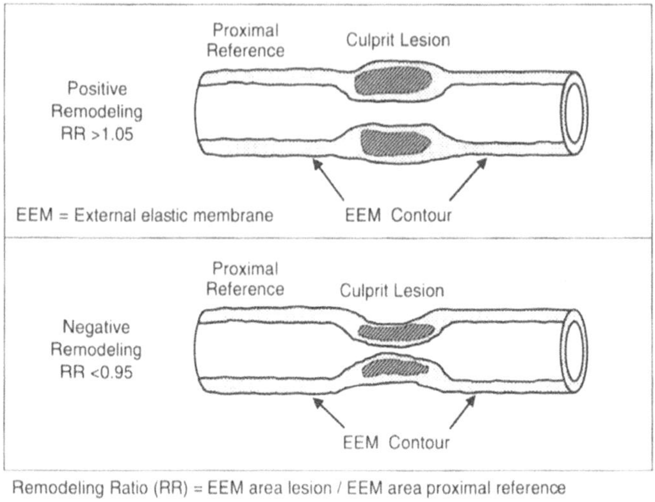

Fig.5: Arterial remodeling describes changes in the EEM area at the lesion site. The direction of remodeling may be related to plaque stability. From [68] with permission.

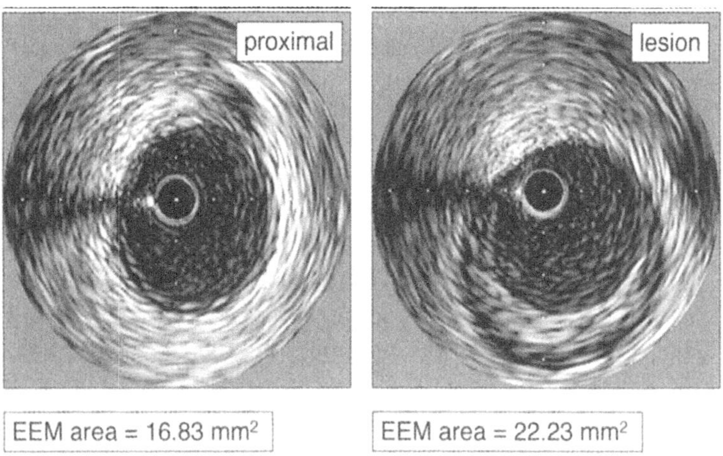

Fig.6: Example of a mildly stenotic lesion with positive remodeling, defined as EEM expansion at the lesion site.

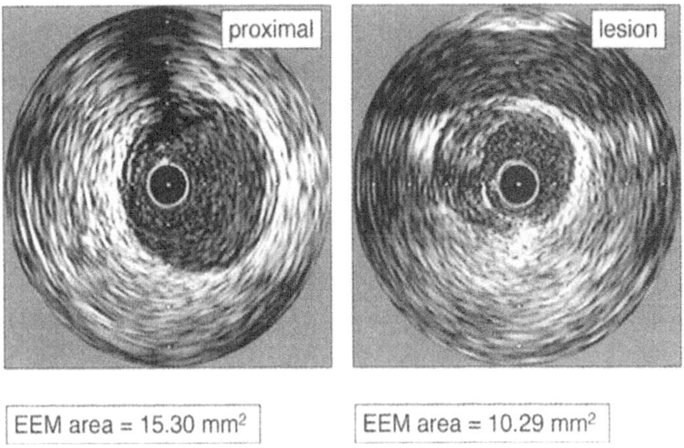

Fig.7: Example of a mildly stenotic lesion with negative remodeling, defined as EEM contraction at the lesion site.

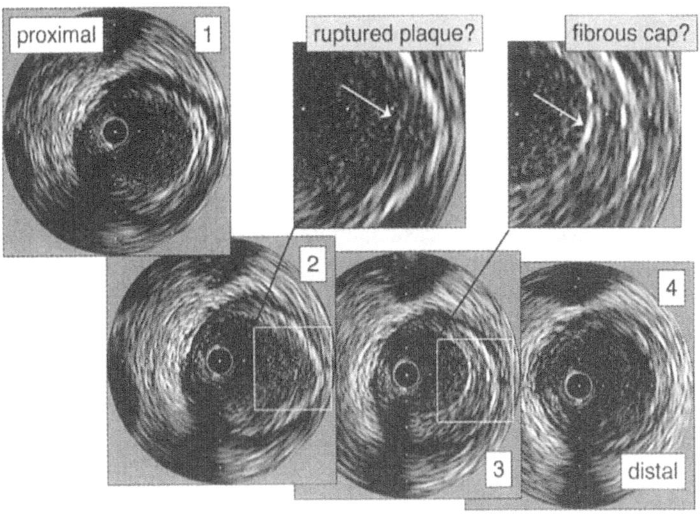

Fig.8: Example of a mildly stenotic lesion with evidence of plaque rupture. The fibrous cap, seen in panel 3, is not present in panel 2. Blood flow in the plaque cavity was evident on the real-time video images.

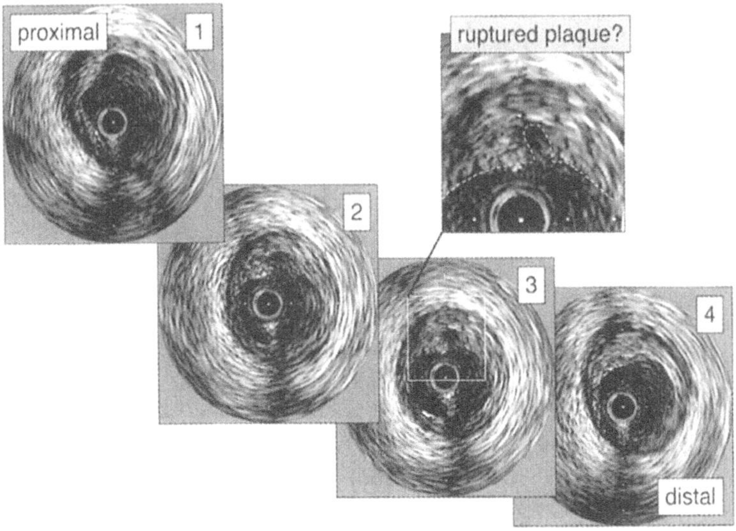

Fig.9: Small cavities (panel 3) can occasionally be observed in mildly stenotic lesions. It is unclear if they reflect plaque rupture.

Rupture

Plaque rupture is frequently found in highly stenotic culprit lesions of patients with acute coronary syndromes (figure 2, panel C). In mildly stenotic lesions an interruption of the fibrous cap is occasionally demonstrated and may be evidence of asymptomatic plaque rupture (figure 8,9). Because a very thin, but intact fibrous cap may not be visible on IVUS, evidence of blood flow in the plaque cavity should be demonstrated. Large plaque cavities, which are evidence of a remote plaque rupture with recanalisation, have also been described [70].

Atheroma Burden and distribution

Atheroma area is determined by subtracting lumen area from EEM area, and therefore includes intima and media. Atheroma area measurements performed in this fashion correlate closely with histology. [71,72]. The integration of consecutive tomographic area measurements along a vessel segment allows the calculation of plaque volume [73,74].

In addition to the area or volume of the plaque, the longitudinal and circumferential distribution is an important descriptor of coronary plaques. Eccentricity and concentricity are two extremes of the circumferential distribution. Eccentric plaques are frequently found at vessel bifurcations opposite to the flow

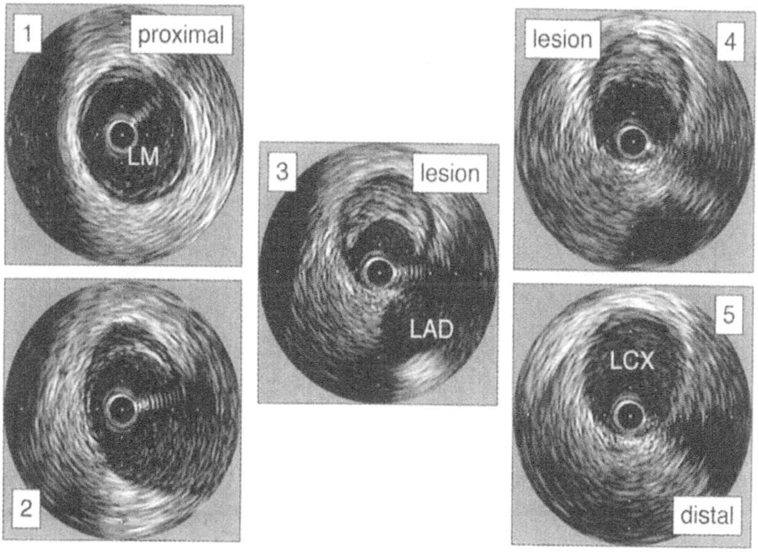

Fig.10: Circumferential plaque distribution is described by the eccentricity of lesions. This example shows a highly eccentric plaque opposite to the flow divider at a vessel bifurcation.

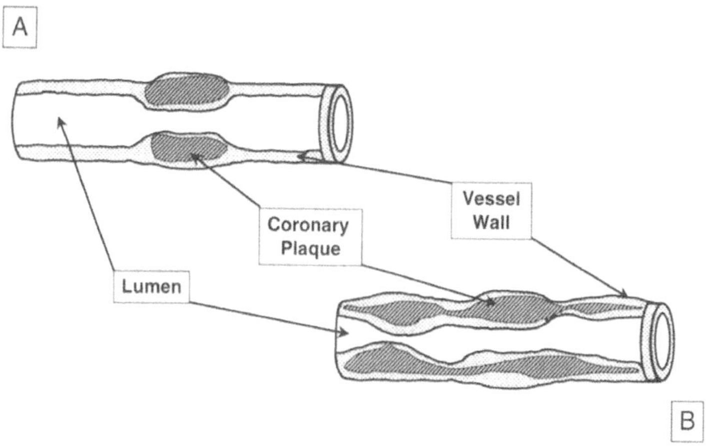

Fig.11: The longitudinal plaque distribution can be described as focal (A) versus diffuse (B) disease. Diffuse disease of the *vessel wall* is frequently demonstrated by intravascular ultrasound in vessels that have focal *luminal* stenosis by angiography.

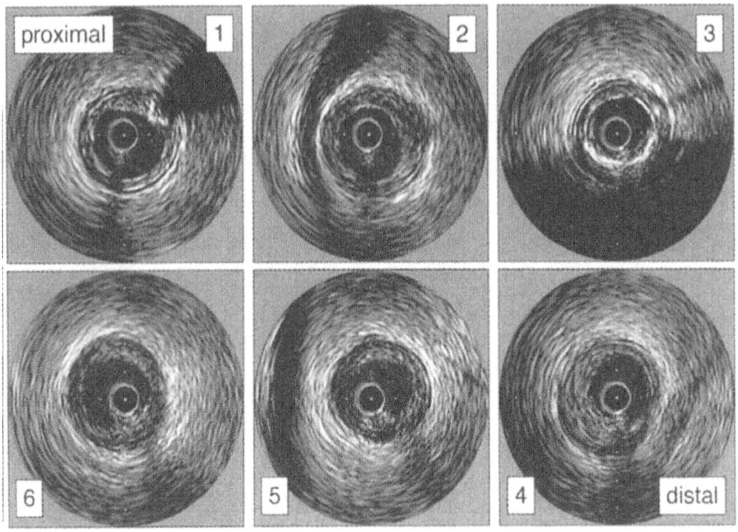

Fig. 12: Example of a vessel segment with diffuse disease. Plaque size and plaque characteristics are non-uniform throughout the segment.

divider [75] (figure 10). The longitudinal extent of disease is described as "diffuse" if the intimal thickness is abnormal at every site within the segment. Focal disease describes a lesion with adjacent normal reference segments (figure 11). IVUS and MRI studies show that, because of arterial remodeling, plaque burden is more diffuse than anticipated based on the luminal dimensions, [76,77].

Limitations of IVUS and Future Developments

Although contemporary ultrasound devices produce remarkably detailed views of the vessel wall, the description of plaque composition relies on simple visual inspection of the video images. The echogenicity and texture of different tissue components may exhibit comparable acoustic properties and therefore appear similar. An important example is the nearly identical appearance of intracoronary thrombus and atheroma with a high lipid content [78,79]. Thus, IVUS can not consistently identify histologic details. Newer developments like radiofrequency analysis (RFA) and elastography will be discussed elsewhere in this monograph. These techniques have not been validated for objective or automated classification of atheromatous lesions but have potential applications for plaque detection [80]. Recent papers describe the reliable detection of lipid pools with RFA [81,82]. Radiofrequency analysis has also shown some promise in differentiating between thrombus and atheroma [83]. Other groups have analyzed the stress distribution across coronary lesions using elastography [84].

Although the safety of IVUS has been demonstrated in several studies, a justified concern is potential injury to the endothelium. This concern in part drives the search for non-invasive imaging techniques like MRI and multi-detector computed tomography [85].

Clinical Observations in Patients with coronary artery disease

Plaque Characteristics in Patients with Acute Coronary Syndromes

Plaque characteristics have been compared in culprit lesion of patients with stable and unstable coronary syndromes. Plaque echolucency, which correlates with the lipid content of plaques, has been associated with the clinical presentation of unstable angina [86-88]. In patients with unstable syndromes a thin fibrous cap and evidence of plaque rupture is often found by IVUS.

Recent studies describe an association between clinical presentation and direction of remodeling [67,68,69,89]. Positive remodeling is associated with unstable lesions, while negative remodeling is more frequently found in stable lesions.

The circumferential distribution of plaque is more often eccentric than concentric in unstable coronary lesions [90]. The longitudinal distribution of plaque along the vessel has not been compared between stable and unstable lesions. However IVUS studies in symptomatic patients often reveal diffuse disease in vessels without significant stenosis. The overall plaque burden in vessels with diffuse disease is larger than in vessels with focal disease. It is an attractive hypothesis that a lager plaque mass may be more "vulnerable" to focal episodes of rupture and thrombosis. This hypothesis is supported by electron-beam CT studies showing a relation between plaque burden and likelihood of new coronary events [91]. However, data from IVUS studies, relating plaque burden and clinical events is not yet available.

Prospective Observation of Asymptomatic Patients in Serial Studies and Future Directions

It is unclear if characteristics found in ruptured lesions reflect those of "vulnerable lesions" before rupture. Therefore the serial observation of mildly stenotic lesions in prospective studies will be necessary. The comparison to clinical symptoms will be important in order to understand characteristics of symptomatic and asymptomatic episodes of rupture. The comparison to clinical symptoms will be important in order to understand characteristics of symptomatic and asymptomatic episodes of rupture (figure 13-15).

A recent intravascular ultrasound study of early, angiographically mildly stenotic lesions describes characteristics of vulnerable plaques before rupture [92]. One hundred fourteen coronary sites without significant stenosis by angiography (<50% diameter stenosis) were identified and examined with intravascular ultrasound at baseline. During 2 years follow-up 12 patients had an acute coronary event caused by a previously examined site. These "vulnerable" sites were compared with lesions that did not cause acute events. At baseline the vulnerable sites were more frequently echolucent and exhibited larger plaque areas with similar lumen areas suggesting positive remodeling.

The progression or regression of plaque burden has been examined with intravascular ultrasound. A small study examined the effect of 3-year treatment with pravastatin or diet in mildly diseased coronary arteries [93]. During follow-up plaque area increased by 41% in the control group but decreased by 7% in the treatment group. Ongoing serial regression-progression trials are the Reversal of Atherosclerosis with Lipitor (REVERSAL) trial, which is comparing 2 lipid-lowering regimens, and the Norvasc for Regression of Manifest Atherosclerotic Lesions (NORMALISE) trial, which is comparing amlodipine, enalapril, and placebo. The primary end points of these trials is plaque burden, however future studies will have to correlate changes in plaque burden to clinical endpoints of ACS.

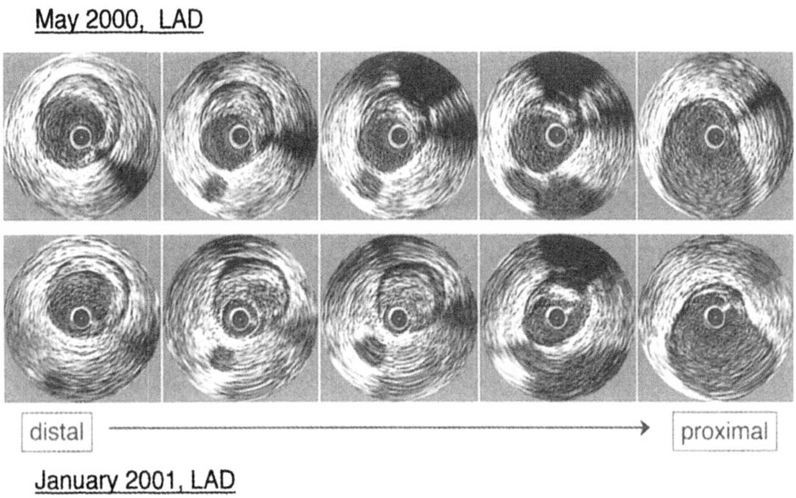

Fig.13: Serial observation of mildly stenotic vessel segments may allow insights into plaque progression, regression, and vulnerability. Figure 13 shows adjacent images of a vessel segment with rapid plaque growth during follow-up.

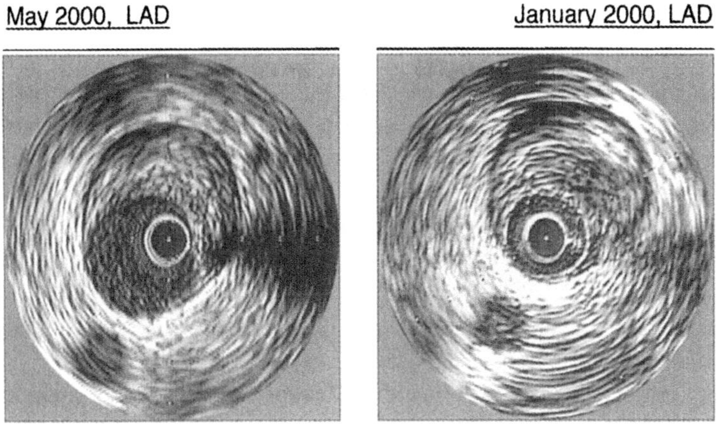

Fig. 14: Figure 14 shows the change at the worst lesion site of the lesion shown in figure 13.

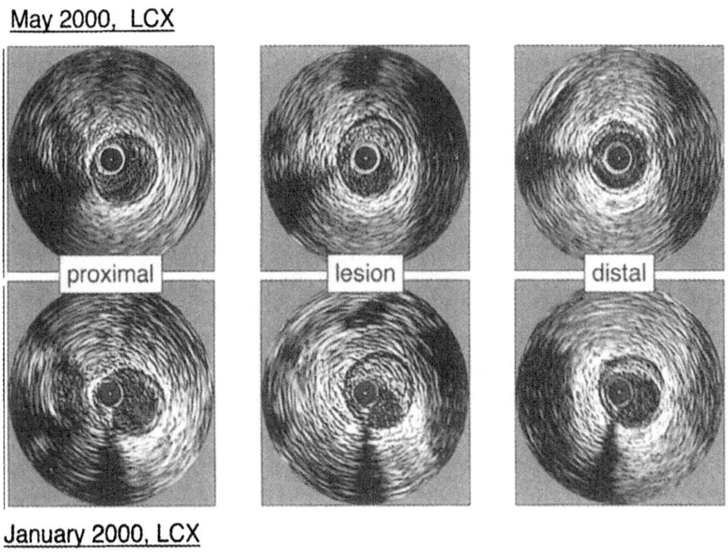

Fig. 15: Figure 15 shows a lesion with minimal change during followup. The complex simultaneous changes of EEM, plaque and lumen size and their relation to plaque vulnerability are incompletely understood.

Conclusions

The vulnerable coronary plaque has a central pathophysiologic role in coronary artery disease and its detection is a major challenge in contemporary cardiology. "Vulnerable" lesions are likely highly dynamic and numerous throughout the coronary tree. Because plaque vulnerability is defined by an event in the future (plaque rupture and thrombosis) the identification of vulnerable lesions will require the prospective observation of large numbers of lesions in large parts of the coronary tree.

Intravascular ultrasound, an established imaging technique, has the ability to observe overall plaque burden and characteristic of individual lesions. However, a combined approach integrating vascular biology, pathology, different imaging techniques and clinical presentation will be necessary. This integrated approach to coronary imaging will allow validation against histology but will also confirm the clinical significance of morphologic findings.

References

1. Newby AC, Libby P, van der Wal A (1999) Plaque instability – the real challenge for atherosclerosis research in the next decade? Cardiovasc Res 41:321-22
2. Hackett D, Davies G, Maseri A (1988) Pre-existing coronary stenoses in patients with first myocardial infarction are not necessarily severe. Eur Heart J 9:1317-1323
3. Giroud D, Li JM, Urban P, et al. (1992) Relation of the site of acute myocardial infarction to the most severe coronary arterial stenosis prior to angiography. Am J Cardiol 69:729-732
4. Ambrose JA, Tannenbaum MA, Alexopoulos D, et al. (1988) Angiographic progression of coronary artery disease and the . development of myocardial infarction. J Am Coll Cardiol 12:56-62
5. Little WC, Constantinescu M, Applegate RJ, et al. (1988) Can coronary angiography predict the site of a subsequent myocardial infarction in patients with mild-to-moderate coronary artery disease? Circulation 78:1157-1166
6. Ross R (1993) The pathogenesis of atherosclerosis: a perspective for the 1990's. Nature 362:801-809
7. Libby P (1995) The molecular bases of the acute coronary syndromes. Circulation 91:2844-2850
8. Falk E, Shah PK, Fuster V (1995) Coronary plaque disruption. Circulation 92:657-671
9. Frenette P, Wagner D (1996) Adhesion molecules. New Engl J Med 334:1526-29
10. Dong ZM, Chapman SM, Brown AA, et al. (1998) The combined role of P- and E-selectins in atherosclerosis. J Clin Invest 102:145-52
11. Faggiotto A, Ross R, Harker L (1984) Studies of hypercholesterolemia in the nonhuman primate. I. Changes that lead to fatty streak formation. Arteriosclerosis 4:323-40
12. Faggiotto A, Ross R (1984) Studies of hypercholesterolemia in the nonhuman primate. II. Fatty streak conversion to fibrous plaque. Arteriosclerosis 4:341-56

13. Qiao JH, Fishbein MC (1991) The severity of coronary atherosclerosis at sites of plaque rupture with occlusive thrombosis. J Am Coll Cardiol 17:1138-42
14. Falk E (1983) Plaque rupture with severe pre-existing stenosis precipitating coronary thrombosis. Characteristics of coronary atherosclerotic plaques underlying fatal occlusive thrombi. Br Heart 50:127-34
15. Glagov S, Weisenberg E, Zarins CK, et al. (1987) Compensatory enlargement of human atherosclerotic coronary arteries. N Engl J Med 316:1371-5
16. Ross R, Glomset JA (1976) The pathogenesis of atherosclerosis I. New Engl J Med 295:369-77
17. Ross R, Glomset JA (1976) The pathogenesis of atherosclerosis II. New Engl J Med 195:420-25
18. Moreno PR, Bernadi VH, Lopez-Cuellar J, et al. (1996) Macrophages, smooth muscle cells and tissue factor in unstable angina. Circulation 94:3090-3097
19. Demer LL (1995) A skeleton in the atherosclerosis closet. Circulation 92:2029-32
20. Fitzpatrick LA, Severson A, Edwards WD, et al. (1994) Diffuse calcification in human coronary arteries. Association of osteopontin with atherosclerosis. J Clin Invest 94:1597-604
21. Huang H, Virmani R, Younis H, et al. (2001) The impact of calcification on the biomechanical stability of atherosclerotic plaques. Circulation 103:1051-1056
22. Strong JP, Malcom GT, McMahan CA, et al. (1999) Prevalence and extent of atherosclerosis in adolescents and young adults: Implications for prevention from pathobiological determinants of atherosclerosis in youth study. JAMA 281:727-35
23. Tutar E, Kapadia SR, Ziada KM, et al. (1999) Coronary atherosclerosis begins at young age: Intravascular ultrasound evidence of disease in individuals <20 years old. Circulation 100,18:I-524abst.2760, abstract
24. Loree HM, Kamm RD, Stringfellow RG, et al.(1992) Effects of fibrous cap thickness on peak circumferential stress in model atherosclerotic vessels. Circulation Research 71:850-858
25. Cheng GC, Loree HM, Kamm RD, et al. (1993) Distribution of circumferential stress in ruptured and stable atherosclerotic lesions. Circulation 87:1179-1187
26. Amento EP, Ehsani N, Palmer H, et al. (1991) Cytokines positively and negatively regulate interstitial collagen gene expression in human vascular smooth muscle cells. Arteriosclerosis 11:1223-30
27. Small DM, Bond MG, Waugh D, et al. (1984) Physicochemical and histological changes in the arterial wall of nonhuman primates during progression and regression of atherosclerosis. J Clin Invest 73:1590-1605
28. Aikawa M, Rabkin E, Okada Y, et al. (1998) Lipid lowering by diet reduces matrix metalloproteinase activity and increases collagen content of rabbit atheroma. Circulation 97:2433-2444
29. Henney AM, Wakeley PR, Davies MJ, et al. (1991) Localization of stromelysin gene expression in atherosclerotic plaques by in sity hybridization. Proc Natl Acad Sci USA 88:8154-58
30. Sukhova GK, Shi GP, Simon DI, et al. (1998) Expression of the elastolytic cathepsins S and K in human atheroma and regulation of their production in smooth muscle cells. J Clin Invest 102,3:576-583
31. Lendon CL, Davies MJ, Born GV, et al. (1991) Atherosclerotic plaque caps are locally weakened when macrophage density is increased. Atherosclerosis 1991;87:87-90

32. Dollery CM, McEwan JR, Henney AM (1995) Matrix metalloproteinases and cardiovascular disease. Circ Res 77:863-68
33. Galis ZS, Sukhova GK, Lark MW, et al. (1994) Increased expression of matrix metalloproteinases and matrix degrading activity in vulnerable regions of human atherosclerotic plaques. J Clin Invest 94:2493-2503
34. Nikkari ST, O'brien KD, Ferguson M, et al. (1995) Interstitial collagenase (MMP1) expression in human carotid atherosclerosis. Circulation 92:1393098
35. Fabunmi RP, Sukhova GK, Sugiyama S, et al. (1998) Expression of tissue inhibitor of metalloproteinases-3 in human atheroma and regulation in lesion-associated cells: a potential protective mechanism in plaque stability. Circ Res 83270-78
36. Gen YJ, Libby P (1995) Evidence for apoptosis in advanced human atheroma. Co localization with interleukin-1 β-converting enzyme. Am J Pathol 147:251-66
37. Han D, Haudenschild C, Hong M, et al. (1995) Evidence for apoptosis in human atherogenesis and in a rat vascular injury model. Am J Pathol 147:267-77
38. Isner JM, Kearney M, Bortman S, et al. (1995) Apotptosis in human atherosclerosis and restenosis. Circulation 91:2703-11
39. Crawford T, Levene CI (1953) Medial thinning in atheroma. J Path Bact 66:19-23
40. Haarer SL, Emig LL, Keiser JA (1998) Vascular remodeling in balloon injured rabbit iliac arteries. Basic Res Cardiol 93:108-115
41. Scott NA, Cipolla GD, Ross CE, et al. (1996) Identification of a potential role for the adventitia in vascular lesion formation after balloon overstretch injury of porcine coronary arteries. Circulation 93:2178-2187
42. Isner JM, Donaldson RF, Fortin AH, et al. (1986) Attenuation of the media of coronary arteries in advanced atherosclerosis. Am J Cardiol 58:937-939
43. Moreno PR, Purushothaman KR, Fuster V (2001) Internal elastic lamina in the process of plaque disruption. JACC 37,supplA:288A, abstract 1258-194
44. Lafont A, Guzman LA, Whitlow PL, et al. (1995) Restenosis after experimental angioplasty. Circ Res 76:996-1002
45. deSmet BJGL, van der Zande J, van der Helm YJM, et al. (1998) Theatheroslerotic yucatan animal model to study the arterial response after balloon angioplasty. Cardiovascular Research 39:224-232
46. Mintz GS, Popma JJ, Pichard AD, et al. (1996) Arterial remodeling after coronary angioplasty: A serial intravascular ultrasound study. Circulation 94:35-43
47. Pasterkamp G, Schoneveld AH, Hijnen DJ, et al. (2000) Atherosclerotic arterial remodeling and the localization of macrophages and matrix metalloproteiases 1,2,and 9 in the human coronary artery. Atherosclerosis 150:245-253
48. Schoenhagen P, Vince DG, Ziada K, et al. (2000) Increased Presence of Matrix-Metalloproteinase 3 in Human Coronary Lesions With Positive Arterial Remodeling. JACC 35,2;supplA:58A,1132-109,abstract
49. Pasterkamp G, Schoneveld AH, van der Wal AC, et al. (1998) Relation of arterial geometry to luminal narrowing and histologic markers for plaque vulnerability: The remodeling paradox. J Am Coll Cardiol 32:655-62
50. Farb A, Burke AP, Tang AL, et al. (1996) Coronary plaque erosion without rupture into a lipid core. Circulation 93:1354-1363
51. van der Wal AC, Becker AE, van der Loss CM, et al. (1994) Site of intimal rupture or erosion of thrombosed coronary atherosclerotic plaques is characterized by an inflammatory process irrespective of the dominant plaque morphology. Circulation 89:36-44

52. Horie T, Sekiguchi M, Hirosawa K (1978) Coronary thrombosis in pathogenesis of acute myocardial infarction. Br Heart J 40:153-161
53. Davies MJ, Thomas A (1984) Thrombosis and acute coronary artery lesions in sudden cardiac ischemic death. N Engl J Med 310:1137-40
54.Tobis JM, Mallery J, Mahon D, et al. (1991) Intravascular ultrasound imaging of human coronary arteries in vivo: analysis of tissue characterizations with comparison to in vitro histological specimens. Circulation 83:913–926
55. Nissen SE, Gurley JC, Grines CL, (1991) Intravascular ultrasound assessment of lumen size and wall morphology in normal subjects and patients with coronary artery disease. Circulation 84:1087-99
56. St Goar FG, Pinto FJ, Alderman EL, et al. (1991) Intravascular ultrasound imaging of angiographically normal coronary arteries: an in vivo comparison with quantitative angiography. J Am Coll Cardiol 18:952–958
57. Gussenhoven EJ, Essed CE, Lancee CT, et al. (1989) Arterial wall characteristics determined by intravascular ultrasound imaging: an in vitro study. J Am Coll Cardiol 14:947–952
58. Fitzgerald PJ, St. Goar FG, Connolly AJ, et al. (1992) Intravascular ultrasound imaging of coronary arteries. Is three layers the norm? Circulation 86:154–158
59. Mintz GS, Douek P, Pichard AD, et al. (1992) Target lesion calcification in coronary artery disease: an intravascular ultrasound study. J Am Coll Cardiol 20:1149 –1155
60. Tuzcu EM, Berkalp B, De Franco AC, et al. (1996) The dilemma of diagnosing coronary calcification: angiography versus intravascular ultrasound. J Am Coll Cardiol 27:832–838
61. Hermiller JB, Tenaglia AN, Kisslo KB, et al. (1993) In vivo validation of compensatory enlargement of atherosclerotic coronary arteries. Am J Cardiol 71:665–668
62. Losordo DW, Rosenfield K, Kaufman J, et al. (1994) Focal compensatory enlargement of human arteries in response to progressive atherosclerosis: in vivo documentation using intravascular ultrasound. Circulation 89:2570–2577
63. Pasterkamp G, Wensing PJ, Post MJ, et al. (1995) Paradoxical arterial wall shrinkage may contribute to luminal narrowing of human atherosclerotic femoral arteries. Circulation 91:1444–1449
64. Mintz GS, Kent KM, Pichard AD, et al. (1997) Contribution of inadequate arterial remodeling to the development of focal coronary artery stenoses: an intravascular ultrasound study. Circulation 95:1791–1798
65. Mintz GS, Popma JJ, Pichard AD, et al. (1996) Arterial remodeling after coronary angioplasty: a serial intravascular ultrasound study. Circulation 94:35 –43
66. Kimura T, Kaburagi S, Tamura T, et al. (1997) Remodeling of human coronary arteries undergoing coronary angioplasty or atherectomy. Circulation 96:475–483
67. Schoenhagen P, Ziada K, Kapadia SR, et al. (2000) Extent and direction of arterial remodeling in stable versus unstable coronary syndromes: an intravascular ultrasound study. Circulation 101:598–603
68. Smits PC, Pasterkamp G, de Jaegere PPT, et al. (1999) Angioscopic complex lesions are predominantly compensatory enlarged: an angioscopic and intracoronary ultrasound study. Cardiovascular Research 41:458-464
69. Filardo SD, Scharzacher SP, Lo ST, et al. (2000) Acute myocardial infarction and vascular remodeling. Am J Cardiol 85:760-762

70. von Birgelen C, Klinkhart W, Mintz GS, et al. (2000) Size of emtied plaque cavity following spontaneous rupture is related to coronary dimensions, not the degree of lumen narrowing. A study with intravascular ultrasound in vivo. Heart 84:483-488
71. Potkin BN, Bartorelli AL, Gessert JM, et al. (1990) Coronary artery imaging with intravascular high-frequency ultrasound. Circulation 81:1575–1585
72. Nishimura RA, Edwards WD, Warnes CA, et al. (1990) Intravascular ultrasound imaging: in vitro validation and pathologic correlation. J Am Coll Cardiol 16:145–154
73. Klingensmith JD, Vince DG, Kuban BD, et al. (2000) Assessment of coronary compensatory enlargement by three-dimensional intravascular ultrasound. The International Journal of Cardiac Imaging 16:87-98
74. Shiran A, Mintz GS, Leiboff B, et al. (1999) Serial volumetric intravascular ultrasound assessment of arterial remodeling in left main coronary artery disease. Am J Cardiol 83:1427-1432
75. Kimura BJ, Russo RJ, Bhargava V, et al. (1996) Atheroma morphology and distribution in proximal left anterior descending coronary artery: In vivo observations. J Am Coll Cardiol 27:825-31
76. Mintz GS, Paanter JA, Pichard AL, et al. (1995) Atherosclerosis in angiographically "normal" coronary artery reference segments. J Am Coll Cardiol 25:1479-85
77. Coulden RA, Moss H, Graves MJ, et al. (2000) High resolution magnetic resonance imaging of atherosclerosis and the response to ballon angioplasty. Heart 83:188-191
78. Bocksch W, Schartl M, Beckmann S, et al. (1995) Intravascular ultrasound imaging in patients with acute myocardial infarction. Eur Heart J 16(Suppl J):46–52
79. Siegel RJ, Ariani M, Fishbein MC, et al. (1991) Histopathologic validation of angioscopy and intravascular ultrasound. Circulation 84:109–117
80. Pasterkamp G, Falk E, Woutman H, et al. (2000) Techniques characterizing the coronary atherosclerotic plaque: influence on clinical decision making? J Am Coll Cardiol 36:13-21
81. Moore MP, Spencer T, Salter DM, et al. (1998) Characterization of coronary atherosclerotic morphology by spectral analysis of radiofrequency signal: in vitro intravascular ultrasound study with histological and radiological validation. Heart 79:459-467
82. Komiyama N, Berry GJ, Kolz ML, et al. (2000) Tissue characterization of atherosclerotic plaques by intravascular ultrasound radiofrequency signal analysis: An in vitro study of human coronary arteries. Am Heart J 140:565-74
83. Bridal SL, Fornes P, Bruneval P, et al. (1997) Parametric (integrated backscatter and attenuation) images constructed using backscattered radio frequency signals (25-56 MHz) from human aortae in vitro. Ultrasound Med Biol 23:215–229
84. de Korte CL, Pasterkamp G, van der Steen AFW, et al. (2000) Characterization of plaque components with intravascular ultrasound elastography in human femoral and coronary arteries in vitro. Circulation 102:617-
85. Schroeder S, Kopp AF, Baumbach A, et al. (2001) Noninvasive detection and evaluation of atherosclerotic coronary plaques with multislice computed tomography. J Am Coll Cardiol 37:1430-1435
86. Hodgson JM, Reddy KG, Suneja R, et al. (1993) Intracoronary ultrasound imaging: correlation of plaque morphology with angiography, clinical syndrome and procedural results in patients undergoing coronary angioplasty. J Am Coll Cardiol 21:35–44

87. Kearney P, Erbel R, Rupprecht HJ, et al. (1996) Differences in the morphology of unstable and stable coronary lesions and their impact on the mechanisms of angioplasty: an in vivo study with intravascular ultrasound. Eur Heart J 17:721–730
88. Erbel R, Ge J, Bockisch A, et al. (1996) Value of intracoronary ultrasound and Doppler in the differentiation of angiographically normal coronary arteries: a prospective study in patients with angina pectoris. Eur Heart J 17:880–889
89. Schoenhagen P, Vince DG, Ziada KM, et al. (2000) The Association between Arterial Expansion (Positive Remodeling) and Unstable Clinical Presentation is Evident in Both Bifurcation and Non-Bifurcation Culprit Lesions: An Intravascular Ultrasound Study. Circulation 102(suppl.II):690, abstract 3340
90. Bocksch WG, Shartl M, Beckmann SH, et al. (1994) Intravascular ultrasound imaging in patients with acute myocardiol infarction: comparison with chronic stable angina pectoris. Coron Artery Dis 5:727-735
91. Janowitz WR, Agatston AS, Viamonte M (1991) Comparison of serial quantitative evaluation of calcified coronary artery plaque by ultrafast computed tomography in persons with and without coronary artery disease. Am J Cardiol 68:1-6
92. Yamagishi M, Terashima M, Awano K, et al. (2000) Morphology of vulnerable coronary plaque: Insights from follow-up of patients examined by intravascular ultrasound before and acute coronary syndrome. J Am Coll Cardiol 35:106-11
93. Takagi T, Yoshida K, Akasaka T, et al. (1997) Intravascular ultrasound analysis of reduction in progression of coronary narrowing by treatment with pravastatin Am J Cardiol 79:1673-1767

Diagnosis of vulnerable plaques in the cardiac catheterization laboratory

Johannes A Schaar, Antonius FW van der Steen, Chourmouzios A. Arampatzis, Rob Krams, Cornelis J Slager, Antoinette G ten Have, Sweder W van de Poll, Frank J Gijsen, Jolanda J Wentzel, Pim J de Feyter, Patrick W Serruys

Erasmus MC Rotterdam, Thoraxcenter, Dr. Molewaterplein 50, Room Ee 23-32
NL 3000 DR Rotterdam, The Netherlands

Abstract: Rupture of vulnerable plaques is the main cause of acute coronary syndrome and myocardial infarction. Identification of vulnerable plaque is therefore essential to enable the development of treatment modalities to stabilize such plaques. Several diagnostic procedures are currently tested to detect vulnerable plaques. Angiography, has a low discriminatory power to identify the vulnerable plaque, but does provide information about the entire coronary system and serves as guide for invasive imaging techniques and therapy. Intracoronary angioscopy offers a direct visualization of the plaque surface and intra-luminal structures like tears and thrombi. However, angioscopy is difficult to perform, invasive and only a limited part of the vessel tree can be investigated. IVUS provides some insight into the composition of coronary plaques. The detection of vulnerable plaques by IVUS is mainly based on a series of case reports with a lack of prospectivity and follow-up. Palpography, an IVUS derived technique, reveals information that is not seen on IVUS. It can differentiate between hard and soft tissue, which may be important for the detection of an instable plaque that is prone to rupture. The clinical value of this technique is currently under investigation. Temperature heterogeneity is an indicator of the metabolic state in a plaque. Preliminary data suggest a coincidence of temperature rise in the plaque and vulnerability. Optical coherence tomography (OCT) can provide images with ultrahigh resolution. Limitations of OCT are the low penetration depth, which hinders studying large vessels, and the light absorbance by blood. Raman spectroscopy can provide quantitative information about the molecular composition of a plaque. Limitations of the technique are the limited penetration depth, the long acquisition time and the absorbance of the light by blood. Near-infrared (NIR) spectroscopy also obtains information on the chemical components of the coronary vessel wall. Future studies will address the question whether NIR spectroscopy is feasible in vivo. High-resolution magnetic resonance imaging (MRI) is a non-invasive modality to characterize atherosclerotic plaques. Due to

the resolution of the technique small plaque structures like fibrous caps cannot yet be assessed using current MR techniques.
Several techniques are currently under development to assess the vulnerable plaque. Some of the techniques show exiting features, but none has proven their value in an extensive in vivo validation and all have a lack of prospective data.

Key words: Vulnerable plaque detection, catheterization laboratory, novel techniques, minimal invasive

Introduction

Rupture of vulnerable plaques is the main cause of acute coronary syndrome and myocardial infarction. Identification of vulnerable plaque is therefore essential to enable the development of treatment modalities to stabilize such plaque. Because myocardial infarction and its consequences are so important, we must investigate options to identify those areas that will be responsible for future events.

A wide variety exists in the stability of coronary atherosclerotic plaques. A plaque may be stable for years, however abrupt disruption of his structure is the main cause of acute coronary syndrome.[1]

This article will discuss the current development of imaging techniques that have the potential to detect vulnerable plaques.

The vulnerable plaque contains certain features that could be diagnosed by various specialized methods. The ideal technique would provide morphological, mechanical and chemical information, however at present, no diagnostic modality providing such all-embracing assessment is available.

The characteristics of vulnerable plaque have been described by numerous reviewers.[2 3] It is important to characterize the following features:

1. size of the lipid core (40% of the entire plaque)
2. thickness of the fibrous cap
3. presence of inflammatory cells
4. amount of remodeling and extent of plaque free vessel wall
5. 3D morphology

Angiography

Coronary angiography has been so far the gold standard to assess the severity of obstructive luminal narrowing. Furthermore it serves as a decision tool to direct therapy such as PTCA or CABG.

Using coronary angiography we can assess the lumen boundaries, but cannot assess information about plaque burden, plaque delineation and plaque components. The predictive power of occurrence of myocardial infarction is rather low since 70% of acute coronary occlusions are in areas that were previously angiographically normal, and only a minority occur where there was severe stenosis.[4] Other studies have affirmed, that the culprit lesion prior to a myocardial infarction has, in 48–78% of all cases, a stenosis smaller than 50%.[5 6 7] Coronary angiograms also often fail to identify the culprit lesion of nontransmural myocardial infarction.[8] The majority of ulcerated plaques are not big enough to be detected by angiography, but can be well assessed pathologically.[9]

Furthermore, we have to take into account that the predictive power of angiography is strongly dependent on the time interval between the angiogram and the myocardial infarction, because both time and interim therapy can influence atherosclerosis. In one study, the angiograms were performed between 1 and 77 months before the event4 and showed that atherosclerosis can be a rapidly progressive process. A recent study evaluated angiograms made one week before acute myocardial infarction showing that signs of thrombosis and rupture were present in a majority of patients.[10] During the year after myocardial infarction, the presence of multiple complex plaques is associated with an increased incidence of recurrent acute coronary syndrome.[11]

Thus, patients with silent non-obstructive coronary atherosclerosis harbor vulnerable plaques that cannot be detected by angiograms, but which are associated with adverse clinical outcomes. If a disrupted ulcerated plaque is seen on angiography, the existence of additional rupture prone plaques is to be expected. Angiography therefore, has a low discriminatory power to identify the vulnerable plaque, but does provide information about the entire coronary system and serves as guide for invasive imaging techniques and therapy.

In the future, our current view of angiography may totally change when non-invasive coronary imaging with MRI or CT will become available for routine use.[12]

Angioscopy

Intracoronary angioscopy offers a direct visualization of the plaque surface and intra-luminal structures like tears and thrombi. It allows assessment of the color of the plaque and thrombus[13] with a sensitivity in detecting such structures higher than when compared to angiography.[14] Angioscopic plaque rupture and thrombus have been shown to be associated with adverse clinical outcomes in patients with complex lesions.[15] Furthermore yellow plaques seem to have an increased instability, in comparison between IVUS and angioscopy.[16] In patients with myocardial infarction all three coronary arteries are widely diseased and have multiple yellow plaques.[17] In a 12-month follow-up study of 157 patients with stable angina, acute coronary syndrome occurred more frequently in patients with yellow plaques than in those with white plaques. These results indicate that acute coronary syndromes occur more frequently in patients with yellow plaques, which can be imaged with angioscopy, but not angiography.[18] However, angioscopy is difficult to perform, invasive and only a limited part of the vessel tree can be investigated. Most importantly, to enable clear visualisation of the vessel wall, the vessel has to be occluded and the remaining blood flushed away with saline, thereby potentially inducing ischaemia. Information regarding the degree of plaque extension into the vessel wall is not provided by angioscopy.

IVUS

Intravascular coronary ultrasound (IVUS) provides real-time high-resolution images of the vessel wall and lumen.[19] The size of IVUS catheters is between 2.9 to 3.5 French. Depending on the distance from the catheter the axial resolution is about 150 microns, the lateral 300 microns. The images appear real time at a frequency of up to 30 frames/sec. Features of the vessel can be detected based on the echogenicity and the thickness of the material. Small structures can be visualized, however only those sized over 160 microns can be estimated accurately. The normal thickness of the media is about 125 -350 μm.

IVUS provides some insight into the composition of coronary plaques. In IVUS images, calcification is characterized by a bright echo signal with distal shadowing which hides plaque components and deeper vessel structures. In comparitive studies between histology and IVUS, plaque calcification can be detected with a sensitivity of between 86 and 97%.[20] [21] The sensitivity to detect micro-calcification ranges around 60%.[22]

In IVUS images lipid depositions are described as echolucent zones and can be detected with a sensitivity of between 78% and 95% and specificity of 30%.[23] [24] This sensitivity is dependent on the amount of lipid and can drop down further if the echo-lucent area is smaller than a quarter of the plaque. Echolucent zones can also be caused by loose tissue and shadowing from calcium, which makes the interpretation of echolucent areas difficult. The sensitivity to differentiate between fibrous and fatty tissue is between 39 and 52%.[25]

The detection of vulnerable plaques by IVUS is mainly based on a series of case reports.[26] [27] The main focus of these reports is the detection of already ruptured plaques. To evaluate the role of IVUS in detecting plaque rupture, a study was performed of 144 patients with angina. Ruptured plaques were characterized by a cavity (echolucent area within the plaque) and a tear of the thin fibrous cap. They were identified in 31 patients of whom 23 (74%) presented with unstable angina. Plaque rupture was confirmed by injecting contrast medium and seeing filling of the plaque cavity on IVUS. Of the patients without plaque rupture (n = 108), only 19 (18%) had unstable angina. The echolucent area (cavity) to total plaque area ratio was larger in the unstable group than in the stable group. The thickness of the fibrous cap in the unstable group was also found to be smaller than in the stable group.[28]

The problem of these studies is the lack of prospectivity and follow up. Only Yamagishi et al. have performed a prospective study with a follow up period of about two years. Large eccentric plaques containing an echolucent zone by IVUS were found to be at increased risk of instability even though the lumen area was preserved at the time of initial study.[29]

It has been demonstrated, that unstable plaques are associated with positive remodeling. Intravascular ultrasound assessment of vascular remodeling may help to classify plaques with the highest probability of spontaneous rupture.[30]

A number of groups have investigated the potential of ultrasound radio-frequency signal analysis for tissue characterization.[31 32 33 34 35 36 37] Many of these studies revealed the potential to identify calcified plaques, but although promising, none has yet produced a technique with sufficient spatial and parametric resolution to identify a lipid pool covered by a thin fibrous cap.

Intravascular elastography/palpography

In 1991, a new technique was introduced to measure the mechanical properties of tissue using ultrasound: elastography.[38] The underlying concept is that upon uniform loading, the local relative amount of deformation (strain) of a tissue is related to the local mechanical properties of that tissue. If we apply this concept to determine the local properties of arterial tissue, blood pressure acts as a stressor. At a given pressure difference, soft plaque components will deform more than hard components. Measurement of local plaque deformation in the radial direction can be obtained with ultrasound. For intravascular purposes, a derivate of elastography called palpography may be a suitable tool.[39] In this approach, one strain value per angle is determined and plotted as a color-coded contour at the lumen vessel boundary. Since radial strain is obtained, the technique may have the potential to detect regions with elevated stress: increased circumferential stress results in an increased radial deformation of the plaque components. In vitro studies with histological confirmation have shown that there are differences of strain normalized to pressure between fibrous, fibro-fatty and fatty components of the plaque of coronary as well as femoral arteries.[40] This difference was mainly evident between fibrous and fatty tissue. The plaque types could not be differentiated by echo-intensity differences on the IVUS echogram.

It is feasible to apply intravascular palpography during interventional catheterization procedures. In a recent study, data were acquired in patients (n=12) during PTCA procedures with echo apparatus equipped with radiofrequency output. The systemic pressure was used to strain the tissue, and the strain was determined using cross-correlation analysis of sequential frames acquired at different pressures. A likelihood function was determined to obtain the frames with minimal motion of the catheter in the lumen since motion of the catheter impairs accuracy of strain estimation. Minimal motion was observed near the end of the passive filling phase. Reproducible strain estimates were obtained within one pressure cycle and over several pressure cycles. Validation of the results was limited to the information provided by the echogram. Significantly higher strain values were found in non-calcified plaques than in calcified plaques.[41] Another in vivo validation study in atherosclerotic Yucatan pigs showed that fatty plaques have an increased mean strain value. High-strain spots were also associated with the presence of macrophages, a further feature of vulnerable plaques.[42]

Palpography reveals information that is not seen on IVUS. To differentiate between hard and soft tissue may be important for the detection of an instable plaque that is prone to rupture. Since palpography is based on clinically available IVUS catheters, the technique can be easily introduced into the catheterization laboratory. By acquiring data at the end of the filling phase, when catheter motion is minimal, the quality and reliability of the palpogram is increased. The clinical value of this technique is currently under investigation.

Thermography

Inflammation produces a temperature rise in the affected tissue. Since atherosclerosis is accompanied by inflammation the hypothesis was created that a temperature rise could be measured at the surface of a plaque. As vulnerable plaque is a very active metabolic area, the hypothesis was extended that vulnerable plaques may have an even higher temperature rise.

Casscells et al. reported that carotid plaques taken at endarterectomy from 48 patients have temperature heterogeneity.[43] The temperature difference between different areas was up to 2.2 °C, and correlated with cell density (R^2= -0.47, p=0.0001). There was a negative correlation between temperature difference and cap thickness (R^2= -0.34, p=0.0001). The same group reported approximately the same in vitro findings in atherosclerotic rabbits.[44] A correlation between temperature rise and macrophage infiltration has also been suggested in an in-vivo rabbit trial.[45] Stefanadis et al. performed studies in humans. Patients with stable angina, unstable angina, and acute myocardial infarction were studied. The thermistor of the thermography catheter has a temperature accuracy of 0.05 °C, a time constant of 300 ms, and a spatial resolution of 0.5 mm. The thermistor of the catheter was driven against the vessel wall by the force of blood flow, without the help of a mechanical device like a balloon. Temperature was constant within the arteries of the control subjects, whereas most atherosclerotic plaques showed higher temperatures compared with healthy vessel wall. Temperature differences between atherosclerotic plaque and healthy vessel wall increased progressively from stable angina to patients with acute myocardial infarction with a maximum difference of 1.5 +/- 0.7 °C.[46] Furthermore patients with a high temperature gradient have a significantly worse outcome than patients with a low gradient.[47] However, this data has yet to be confirmed prospectively in other centers, and the influence of parameters such as coronary blood flow or catheter design has to be studied in the future.

Optical Coherence Tomography

Optical Coherence Tomography (OCT) can provide images with ultrahigh resolution. The technique measures the intensity of back-reflected light in a similar way as IVUS measures acoustic waves.[48] It is an invasive technique with a catheter advanced over a 0.014'' wire. With a Michelson interferometer light is split into two signals. One is sent into the tissue and the other to a reference arm with a mirror. Both signals are reflected and cross correlated by interfering the light beams. To achieve cross correlation at incremental penetration depths in the tissue, the mirror is dynamically translated. The intensity of the interfering signals at a certain mirror position represents backscattering at a corresponding depth. High-resolution images with a resolution ranging from 4–20 μm can be achieved[49] with a penetration depth up to 2 mm. Images can be acquired real time at 15 frames/sec.

Early attempts were made to validate OCT using histology. A lipid pool generates decreased signal areas, and a fibrous plaque produces a homogenous signal rich lesion.[50] In vitro comparison of OCT with IVUS demonstrated superior delineation by OCT of structural details like thin caps or tissue proliferation.[51]

Limitations of OCT are the low penetration depth, which hinders studying large vessels, and the light absorbance by blood which currently needs to be overcome by saline infusion or balloon occlusion with associated potential for ischemia. Special techniques like index matching may improve imaging through blood.[52]

Raman Spectroscopy

Raman spectroscopy is a technique that can characterize the chemical composition by utilizing the Raman effect.[53] The Raman effect is created when incident light (wave length 750 – 850 nm) excites molecules in a tissue sample, which backscatter the light while changing wave length. This change in wave length is the Raman effect.[54] The wave length shift and the signal intensity are dependent on the chemical components of the tissue sample. Due to this unique feature, Raman spectroscopy can provide quantitative information about the molecular composition of the sample.[55] The spectra obtained from tissue require post-processing to differentiate between plaque components.

Even in the presence of blood, Raman spectra have been shown to be obtainable in vivo from the aortic arch of sheep.[56] In a study using mice that received a high-fat/high-cholesterol diet for 0, 2, 4, or 6 months, Raman spectroscopy showed good correlation between cholesterol accumulation and total serum cholesterol exposure (r approximately 0.87, $P<0.001$). In female mice (n=10) that were assigned to an HFC diet, with or without 0.01% atorvastatin, a strong reduction in cholesterol accumulation (57%) and calcium salts (97%) ($P<0.01$) was demonstrated in the atorvastatin-treated group. Raman spectroscopy can therefore be used to quantitatively study the size and distribution of depositions of cholesterol and calcification.[57]

Limitations of the technique are the limited penetration depth (1 – 1.5 mm), the long acquisition time and the absorbance of the light by blood. Raman spectroscopy gives no geometrical information.

Near-Infrared Spectroscopy

Near-infrared (NIR) spectroscopy also obtains information on the chemical components of the coronary vessel wall. Molecular vibrational transitions measured in the NIR region (750 – 2500 nm) give qualitative and quantitative results on plaque composition. NIR spectroscopy sensitivity and specificity for the histological features of plaque vulnerability were 90% and 93% for lipid pool, 77% and 93% for thin cap, and 84% and 89% for inflammatory cells.[58] A differentiation between vulnerable and non-vulnerable carotid plaques could be achieved ex vivo.[59] Future studies will address the question whether NIR spectroscopy is feasible in vivo. Problems like acquisition time, blood scattering, influence of pH and temperature must be addressed.

Magnetic Resonance Imaging

High-resolution magnetic resonance imaging (MRI) is a non-invasive modality to characterize atherosclerotic plaques. Combining information from T1 and T2-weighted imaging can permit in vitro identification of the atheromatous core, collageneous cap, calcification, media, adventitia, and perivascular fat.[60] In a small number of patients (n=6) a matching between in vivo and in vitro measurements of carotid arteries was seen.[61] Yuan et al. determined the accuracy of in vivo MRI for measuring the cross-sectional maximum wall area of atherosclerotic carotid arteries in a group of 14 patients undergoing carotid endarterectomy. The authors showed a strong correlation between in vivo and in vitro measurements.[62] Although not perfect, it may be possible to identify carotid plaques at high risk for stroke using MRI.[63] Images of carotid arteries can be further improved using a coil placed close to the carotid artery at the surface of the neck.[64] An in-plane resolution of 0.4 x 0.4 mm and a slice thickness of 3 mm may allow an assessment of fibrous cap thickness and integrity.[65]

Imaging of coronary arteries with MRI is more difficult than imaging carotid plaques since cardiac and respiratory motion, the small plaque size, and the location of the coronary arteries can cause acquisition problems. Nevertheless, high resolution MRI of the human coronary wall of angiographically normal and abnormal vessels has been shown to be feasible. In a study by Botnar et al. the coronary wall thickness and wall area were significantly enlarged in patients with coronary artery disease demonstrated by angiography.[66] Small plaque structures like fibrous caps cannot yet be assessed using current MR techniques. Thinner slices and higher in plane resolution are needed to better delineate coronary plaques.

ANGUS and Shear Stress

High resolution reconstruction of 3D coronary lumen and wall morphology is obtained by combining angiography and IVUS.[67] Briefly a biplane angiogram of a sheath-based IVUS catheter taken at end-diastole allows reconstruction of the 3D pullback trajectory of the catheter. Combining this path with lumen and wall information derived from IVUS images that are successively acquired during catheter pullback at end-diastole gives accurate 3D lumen and wall reconstruction with resolution determined by IVUS. Filling the 3D lumen space with a high resolution 3D mesh allows calculation of the detailed blood velocity profile in the lumen.[68]

For this purpose absolute flow and blood viscosity need to be provided as boundary conditions. From the blood velocity profile local wall shear stress on the endothelium can be accurately derived. Wall shear stress is the frictional force, normalized to surface area that is induced by the blood passing the wall. Although from a mechanical point of view shear stress is of a very small magnitude compared to blood pressure induced tensile stress it has a profound influence on vascular biology[69] and explains the localization of atherosclerotic plaque in the presence of systemic risk factors.[70] Many of these biological processes also influence the stability of the vulnerable plaque including inflammation, thrombogenicity, vessel remodeling, intimal thickening or regression and smooth muscle cell proliferation. Therefore, the study of this parameter as derived by image based modeling is of utmost importance.

Conclusion

Assessment of atherosclerosis by imaging techniques is essential for in vivo identification of vulnerable plaques. Several invasive and noninvasive imaging techniques are currently in development.

OCT has the advantage of high resolution, thermography measures metabolism and NIR spectroscopy obtains information on chemical components. IVUS and IVUS-palpography are easy to perform and assess morphology and mechanical instability. Shear stress is an important mechanical parameter deeply influencing vascular biology. MRI and CT have the advantage of non-invasive imaging.

Nevertheless all techniques are still under development and at present, none of them can identify a vulnerable plaque alone or predict its further development. This is related to fundamental methodological insufficiencies that may be resolved in the future. From a clinical point of view, most techniques currently assess only one feature of the vulnerable plaque. Thus the combination of several modalities (see figure 1)[71] will be of importance in the future to ensure a high sensitivity and specificity in detecting vulnerable plaque.

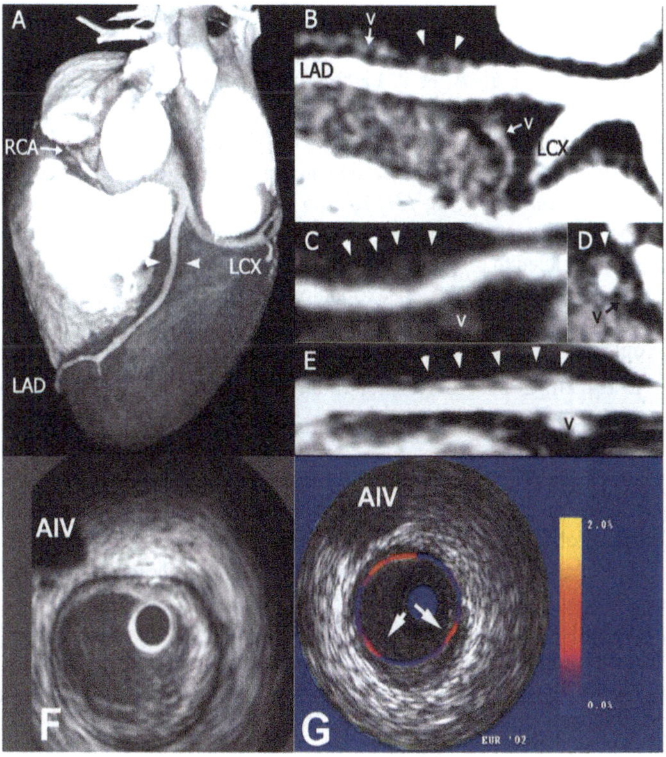

Figure 1: Non-invasive coronary imaging, using a 16-slice spiral computed tomography scanner (MSCT) (Sensation 16, Siemens AG, Forchheim, Germany) suggested a non-obstructive lesion in the mid-LAD (A, arrowhead), which was confirmed with coronary angiography. The attenuation value of the plaque was measured as 80 Hounsfield Units (HU), suggesting a mixed plaque composition without calcification (C, D, arrowheads). The entire segment can be shown in a single plane by means of vessel tracking (E, arrowhead). The great cardiac vein can be differentiated from the plaque by the higher and homogeneous attenuation of the venous lumen (v). Palpography (G) delivers strain information of this plaque's surface. At the right part a scale ranging from 0% (blue) to 2% (yellow) characterizes the strain pattern. The strain images are colour-coded, blue indicates stiff (low strain) material and red indicates softer (higher strain) material. In the cross section , which is identically in the same position with the IVUS image (F), an eccentric soft plaque is visible with shoulders of high strain (arrows) either side of the otherwise stable cap. Between 22.00 and 24.00 hrs the palpogram appears to show an area of high strain, however, this is caused by the nearby cardiac vein (AIV).

References

1 Libby P. Molecular bases of the acute coronary syndromes. Circulation 1995;91:2844-50.

2 Falk E. Stable versus unstable atherosclerosis: clinical aspects. Am Heart J 1999; 138:S421-5

3 Virmani R, Burke AP, Farb A, et al. Pathology of the unstable plaque. Prog Cardiovasc Dis 2002; 44: 349-56.

4 Little WC, Constantinescu M, Applegate RJ, et al. Can coronary angiography predict the site of a subsequent myocardial infarction in patients with mild-to-moderate coronary artery disease? Circulation 1988; 78: 1157-66.

5 Nobuyoshi M, Tanaka M, Nosaka H, et al.Progression of coronary atherosclerosis: is coronary spasm related to progression? J Am Coll Cardiol 1991; 18: 904-10.

6 Ambrose JA, Tannenbaum MA, Alexopoulos D, et al. Angiographic progression of coronary artery disease and the development of myocardial infarction. J Am Coll Cardiol 1988;12: 56-62.

7 Giroud D, Li JM, Urban P, et al. Relation of the site of acute myocardial infarction to the most severe coronary arterial stenosis at prior angiography. Am J Cardiol 1992; 69: 729-32.

8 Kerensky RA, Wade M, Deedwania P, et al. Revisiting the culprit lesion in non-Q-wave myocardial infarction. Results from the VANQWISH trial angiographic core laboratory. J Am Coll Cardiol 2002; 39: 1456-63.

9 Frink RJ. Chronic ulcerated plaques: new insights into the pathogenesis of acute coronary disease. J Invasive Cardiol 1994; 6: 173-85.

10 Ojio S, Takatsu H, Tanaka T, et al. Considerable time from the onset of plaque rupture and/or thrombi until the onset of acute myocardial infarction in humans: coronary angiographic findings within 1 week before the onset of infarction. Circulation 2000; 102: 2063-9.

11 Goldstein JA, Demetriou D, Grines CL, et al. Multiple complex coronary plaques in patients with acute myocardial infarction. N Engl J Med 2000; 343: 915-22.

12 Kim WY, Danias PG, Stuber M, Coronary magnetic resonance angiography for the detection of coronary stenoses. N Engl J Med 2001; 345: 1863-9.

13 Mizuno K, Satomura K, Miyamoto A, et al. Angioscopic evaluation of coronaryartery thrombi in acute coronary syndromes. N Engl J Med 1992; 326: 287-91.

14 Sherman CT, Litvack F, Grundfest W, et al. Coronary angioscopy in patients with unstable angina pectoris. N Engl J Med 1986 Oct 9;315(15):913-9.

15 Feld S, Ganim M, Carell ES, et al. Comparison of angioscopy, intravascular ultrasound imaging and quantitative coronary angiography in predicting clinical outcome after coronary intervention in high risk patients. J Am Coll Cardiol 1996; 28: 97-105.

16 Takano M, Mizuno K, Okamatsu K, et al. Mechanical and structural characteristics of vulnerable plaques: analysis by coronary angioscopy and intravascular ultrasound. J Am Coll Cardiol 2001; 38: 99-104.

17 Asakura M, Ueda Y, Yamaguchi O, et al. Extensive development of vulnerable plaques as a pan-coronary process in patients with myocardial infarction: an angioscopic study. J Am Coll Cardiol 2001; 37: 1284-8.

18 Uchida Y, Nakamura F, Tomaru T, Morita T, Oshima T, Sasaki T, Morizuki S, Hirose J. Prediction of acute coronary syndromes by percutaneous coronary angioscopy in patients with stable angina. Am Heart J 1995; 130: 195-203.

19 Bom N, Li W, van der Steen AF, Lancee CT, Cespedes EI, Slager CJ, deKorte CL. Intravascular imaging. Ultrasonics 1998; 36: 625-8.
20 Di Mario C, The SH, Madretsma S, et al. Detection and characterization of vascular lesions by intravascular ultrasound: an in vitro study correlated with histology. J Am Soc Echocardiogr 1992; 5: 135-46.
21 Sechtem U, Arnold G, Keweloh T, et al. In vitro diagnosis of coronary plaque morphology with intravascular ultrasound: comparison with histopathologic findings. Z Kardiol 1993; 82: 618-27.
22 Friedrich GJ, Moes NY, Muhlberger VA, et al. Detection of intralesional calcium by intracoronary ultrasound depends on the histologic pattern. Am Heart J 1994; 128: 435-41.
23 Potkin BN, Bartorelli AL, Gessert JM, et al. Coronary artery imaging with intravascular high-frequency ultrasound. Circulation 1990; 81: 1575-85.
24 Rasheed Q, Dhawale PJ, Anderson J et al. Intracoronary ultrasound-defined plaque composition: computer-aided plaque characterization and correlation with histologic samples obtained during directional coronary atherectomy. Am Heart J 1995; 129: 631-7.
25 Hiro T, Leung CY, Russo RJ, et al. Variability of a three-layered appearance in intravascular ultrasound coronary images: a comparison of morphometric measurements with four intravascular ultrasound systems. Am J Card Imaging 1996; 10: 219-27.
26 Ge J, Haude M, Gorge G, et al. Silent healing of spontaneous plaque disruption demonstrated by intracoronary ultrasound. Eur Heart J 1995; 16: 1149-51.
27 Jeremias A, Ge J, Erbel R. New insight into plaque healing after plaque rupture with subsequent thrombus formation detected by intravascular ultrasound. Heart 1997; 77: 293.
28 Ge J, Chirillo F, Schwedtmann J, et al. Screening of ruptured plaques in patients with coronary artery disease by intravascular ultrasound. Heart 1999; 81: 621-7
29 Yamagishi M, Terashima M, Awano K, et al. Morphology of vulnerable coronary plaque: insights from follow-up of patients examined by intravascular ultrasound before an acute coronary syndrome. J Am Coll Cardiol 2000; 35: 106-11
30 von Birgelen C, Klinkhart W, Mintz GS, et al. Plaque distribution and vascular remodeling of ruptured and nonruptured coronary plaques in the same vessel: an intravascular ultrasound study in vivo. : J Am Coll Cardiol 2001; 37: 1864-70
31 Nair A, Kuban BD, Obuchowski N, et al. Assessing spectral algorithms to predict atherosclerotic plaque composition with normalized and raw intravascular ultrasound data. Ultrasound Med Biol 2001; 10: 1319-31.
32 Landini L, Sarnelli R, Picano E, et al. Evaluation of frequency dependence of backscatter coefficient in normal and atherosclerotic aortic walls. Ultrasound Med Biol 1986; 5: 397-401.
33 Wilson LS, Neale ML, Talhami HE et al. Preliminary results from attenuation-slope mapping of plaque using intravascular ultrasound. Ultrasound Med Biol 1994; 20:529-42.
34 Jeremias A, Kolz ML, Ikonen TS, et al. Feasibility of in vivo intravascular ultrasound tissue characterization in the detection of early vascular transplant rejection. Circulation 1999; 100: 2127-30.

35 Wickline SA, Miller JG, Recchia D, et al. Beyond intravascular imaging: Quantitative ultrasonic tissue characterization of vascular pathology. IEEE Ultrasonics Symposium 1994; 3: 1589-97.
36 Bridal SL, Beyssen B, Fornes P, Julia P, Berger G. Multiparametric attenuation and backscatter images for characterization of carotid plaque. Ultrason Imaging 2000; 22: 20-34.
37 Spencer T, Ramo MP, Salter DM et al. Characterisation of atherosclerotic plaque by spectral analysis of intravascular ultrasound: an in vitro methodology. Ultrasound Med Biol 1997;23: 191-203.
38 Ophir J, Cespedes I, Ponnekanti H, et al. Elastography: a quantitative method for imaging the elasticity of biological tissues. Ultrason Imaging 1991; 13: 111-34.
39 Doyley MM, Mastik F, de Korte CL, et al. Advancing intravascular ultrasonic palpation toward clinical applications. Ultrasound Med Biol 2001; 27: 1471-80.
40 de Korte CL, Pasterkamp G, van der Steen AF, et al. Characterization of plaque components with intravascular ultrasound elastography in human femoral and coronary arteries in vitro. Circulation 2000; 102: 617-23.
41 de Korte CL, Carlier SG, Mastik F, et al. Morphological and mechanical information of coronary arteries obtained with intravascular elastography; feasibility study in vivo. Eur Heart J 2002; 23: 405-13.
42 de Korte CL, Sierevogel MJ, Mastik F, Strijder C, et al. Identification of atherosclerotic plaque components with intravascular ultrasound elastography in vivo: a Yucatan pig study. Circulation 2002; 105: 1627-30.
43 Casscells W, Hathorn B, David M, et al. Thermal detection of cellular infiltrates in living atherosclerotic plaques: possible implications for plaque rupture and thrombosis. Lancet 1996; 347: 1447-51.
44 Casscells W, David M, Bearman G et al. Thermography. In: The Vulnerable Atherosclerotic Plaque. Edited by V. Fuster, Futura Publishing Company, 1999: 231-42
45 Verheye S, De Meyer GR, Van Langenhove G et al. In vivo temperature heterogeneity of atherosclerotic plaques is determined by plaque composition. Circulation 2002; 105: 1596-601.
46 Stefanadis C, Diamantopoulos L, Vlachopoulos C et al. Thermal heterogeneity within human atherosclerotic coronary arteries detected in vivo: A new method of detection by application of a special thermography catheter. Circulation 1999; 99:1965-71.
47 Stefanadis C, Toutouzas K, Tsiamis E, et al. Increased local temperature in human coronary atherosclerotic plaques: an independent predictor of clinical outcome in patients undergoing a percutaneous coronary intervention. J Am Coll Cardiol 2001;37: 1277-83.
48 Huang D, Swanson EA, Lin CP, et al. Optical coherence tomography. Science 1991; 254: 1178.
49 Boppart SA, Bouma BE, Pitris C, et al. In vivo cellular optical coherence tomography imaging. Nat Med 1998; 4: 861-5.
50 Jang IK, Bouma BE, Kang DH, et al. Visualization of coronary atherosclerotic plaques in patients using optical coherence tomography: comparison with intravascular ultrasound. J Am Coll Cardiol 2002 Feb 20;39(4):604-9
51 Brezinski ME, Tearney GJ, Weissman NJ, et al. Assessing atherosclerotic plaque morphology: comparison of optical coherence tomography and high frequency intravascular ultrasound. Heart 1997; 77: 397-403.

52 Brezinski M, Saunders K, Jesser C, et al. Index matching to improve optical coherence tomography imaging through blood. Circulation 2001; 103: 1999-2003.
53 Baraga JJ, Feld MS, Rava RP. In situ optical histochemistry of human artery using near-infrared Fourier transform Raman spectroscopy. Proc Natl Acad Sci U S A. 1992;89: 3473-3477.
54 Van de Poll SWE, Motz JT, Kramer JR. Prospects of Laser Spectroscopy to detect vulnerable plaque. In: Cardiovascular plaque rupture. Ed: Brown DL, Dekker, New York, 2002.
55 Hanlon EB, Manoharan R, Koo TW, et al. Prospects for in vivo Raman spectroscopy. Phys Med Biol 2000; 45: R1-59.
56 Buschman HP, Marple ET, Wach ML, et al. In vivo determination of the molecular composition of artery wall by intravascular Raman spectroscopy. Anal Chem 2000; 72:3771-5.
57 van De Poll SW, Romer TJ, Volger OL, et al. Raman spectroscopic evaluation of the effects of diet and lipid-lowering therapy on atherosclerotic plaque development in mice. Arterioscler Thromb Vasc Biol 2001; 21: 1630-5.
58 Moreno PR, Lodder RA, Purushothaman KR, et al. Detection of lipid pool, thin fibrous cap, and inflammatory cells in human aortic atherosclerotic plaques by near-infrared spectroscopy. Circulation 2002; 105: 923-7.
59 Wang J, Geng YJ, Guo B, etal. Near-infrared spectroscopic characterization of human advanced atherosclerotic plaques. J Am Coll Cardiol 2002; 39: 1305-13.
60 Toussaint JF, Southern JF, Fuster V, et al. T2-weighted contrast for NMR characterization of human atherosclerosis. Arterioscler Thromb Vasc Biol 1995; 15: 1533-42.
61 Toussaint JF, LaMuraglia GM, Southern JF, et al. Magnetic resonance images lipid, fibrous, calcified, hemorrhagic, and thrombotic components of human atherosclerosis in vivo. Circulation 1996; 94: 932-8.
62 Yuan C, Beach KW, Smith LH Jr, et al. Measurement of atherosclerotic carotid plaque size in vivo using high resolution magnetic resonance imaging. Circulation 1998; 98: 2666-71.
63 Yuan C, Mitsumori LM, Beach KW, et al. Carotid atherosclerotic plaque: noninvasive MR characterization and identification of vulnerable lesions. Radiology 2001; 221: 285-99.
64 Hayes CE, Mathis CM, Yuan C. Surface coil phased arrays for high-resolution imaging of the carotid arteries. J Magn Reson Imaging 1996; 6: 109-12.
65 Hatsukami TS, Ross R, Polissar NL, et al. Visualization of fibrous cap thickness and rupture in human atherosclerotic carotid plaque in vivo with high-resolution magnetic resonance imaging. Circulation 2000; 102: 959-64.
66 Botnar RM, Stuber M, Kissinger KV, et al. Noninvasive coronary vessel wall and plaque imaging with magnetic resonance imaging. Circulation 2000; 102: 2582-7.
67 Slager, CJ, Wentzel JJ, Schuurbiers JCH et al., True 3dimensional reconstruction of coronary arteries in patients by fusion of angiography and IVUS (ANGUS) and its quantitative validation. Circulation, 2000. 102: 511-6
68 Thury A, Wentzel JJ, Schuurbiers JC, Ligthart JM, Krams R, de Feyter PJ, Serruys PW, Slager CJ. Prominent role of tensile stress in propagation of a dissection after coronary stenting: computational fluid dynamic analysis on true 3d-reconstructed segment. Circulation. 2001; 104: E53-4.

69 Malek, A.M., S.L. Alper, and S. Izumo, Hemodynamic shear stress and its role in atherosclerosis. JAMA 1999; 282: 2035-42.
70 Asakura, T. and T. Karino, Flow patterns and spatial distribution of atherosclerotic lesions in human coronary arteries. Circ Res, 1990. 66: 1045 -66.
71 Arampatzis CA, Ligthart JMR, Schaar JA et al. Detection of avulnerable coronary plaque: a treatment dilemma. Circulation 2003 in press.

Part 4

Carotid Scanning

Conventional and compound scanning of the Carotid artery

J. E. Wilhjelm

Center for Arteriosclerosis Detection with Ultrasound (CADUS), Ørsted•DTU, DK-2800 Kongens Lyngby, Denmark

Summary: Plaque in the carotid artery is typically formed at the bifurcation in the neck, but in contrast to most other vessels in the body, rather than the reduction in lumen diameter it is risk of embolization of small blood vessels in the brain, *i.e.* the risk of stroke, that is the primary cause for concern. The carotid artery can be visualized with extra-vascular ultrasound and the main parameters of this scanning situation is considered in this chapter, including aspects of the degradation due to the medium between transducer and plaque. The ultrasound images are affected by variations in gray scale value due to speckle and angle-dependence and due to intervening anisotropic tissue. Spatial compound imaging is introduced and it is shown with in vitro and in vivo examples how the speckle can be reduced this way. Error sources and potential problems are discussed. For the in vitro investigations, it is shown how the tissue types in the plaque can be determined by scanning the plaque casted into an agar block and subsequently cutting this block at given calibration marker locations. Both anatomical and histological images can then be obtained at the same scan plane as that of the ultrasound image. Finally, future techniques for classification of plaque tissue with extra-vascular ultrasound is discussed.

Key words: Carotid artery, Spatial compounding, Speckle reduction, Plaque characterization, Angle-dependence

1 Atherosclerosis in the carotid arteries

Plaque deposits in the carotid arteries are typically confined to the region of bifurcation of the common carotid artery. As in other parts of the body, plaque in the carotid artery can obstruct the blood flow and it can produce thrombi (a small fragment such as a blood clot formed at the surface of the plaque or part of the plaque). However, in contrast to other parts of the body, the obstruction of the blood flow it not necessarily as critical given that the primary organ supplied via these arteries - the brain - obtain blood from four arteries in parallel: The two carotid arteries in the front of the neck and the two vertebral arteries in the back of the neck. Thus, the main problem is the emboli which can obstruct small vessels in the brain and thereby cause ischemia with subsequent damage of brain tissue. Such an attack, commonly termed *stroke*, occurs for about two people a year in a population of 1 000 in the industrialized countries. The mortality is 30%, making it the third most common cause

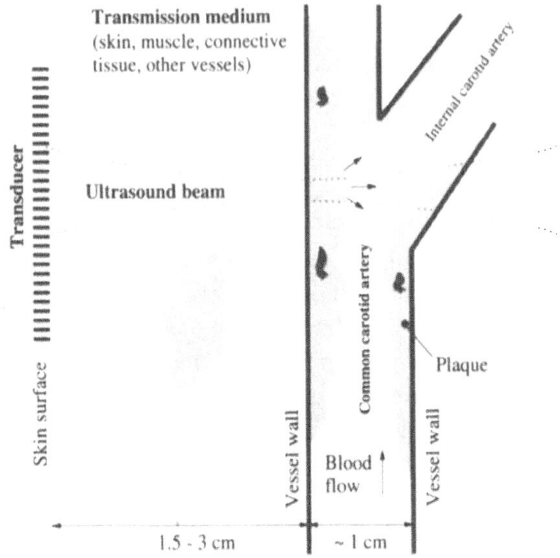

Fig. 1. Schematic cross-sectional view of non-invasive scanning of the carotid artery. The transducer will typically be a linear array.

of death. [1] Of the surviving 70%, many experience permanent disabilities. Thus, prevention is important. Recent studies have shown that the risk of stroke is reduced by more than 50% by surgical removal (endarterectomy) of carotid plaques that caused diameter reductions of more than 70% [2,3] assessed angiographically. On the other hand, as not all carotid plaques are vulnerable and thereby associated with high stroke risk, and as carotid endarterectomy carries a certain risk for the patient and is expensive, optimized selection of patients for operation is crucial.

The previous chapters in this book have explained how IVUS can be used for characterization of plaque in most of the vessels in the body. However, IVUS scanning of the carotid arteries would involve some problems: *i*) a catheter manipulated close to the plaque could in principle cause production of emboli, that in turn could cause stroke during such a diagnostic procedure *ii*) an invasive diagnosis is several times more expensive than a non-invasive and *iii*) there is always a larger risk involved with invasive procedures. However, the use of IVUS scanning has been reported in connection with carotid stent placement. [4,5]

2 The scanning situation

A cross-sectional view of a typical scanning situation is sketched in Figure 1. A transducer frequency of 7.5 MHz is typically applied. At c = 1540 m/s, this gives λ = 0.2 mm. The typical length of the plaque is 10 - 25 mm (50λ - 120λ) while the thickness of the plaque (wall to lumen) is typically 1 - 3 mm (5λ - 15λ). [6] However, total occlusion can also occur.

The plaque is typically located at a depth around 20 mm (100λ) from the skin surface (Danish patients with median age 63 years, range 35 - 73 years [6]). Thus the ultrasound energy must travel at least 40 mm round trip though the medium between transducer and plaque, which consists of skin, subcutaneous fat, various layers of muscle, other vessels and connective tissue. This inhomogenous coupling medium will have some degrading effects on the echo from the plaque, compared to scanning through a more homogenous medium (at 7.5 MHz), such as blood. Specifically, the different tissues exhibits variations in speed of sound, density and attenuation.

The variation in speed of sound distorts the emitted as well as the echoed wavefront (phase aberration) which in turn results in phase cancellation of the echoed wavefront at the transducer surface.

The attenuation varies from about negligible in blood to a variable higher level in the remaining tissue, both due to the interfaces between the tissues and the variation in attenuation through a given tissue type. As an example on the latter, anisotropic tissues, such as muscle, seems to infer a stronger attenuation for ultrasound propagation *along* the fibres, than propagation *perpendicular* with the fibres. Specifically, for cardiac muscle, an in vitro investigation reports the attenuation to 0.47 dB/mm and 0.25 dB/mm, respectively (at 7.5 MHz). [7] Thus, for a 5 mm layer of muscle this can, in principle, give a variation of 2.2 dB in total round-trip attenuation. The mechanism of this angle-dependence of attenuation is not fully understood, and it is interesting to note that apparently, the largest attenuation is obtained

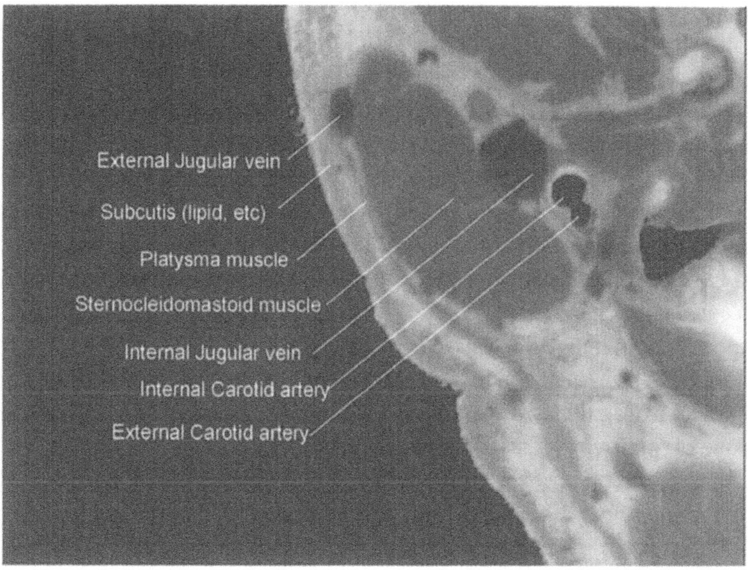

Fig. 2. Example of cross-sectional view of the overlying tissues, when scanning the carotid artery. Macroscopic image from the Visible Human Project (U.S. National Library of Medicine).

at the same angle of propagation relative to the fibres, where the level of backscattering is the *smallest.* [8,9,10]

Thus, the echo signal from a plaque region is not only due to the plaque material but also (to a yet unknown degree) to the inhomogenous medium in between plaque and skin.

In most ultrasound scanners, the attenuation compensation (Time-Gain control) is constant and about 0.1 dB/MHz per millimeter travelled. Thus, for a depth of 20 mm imaged at 7.5 MHz, the total attenuation compensation will be about 30 dB.

A specific example of the content of the inhomogenous medium between transducer and plaque can be obtained from a cross-sectional gross-anatomical picture of the neck such as those available from the Visible Human Project, as shown in Figure 2. Specifically, this picture shows the internal and external carotid artery at the bifurcation situated behind or next to other vessels, muscles, connective tissues, subcutaneous lipid and skin. The scanning of the carotid artery are often conducted through the sternocleidomastoid muscle more or less perpendicular to the muscle fibres.

In addition to the above, the echo from the plaque itself is influenced by additional factors, such as speckle noise and angle-dependence from the anisotropic part of the plaque. This will be further elaborated on in Section 4.

3 Multi-angle spatial compounding

3.1 Background

A possible way to reduce the influence of some of the above mentioned problems is spatial compound imaging which can reduce the angle-dependence and the speckle [11] noise. [12, 13, 14, 15, 16] Spatial compounding has been known for long, but the clinical use has been quite limited, except for the use in static B-mode scanners up to the beginning of the 80's. Even though a few spatial compound imaging systems have been constructed in the early 80's [17, 18, 14] subsequent publications only consider such systems theoretically. [19, 13, 20] A few exceptions used spatial compounding to image bone structures. [21, 22]

Within CADUS, a digital off-line ultrasound scanner for Multi-Angle Compound Imaging (MACI), denoted Xtra, was build by SK Jespersen *et al.* in 1997 with the purpose of improving visualization of vascular tissue. [12]

Concurrent with our work, the company ATL (Bothell, WA, USA) developed a similar system operating in real-time (called Sono-CT™ and incorporated into the HDI 5000 system), which was introduced in the fall of 1999 [23]. This system can make spatial compound images based on up to 9 single-angle images, while the largest beam angle is up to ±20°.

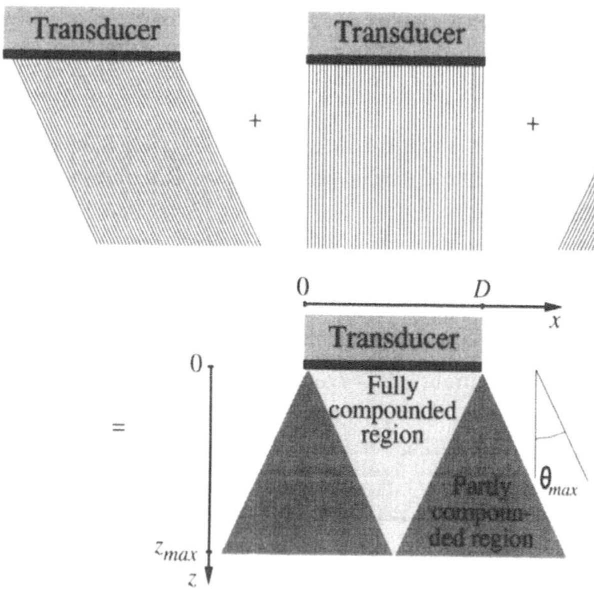

Fig. 3. Principle in spatial compound imaging. In this example, images are recorded with three different scanning angles, and then averaged to form the final image.

3.2 Principle

The MACI system makes use of a conventional low-pitch linear array transducer. The active aperture is selected as a subset of the elements and this aperture is then operated in phased array mode, to create an ultrasound beam with a given angle, θ_i. The aperture is then moved along the entire array to record an image at that angle. This operation is repeated for all desired beam angles, θ_i, where $i = 1, 2, ..., N_\theta$. The final compound image is created by forming the average of the recorded single-angle images. Figure 3 sketches the location of the scan lines used in this procedure when $N_\theta = 3$. It can be observed that the fully compounded region is a triangle. From the geometry, the depth of this region can be found to: [12]

$$z_{max}(D, \theta_{max}) = \frac{D}{(2\tan(\theta_{max}))} \tag{1}$$

where D is the total length of the array and θ_{max} is half the angular span. As an example, consider $D = 40$ mm and $\theta_{max} = 21°$, which gives $z_{\max} = 52$ mm.

3.3 Experimental investigations

3.3.1 Experimental scanner

Using a custom-made experimental ultrasound system for scanning plaque in vitro provides some valuable insight into the behavior and mechanism of conventional and spatial compound scanning.

Being an experimental off-line system, the hardware of the Xtra system consists of three major components: *i)* A 192-element linear transducer array with center frequency 7.5 MHz, pitch 208 μm (distance between centres of adjacent transducer elements), transverse acoustical focal distance of 16 mm and array length of $D = 40$ mm. *ii)* A control computer containing a PCI-based 12 bit AD-card and a timing controller (real-time scan controller, RTSC) for the ultrasound system. *iii)* A 19" rack containing the custom designed transmit and receive system hardware, power supply and transducer connector.

The real-time transmit system contains 64 parallel transmitters with fully programmable transmit delays (0 - 34 μs) and apodization control (25 - 0 dB). The delay resolution of the transmit system is 16.7 ns. The receive system consists of 64 parallel pre-amplifiers and a multiplexer which selects one of the receive channels at a time which subsequently passes a time gain compensation (TGC) amplifier and an anti-aliasing lowpass filter before it is digitized in the 12 bit AD-converter located in the control computer (sampling frequency up to 60 MHz).

The system software consists of three main parts: *i)* Creation of setup data for the entire system, *ii)* Control of hardware and recording of the single-element signals and *iii)* processing of the received signals. The digital signal processing performed on the received signals consist of dynamic beamforming, bandpass filtering, envelope detection, scan conversion, compounding, logarithmization, scaling and display. [12]

3.3.2 Creating a reference for comparison

In order to evaluate ultrasound images and evaluate different imaging techniques, a reference that reveals tissue shape, size and type is needed. This can be obtained with the following technique.

Formalin fixed carotid atherosclerotic plaques, removed during prophylactic carotid endarterectomy, were cast into a rectangular agar block. Such a tissue sample normally contains the atherosclerotic plaque surrounded by intima and some part of the media. By supporting the plaque on tiny wires during casting, the plaque was hold in the middle of the agar block. A lid with two rows of small rectangular openings was placed on top of the liquid agar. When the liquid agar had hardened, the lid was removed, and the upper surface of the agar block then contained two parallel rows of calibration markers, both running parallel with the "long axis" of the plaque. These calibration markers can then be visualized in the ultrasound images and they can also be used as guides for slicing the agar block with plaque after scanning. Figure 4 provides a photograph of the agar block where the left part of the block has been cut away thereby showing tissue cut free at the left end of the block. The two bands of calibration markers were located so that they were at the right and left sides of the ultrasound images, in order to avoid variation in the interference on the echo signals from the plaque due to the water-agar interface.

The distance between two adjacent calibration markers was selected to 5 mm. Slicing was done both at the center of a marker and between two markers. Thus the distance between slices was 2.5 mm. After cutting each slice, a macroscopic photograph was taken of the slice. There were typically 8 to 16 slices per plaque.

Each slice could then be further trimmed and placed in a tissue cassette for subsequent histological analysis.

3.3.3 In vitro images of carotid plaque

Prior to cutting the agar block, the Xtra system was used to record ultrasound images at parallel cross-sectional scan planes covering the entire plaque. The separation between scan planes was 0.5 mm. This allowed to precisely identify the rows of calibration markers on the top of the agar block and thereby identify the ultrasound images that were associated with anatomical and histological images. Seven

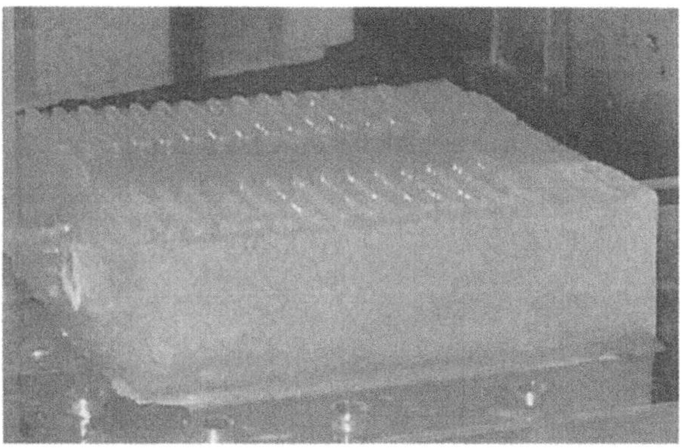

Fig. 4. Carotid atherosclerotic plaque casted into a rectangular agar block with calibration markers on the top.

single-angle images were recorded at each scan plane using the angles -21°, -14°, -7°, 0°, 7°, 14° and 21°. The scanning took place in pure demineralized water at ~20°.

A typical set of single-angle images and corresponding compound image are given in Figure 5. The conventional B-mode image corresponds to the 0° image, while the MACI image was obtained from averaging of all seven single-angle images.

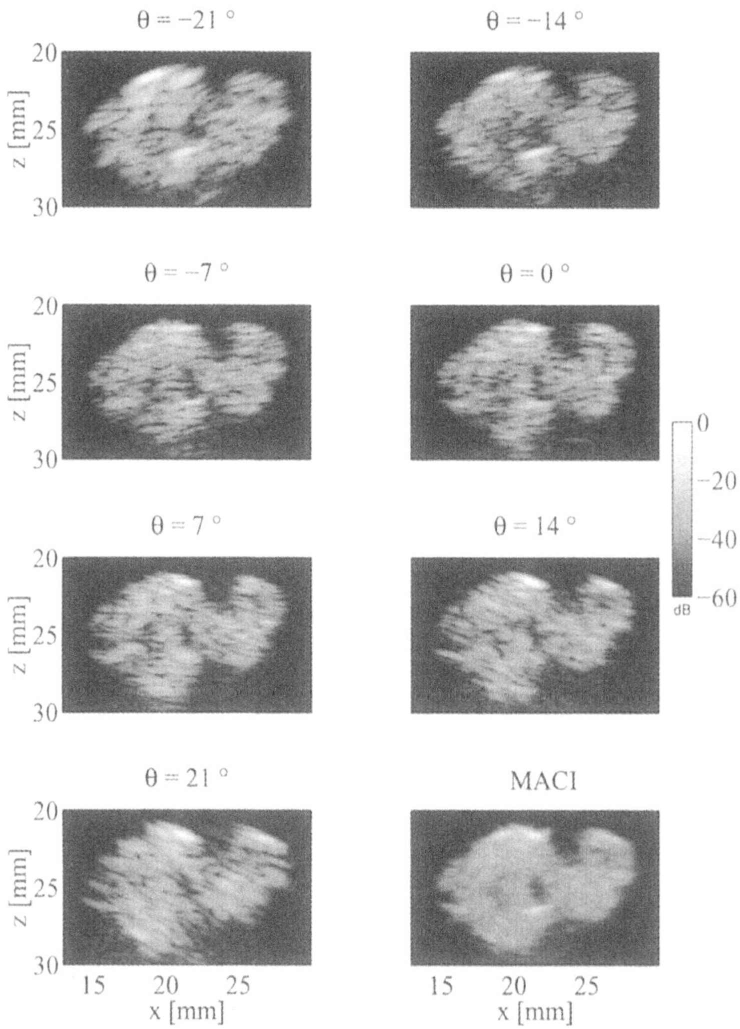

Fig. 5. Example of single-angle images and MACI image of a carotid atherosclerotic plaque scanned in vitro with the Xtra system.

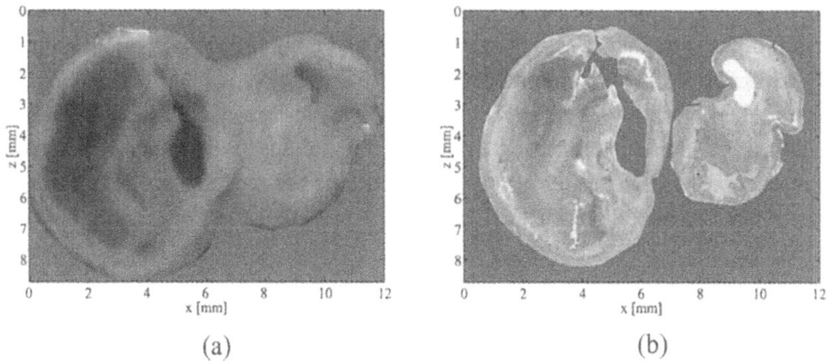

Fig. 6. The macroscopic (anatomical) image (a) and the histological image (elastin-trichrome) (b) corresponding to the ultrasound scan plane in Figure 5.

These images are displayed in a coordinate system where (0,0) correspond to the center of the leftmost transducer element.

Figure 6a shows the corresponding macroscopic photograph with metric axes. The left part of the image show the atherosclerotic plaque from the internal carotid artery surrounded by the intima and part of the media layer. The right part of the image shows plaque from the smaller external carotid artery with intima and some part of media. The narrow opening at the upper part of the plaque from the internal carotid artery is the surgical cut into the residual lumen made during the removal of the plaque. Figure 6b shows the corresponding histological image which reveals a large pool of lipid with blood elements in the central part of the internal carotid artery. The tissue region on the histological image is somewhat smaller than that on the anatomical image due to shrinkage in the histological process.

In the ultrasound images in Figure 5, notice how the speckle phenomena gives the individual single-angle images a pronounced mottled appearance, yet this is strongly suppressed in the MACI image accompanied by only a 30% increase in resolution size (mainly in the vertical image direction [24]). With the MACI approach, it may be easier to assess *e.g.* any possible residual lumen in an atherosclerotic artery, as can be seen when comparing the 0° image with the MACI image in Figure 5, relative to the reference in Figure 6a The center region of the plaque ($x \cong 16$ - 21 mm, $z \cong 24$ - 27 mm in Figure 5) - which is identified as lipid with blood from Figure 6b - appears somewhat darker than the remaining part, which is consistent with the observation of low echogenicity from lipid regions [25]. At the upper interface between agar and plaque from the internal carotid artery, the location of the bright echo changes along the interface with scanning angle ($x \cong 18$ mm, $z \cong 22$ mm). At ($x \cong 22$ mm, $z \cong 26$ mm) another strong reflection can be identified at the bottom of the residual lumen. As the reflection is strongest at $\theta = -14°$, this structure might have some preferred orientation perpendicular to the beam angle $\theta = -14°$. However, this - apparently strong - reflector is difficult to fully identify in Figure 6a, maybe because it is located out of the photographic plane of Figure 6a and b (the -3 dB width of the ultrasound beam perpendicular to the scan plane is about 0.8 mm, while the thickness of the histological image is 0.05 mm). Another possible reason could be that the bottom of the residual lumen happens to be perpendicular to the $\theta = -14°$ beam in the image plane and in a plane perpendicular to the image plane and comprising the $\theta = -14°$ beam. Nevertheless, both of these echogenic regions are more uniformly visualized on the MACI image.

3.4 Clinical investigations

While in vitro investigations allow comparisons to reference images and thereby allows for quantitative evaluation of the imaging technique, in vivo investigations provides realistic comparison between conventional and compound images. Figure 7 shows a conventional B-mode image and the corresponding spatial compound image of an atherosclerotic lesion, recorded with the HDI 5000 system. The B-mode image is extracted from the series of seven single-angle image recorded for each compound image. Note the artifacts outside the region of interest at especially the lower right part of the image due to the application of multiple scan angles. On these static images, it can be observed that the lumen is blacker and the plaque and vessel wall outlines are better defined in the compound image, compared to the conventional image. The improvement seen in this example gets even more pronounced for anechoic plaque. [26]

Another example is given in Figure 8 which shows a cross-sectional view of the brachial artery, recorded for the purpose of measuring changes in lumen diameter. The outline of the artery lumen, as well as other items in the image is more well-defined on the compound image. Specifically, while the upper and lower artery wall

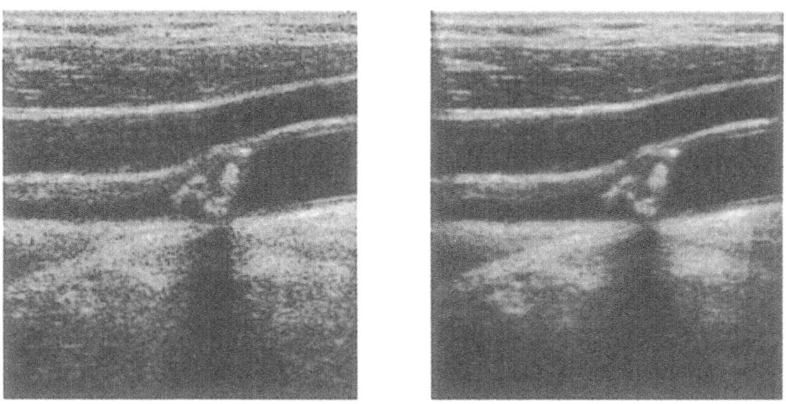

(a) (b)

Fig. 7. Example of conventional B-mode (0° image) (a) and spatial compound image (b) from the same recording of a carotid plaque in vivo with the ATL HDI 5000 system. Courtesy of S Kofoed MD, KAS Gentofte, Denmark.

is well defined in both images, the artery wall between 1 and 5 o'clock and between 7 and 11 o'clock is much more well-delineated in the compound image, compared to the conventional image.

3.5 Advantages and limitations of spatial compounding

When comparing the conventional B-mode images (0° images) to the multi-angle compound images, it is observed that for both in vitro and in vivo, the MACI method features images with reduced angle-dependence, reduced random variation and improved delineation of the plaque outline. These improvements are due to the fact that *i*) more scan lines are perpendicular or nearly perpendicular to the tissue interfaces or fibres and *ii*) because spatial compounding reduces the variation in the image due to speckle.

Despite the encouraging images obtained with MACI, it should be pointed out that a number of factors influence the achievable image quality in the present implementation. Three of them are based on the fact that single images are recorded and then added [12]:

1. Errors in the *mean speed of sound* assumed by the scanner will cause blurring of the resulting compound image, because the individual single-angle images will be affected by a range and steering angle error causing the images recorded from different directions to be misaligned when added.
2. Image blurring may also be caused by tissue movement: if the individual single-angle images are recorded consecutively, these might not align exactly when added to form the compound image. This has been studied previously [12], and an estimate based on typical movement data for the carotid artery

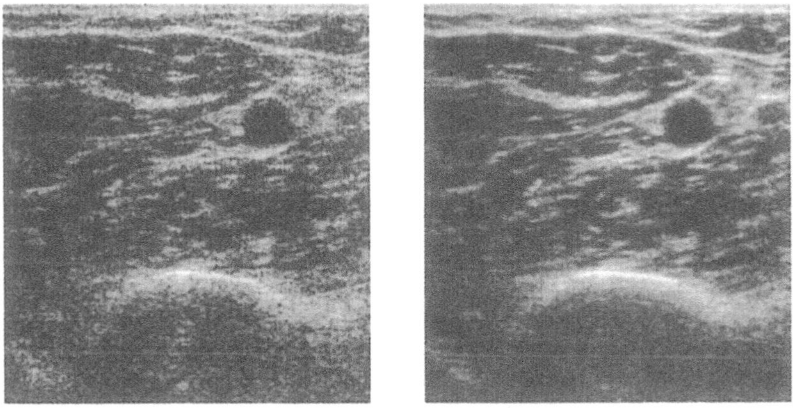

(a) (b)

Fig. 8. Example of conventional B-mode (0° image) (a) and spatial compound image (b) from the same recording of the brachial artery with the ATL HDI 5000 system. Courtesy of J Bismuth MD, KAS Gentofte, Denmark.

gave a maximal displacement that was approximately half of the -6 dB axial width of the point spread function.

3. Spatial variation in the speed of sound of the tissue being scanned may also cause blurring, due to local misalignment between the individual single-angle images.

The combined effect of these factors can, however, only be assessed with a system that is capable of scanning real-time *in vivo* and deliver digital un-logarithmiticed single-angle images for further analysis. The fourth important limitation is technological in nature:

4. When ultrasound images are recorded with a beam angle different from zero, grating lobes might occur, depending on how fine a pitch is used in the linear array transducer. Too large a pitch, and thereby too high grating lobe levels will severely reduce the contrast resolution of the individual single-angle images for high beam angles and thereby also - to a lesser degree - the final compound image. [12] Thus, a consequence can be that the brightness of anechoic *(e.g.* fatty) regions will be mainly governed by the level of the grating lobes. Because the levels of the grating lobes decrease with pitch, a possible solution is to decrease the pitch. However, if the length of the transducer array is to be kept constant (to avoid reducing the fully compounded region), this can only be done by increasing the number of transducer elements and thereby, if the size of the transmit/receive aperture is to be kept constant as well, the number of channels in the ultrasound system.

Theoretically, factors one and two could be handled by using synthetic aperture imaging, involving parallel recording of all (several hundreds) transducer signals. The signals should be stored in memory and subsequently beamformed and scan converted. This would reduce the movement artifacts, and if sufficient computational power was available, the image formation could be done with different assumed sound velocities, and the sound velocity yielding the "sharpest" image would then be assumed to be the "correct" sound velocity.

4 Plaque classification - the challenges of future clinical methods

Not all carotid plaques seems to have a high likelihood of causing dangerous neurological events, and it is believed that plaque composition (especially the relative content of lipids) in addition to size is a predominant factor as to whether a plaque is benign or malignant. [27,28] In principle, the relative content of lipid can be revealed by extravascular ultrasound as shown in several newer carotid studies based on *large pools* of patients. [1,29,30,31,32] Also relations between ultrasonographic appearance and hemorrhage has been reported. [33,34,35,36] However, the ability to accurately unveil the composition of the plaque for the *individual patient* has yet to be shown.

Relations between echogenicity and plaque materials has also been demonstrated in vitro for frequencies up to 10 MHz (for higher frequencies, see [45]). Specifically, two studies [25, 37] have evaluated the relation between plaque tissue type and integrated backscatter (IBS). The IBS was 24 dB larger for calcified material than for fatty material at normal incidence. Likewise angle-dependence of IBS showed largest variation for calcified tissue (20 dB) and smallest for fatty (8 dB). [25]. Differentiating young plaques from old ones have been investigated as well [38], just as attenuation (transmission measurements) has been investigated for classification purposes [39].

To derive quantitative tissue parameters from ultrasound echo signals is a difficult task, because - generally speaking - the 1D electrical signal associated with a given scan line is the result of not only the 3D distribution of scatterers and the geometry and acoustic properties of tissues and tissue interfaces, but also the field pattern and frequency response of the given transducer and constructive/destructive interference in the backscattered field at the surface of the receiving transducer. While the first couple of factors are only indirectly related to tissue properties the latter factors are not at all related to the tissue properties. This may have a varying and sometimes strong influence on the electrical echo signals.

As an example, the (electrical) echo signal due to insonification of an interface between two tissue types depends on the average specific acoustic impedance differences between the two tissues constituting the interface, but also strongly on the surface characteristics and orientation of the interface relative to the transducer as well as on transducer geometry. The change is manifested in both amplitude and signal waveform. [40] Several authors [41, 42] have noted the angle-dependence for vascular tissue, even though their in vivo observations always will be influenced by the change in the propagation path that accompany a change in scanning angle.

For the purpose of quantitative analysis of ultrasound data, MACI has potentials for overcoming some of the above problems. Since more independent data are recorded from the same tissue region, the speckle is reduced and estimates of parameters such as mean echogenicity of isotropic tissue should therefore be correspondingly more robust, compared with conventional B-mode imaging. For isotropic tissue (probably such as fibrous tissues) there might be a (more) deterministic angle-dependence, which in principle could be exploited.

Finally, as stated in the beginning of this chapter, the change in the ultrasound field due to the inhomogenous medium between the transducer and the target tissue must be considered. If the effect of the medium between transducer and plaque is on the same order of magnitude as the variation in the echo signals due to variation in tissue types, then the influence of the transmission medium on the echo signal must be removed. Otherwise, simple parameters such as integrated backscatter might not be useful in classifying tissue types.

An approach to solve this problem is to use what appears to be the only reference tissue close to the plaque, the blood. This has been suggested by Pedersen [43, 44] in a study where it was investigated to which degree a statistically stable estimate of the backscattering from blood can be measured. The influence of the inhomogenous medium will then be nearly the same for the estimate of the backscattering from blood as it will be for a small plaque region located slightly further away along the ultrasound beam. By calculating the integrated backscatter of these two targets, a more absolute measure for the backscattering from a small plaque region can be obtained, as has been demonstrated by in vitro studies.[44]

5 References

1. Lorenzen I, Bendixen G, Hansen NE (1999) Medicinsk kompendium, part 2, ed 15, Nyt Nordisk forlag Arnold Busck, Copenhagen

2. European Carotid Surgery Trialist' Collaborative Group (ECST). MRC European Carotid Surgery Trial: Interim results of symptomatic patients with severe (70-99%) or with mild (0-29%) carotid stenosis (1991). The Lancet:337:1235-1243

3. North American Symptomatic Carotid Endarterectomy Trial Collaborators (NASCET) (1991) Beneficial effect of carotid endarterectomy in symptomatic patients with high-grade carotid stenosis. New Engl. J. of Med:325:7:445-453

4. Reid DB, Diethrich EB, Marx P, Wrasper R. (1996) Intravascular ultrasound assessment in carotid interventions. J Endovasc Surg:3:2:203-10

5. Wilson EP, White RA, Kopchok GE (1996) Utility of intravascular ultrasound in carotid stenting. J Endovasc Surg:3:1:63-68

6. Wilhjelm JE, Grønholdt M-LM , Wiebe B, Jespersen SK, Hansen LK & Sillesen H (1998) Quantitative Analysis of Ultrasound B-mode Images of Carotid Atherosclerotic Plaque: Correlation with Visual Classification and Histological Examination. IEEE Transactions on Medical Imaging, 17:6: 910 - 922

7. Mottley JG, Miller JG (1988) Anisotropy of the ultrasonic attenuation in soft tissues: Measurements in vitro. J. Acoust. Soc. Am., 88:3: 1203-1210

8. Mottley JG, Miller JG (1988) Anisotropy of the ultrasonic backscatter of myocardial tissue: I. Theory and measurements in Vitro. J. Acoust. Soc. Am., 83:2: 755-761

9. Madaras EI, Perez J, Sobel BE, Mottley JG, Miller JG (1988) Anisotropy of the ultrasonic backscatter of myocardial tissue: II. Measurements in vivo. J. Acoust. Soc. Am., 83:2: 762-769

10. Aygen M, Popp RL (1987) Influence of the Orientation of Myocardial Fibers on Echocardiographic Images. The American Journal of Cardiology:60:147-152

11. Burckhardt CB (1978) Speckle in Ultrasound B-Mode Scans. IEEE Transactions on SU, 25:1: 1-6

12. Jespersen SK, Wilhjelm JE, Sillesen H (1998) Multi-Angle Compound Imaging. Ultrasonic Imaging, 20: 81-102

13. O'Donnell M, Silverstein SD (1988) Optimum Displacement for Compound Image Generation in Medical Ultrasound. IEEE Transactions on UFFC, 35:4: 470-476

14. Shattuck DP, Ramm OTv (1982) Compound Scanning with a Phased Array. Ultrasonic Imaging, 4: 93-107

15. Trahey GE, Smith SW, Ramm OTv (1986) Speckle Pattern Correlation with Lateral Aperture Translation: Experimental Results and Implications for Spatial Compounding. IEEE Transactions on UFFC, 33:3: 257-264

16. Wagner RF, Insana MF, Smith SW (1988) Fundamental Correlation Lenghts of Coherent Speckle in Medical Ultrasonic Images. IEEE Transactions on UFFC, 35:1: 34-44

17. Berson M, Roncin A, Pourcelot L (1981) Compound Scanning with an Electrically Steered Beam. Ultrasonic Imaging, 3: 303-308

18. Carpenter DA, Dadd MJ, Kossoff G (1980) A Multimode Real Time Scanner. Ultrasound in Med. & Bio., 6: 279-284

19. Anderson ME, Soo MS, Bentley RC, Trahey GE (1997) The detection of breast microcalcifications with medical ultrasound. J. Acoust. Soc. Am., 101:1: 29-39

20. Shankar PM, Newhouse VL: Speckle Reduction with Improved Resolution in Ultrasound Images. IEEE Transactions on SU, 32:4:537-543

21. Hernandez A, Basset O, Chirossel P, Gimenez G (1996) Spatial Compounding in Ultrasonic Imaging Using an Articulated Scan Arm. Ultrasound in Med. & Biol., 22: 229-238

22. Daniel F. Leotta, Roy W. Martin (2000) Three-dimensional ultrasound imaging of the rotator cuff: spatial compounding and tendon thickness measurement. Ultras. in Med. & Biol.:26:4: 509-525

23. Entrekin RR, Jago JR, Kofoed SC: Real-time spatial compound imaging: Technical performance in vascular applications (2000). Acoustical Imaging: 25:331-342

24. Jespersen SK, Wilhjelm JE, Sillesen H (2000) In vitro Spatial Compound Scanning for Improved Visualization of Atherosclerosis. Ultrasound in Med. & Biol: 26:8:1357-1362

25. Picano E, Landini L, Distante A, Salvadori M, Lattanzi F, Masini M, L'Abbate A (1985) Angle dependence of ultrasonic backscatter in arterial tissues: a study in vitro. Circulation, 72:3: 572-576

26. Kofoed SC, Grønholdt MLM, Wilhjelm JE, Bismuth J, Sillesen H: Real-time spatial compound imaging improves reproducibility in the evaluation of atherosclerotic carotid plaques. Ultrasound in Medicine and Biology, "in press".

27. Elatrozy T, Nicolaides A, Tegos T, Griffin M (1998) The objective characterisation of ultrasonic carotid plaque features. Eur J Vasc Endovasc Surg:16:223-230

28. Polak JF, Shemanski L, O'Leary DH, Lefkowitz D, Price TR, Savage PJ, Brant WE, Reid C (1998) Hypoechoic plaque at US of the carotid artery: an independent risk factor

for incident stroke in adults aged 65 years or older. Cardiovascular Health Study. Radiology:208:649-654

29. El-Barghouty NM, Levine T, Ladva S, Flanagan A, Nicolaides A (1996) Histological verification of computerised carotid plaque characterisation. Eur. J. Vasc. Endovasc. Surg., 11: 414-416

30. ECPSG: Carotid Artery Plaque Composisiton - Relationship to Clinical Presentation and Ultrasound B-mode Imaging (1995). Eur J Vasc Endovasc Surg: 10, 23-30.

31. Hatsukami TS, Thackray BD, Primozich JF, Ferguson MS, Burns DH, Beach KW, Detmer PR, Alpers C, Gordon D, Strandness Jr DE (1994) Echolucent regions in carotid plaque: Preliminary analysis comparing three-dimensional histologic reconstructions to sonographic findings. Ultrasound in Med. & Biol:20:8:743-749

32. Urbani MP, Picano E, Parenti G, Mazzarisi A, Fiori L, Paterni M, Pelosi G, Landini L (1993) In Vivo Radiofrequency-Based Ultrasonic Tissue Characterization of the Atherosclerotic Plaque. Stroke:24:10:1507-1512

33. Bluth EI, Kay D, Merritt: CRB, Sullivan M, Farr G, Mills, NL, Foreman M, Sloan K, Schlater M, Stewaert J (1986) Sonograhic Characterization of carotid plaque: Detection of hemorrhage. ARJ 146:1061-1065

34. Gray-Weale AC, Graham JC, Burnett JR, Byrne K, Lusby RJ (1988) Carotid artery atheroma: Comparison of preoperative B-mode ultrasound appearance with carotid endarterectomy specimen pathology. J Cardiovasc Surg: 29: 676-681

35. O'Donnell Jr TF, Erdoes L, Mackey WC, McCullough J, Shepard A, Heggerick P, Isner J, Callow AD (1985) Correlation of B-Mode Ultrasound Imaging and Arteriography with Pathologic Findings at Carotid Endarterectomy. Arch Surg:120:443-449.

36. Weinberger J, Marks SJ, Gaul JJ, Goldman B, Schanzer H, Jacobsen J, Dikman S (1987) Atherosclerotic Plaque at the Carotid Artery Bifurcation. J Ultrasound Med:6: 363-366

37. Barzilai B, Saffitz JE, Miller JG, Sobel BE (1987) Quantitative Ultrasonic Characterization of the Nature of Atherosclerotic Plaques in Human Aorta . Circulation Research, 60:3:459-463

38. Goes E, Janssens W, Maillet B, Freson M, Steyaert L, Osteaux M (1990) Tissue Characterization of Atheromatous Plaques: Correlation Between Ultrasound Image and Histological Findings. J Clin. Ultrasound: 18:8:611-617

39. Picano E, Landini L, Distante A, Benassi A, Sarnelli R, L'Abbate A (1985) Fibrosis, Lipids, and Calcium in Human Atherosclerotic Plaque. Circulation Research:56:4:556-562

40. Wilhjelm JE, Pedersen PC Jacobsen SM (2001) The Influence of Roughness, Angle, Range and Transducer Type on the Echo Signal from Planar Interfaces. IEEE Trans. on Ultrasonics, Ferroelectrics, and Freq. Cont., 48:2: 511 - 521

41. Beach KW, Primozich JF, Strandness Jr DE (1994) Pseudocolor B-mode Arterial Images to Quantify Echogenicity of Atherosclerotic Plaque. Ultrasound in Med. & Biol.20: 8: 731-742

42. Labs KH, Jäger KA, Fitzgerald DE, Woodcock JP, Neuerburg-Heusler: Diagnostic vascular ultrasound, Edward Arnold, London, 1992

43. Pedersen PC (1997) Ultrasound technique for arterial plaque characterization based on absolute backscatter measurements. Ultrasonic Imaging, 19:1: 46-47

44. Pedersen PC and Cakareski Z (1998) Determination of ultrasound backscatter level of vascular structures, with application to arterial plaque characterization. 24th International Symposium on Acoustical Imaging, Santa Barbara, CA, Sept. 23 - 25, pp. 273 - 280

45. de Kroon, M.G.M. van der Wal, L.F. Gussenhoven, W.J. Bom, N.: Angle-dependent backscatter from the arterial wall. Ultrasound in Med and Biol. 1991, 17:2 , pp. 121-126.

3D Imaging of the Carotid Arteries

Aaron Fenster, PhD, Dónal B. Downey MD

The John P. Robarts Research Institute
100 Perth Drive, London, ON, N6G 4N9 Canada
and the Department of Diagnostic Radiology & Nuclear Medicine
The University of Western Ontario, London, Canada

Summary: Although ultrasonography is an important cost-effective imaging modality, technical improvements are needed before its full potential is realized for accurate and reproducible monitoring of carotid disease and plaque burden. 2D viewing of 3D anatomy, using conventional ultrasonography limits our ability to quantify and visualize carotid disease and is partly responsible for the reported variability in diagnosis and monitoring of disease progression. Efforts of investigators have focused on overcoming these deficiencies by developing 3D ultrasound imaging techniques that are capable of acquiring B-mode, color Doppler and power Doppler images of the carotid arteries using existing conventional ultrasound systems, reconstructing the information into 3D images, and then allowing interactive viewing of the 3D images on inexpensive desktop computers. In addition, the availability of 3D ultrasound images of the carotid arteries has allowed the development of techniques to quantify plaque volume and surface morphology as well as allowing registration with other 3D imaging modalities. This paper describes 3D ultrasound imaging techniques used to image the carotid arteries and summarizes some of the developments aimed at quantifying plaque volume and morphology.

Key words: Carotid arteries, 3D ultrasonography, carotid disease, 3D imaging

Introduction

Determining carotid stenosis severity is important in identifying patients who would benefit from carotid endarterectomy. However, identification of vulnerable plaques in the carotid arteries that are likely to rupture leading to throbogenic events is our next challenge. Research related to the latter issue is now active and involves multiple disciplines including imaging. Imaging research is being transformed from imaging of the carotids for stenosis severity measurements to the development of techniques to visualize the vessel wall, plaque surface features and plaque composition. With improved strategies to treat atherosclerosis non-surgically, sensitive non-invasive 3D imaging techniques allowing direct plaque visualization are becoming more important in serial monitoring of disease progression and regression.

The utility of ultrasonography (US) in diagnosis and assessment of carotid disease is well established. Because it is non-invasive and due to the continuous improvements in image quality, US is progressively achieving a greater role because it can provide direct views of the vessel wall and plaque. Although carotid stenosis severity assessment has been an important focus, viewing the vessel wall and plaque is allowing investigators to focus on quantitative assessment of plaque morphology, volume, ulcerations, and composition.

Use of conventional 2D US has resulted in mixed results in attempts to quantify plaque morphology and composition and correlate these with risk of stroke. Some investigators reported a high accuracy for identifying features such as intraplaque hemorrhage,[1,2] while others found only moderate sensitivity and specificity.[3] Some of these discrepancies related to lack of standardization and variability of the conventional US exam. However, it is generally agreed that improved US techniques are required to allow accurate quantification and characterization of plaques. Improvements that hold promise are compound imaging[4] providing improved visualization of the plaque, coupled with 3D US techniques[5-8] overcoming limitations of 2D US discussed below.

Limitations of conventional US imaging

In a conventional US examination, the transducer is manipulated to obtain a series of 2D images, which are mentally combined by the operator to form a subjective impression of the 3D anatomy and pathology. The success of the procedure is thus largely dependent on the skill and experience of the operator in performing these tasks. This is a sub-optimal approach, due to the following limitations:

(a) Conventional US images are 2D; hence, the diagnostician must mentally transform multiple images to develop a 3D impression of the complex three-dimensional structure of the vessel and plaque. This may lead to variable and incorrect quantification of plaque morphology, size and extent.

(b) It is difficult to localize the thin 2D ultrasound image plane in the organ and reproduce it at a later time, making conventional US non-optimal for quantitative prospective or follow-up studies, particularly where small changes are being followed over the course of time.

(c) The patient's anatomy sometimes restricts the image angle, resulting in inaccessibility of the optimal image plane necessary for diagnosis of carotid disease and assessment of plaque morphology.

3D ultrasound imaging

It has been hypothesized that 3D US with visualization of the plaque surface, volume measurement and plaque characterization would provide the capability to identify vulnerable plaques.[9-11] In addition, 3D imaging of the plaque may also allow quantitative monitoring of plaque progression and regression (changes in volume and morphology), provide important information about the plaque's response to therapy (e.g. lipid lowering, diet) and natural history of plaques.[12,13]

In the following sections, we discuss and provide examples of 3D ultrasound imaging tools, which allow measurement and visualization of plaque volume, surface features, and composition, and provide detailed information on changes in carotid plaques. We also discuss how this information can be correlated to 3D MR images of carotid plaques.

3D ultrasound scanning techniques

Most 3D US system approaches make use of conventional 1D US transducers producing 2D US images and different methods to locate the position of 2D images within the volume under investigation. Over the past 2 decades, two different 3D ultrasound imaging approaches have been utilized in imaging the carotid arteries: mechanical linear scanners, sensed free-hand techniques. These approaches are summarized in the subsequent sections.

Mechanical linear 3D US scanning

In this approach, a motorized mechanism translates the transducer linearly across the patient's skin, while the 2D images are acquired at regular spatial intervals.

Thus, the acquired images are uniformly spaced and parallel (see Fig. 1). The ability to vary the translation speed and sampling interval is necessary to match the sampling rate to the frame rate of the ultrasound machine, and to match the sampling interval to (half) the elevational resolution of the transducer.[14,15]

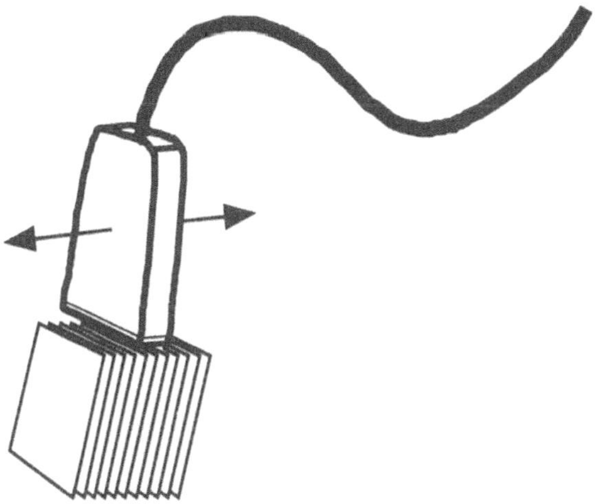

Fig. 1. Schematic diagram of the mechanical linear scanning approach, in which a series of parallel 2D images are collected and used to reconstruct the 3D image.

The simple predefined geometry of the acquired 2D image planes allows for a fast and accurate 3D image reconstruction.[16] Thus, with this approach, a 3D image can be obtained immediately after performing the scan, and its resolution can be optimized. Because the 3D image is produced from the series of conventional 2D images, its resolution will not isotropic. In the direction parallel to the acquired 2D image planes, it will equal that of the original 2D images, but in the direction of the 3D scan, it will be equal to the elevational resolution of the transducer. Since the 3D image resolution will generally be the worst in the scanning direction, a transducer with good elevational resolution should be used.

Typically, 200 images (336 x 352 pixels each) are acquired at 0.5 mm intervals in a time that depends on the ultrasound machine frame rate and whether cardiac gating is used. Scanning parameters are typically adjusted depending on the type of acquisition. For example: 1) 3D B-mode - typically 2 or 3 focal zones are used resulting in about 15 frames/sec and a 3D scan of 13 sec for 200 images; 2) 3D power Doppler – increased persistence and no cardiac gating, resulting in an acquisition rate of 7.5 images/sec and a 3D scan of 20 sec for 150 images.

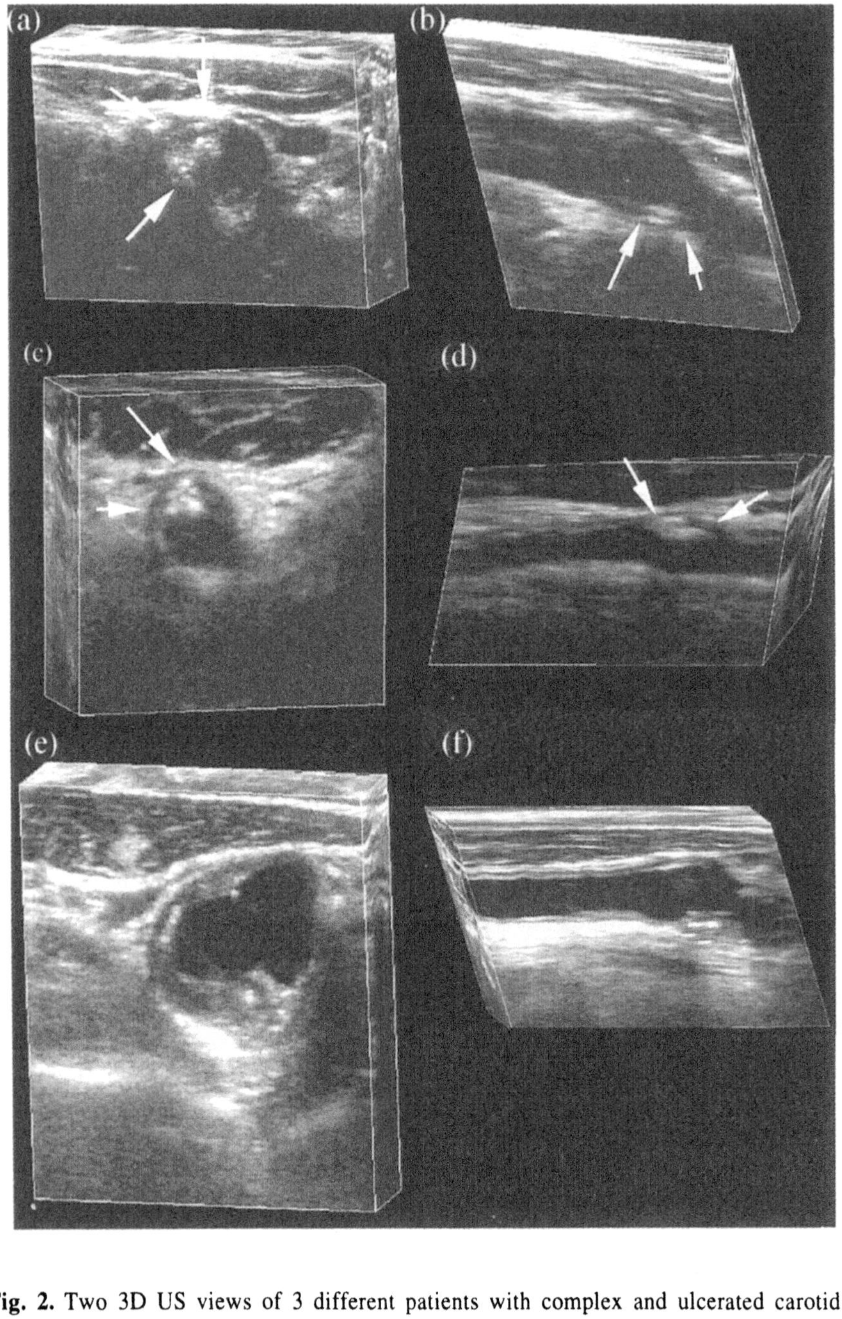

Fig. 2. Two 3D US views of 3 different patients with complex and ulcerated carotid plaques. For each patient, a transverse and longitudinal view is shown side-by-side.

The linear scanning approach has been successfully implemented in many vascular imaging applications, using B-mode images, color Doppler images of the carotid arteries,[17-21] vascular test phantoms,[22-24] and power Doppler images.[17,22] An example of a mechanically linearly scanned 3D ultrasound image of a patient's carotid arteries with a complex plaque made is shown in Fig. 2.

Free-hand 3D scanning with magnetic field sensors

Although the mechanical scanning approach offers short imaging times, high quality 3D images, and fast reconstruction times, the devices are bulky and inconvenient to use at times. To overcome this problem, free-hand scanning techniques that do not require a motorized fixture have been developed by many investigators.[25-27] In these approaches, a sensor is attached to the transducer to measure its position and orientation. Thus, the transducer is manipulated in the usual manner over the anatomy to be imaged, while 2D images are acquired by a computer together with their positions and orientations. This information is then used to reconstruct the 3D image. Since the relative locations of the acquired 2D images are not predefined, the operator must ensure that the set of 2D images has no significant gaps.

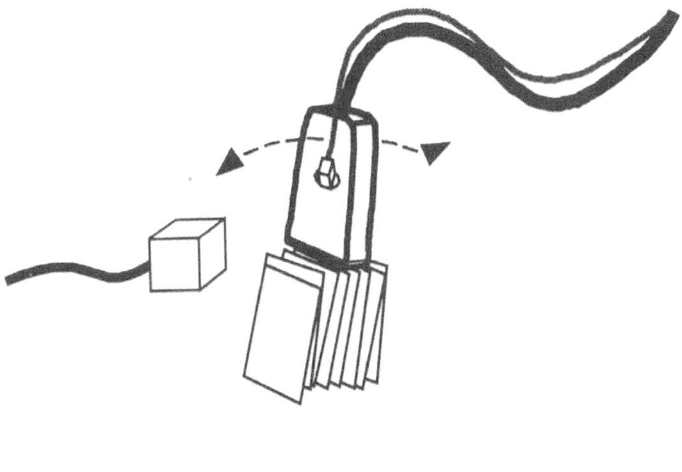

Fig. 3. Schematic diagram showing free-hand scanning approach using magnetic field sensing. Since the arrangement of the acquired 2D images has no regular geometry, the position and orientation of each plane must be known accurately

Although several free-hand scanning approaches have been developed, (e.g., articulated arms, acoustic sensing, magnetic field sensing, and image-based sensing) the most popular makes use of a 6 degree-of-freedom magnetic field sensor. In this approach, a transmitter is used to produce a spatially varying magnetic field, and a small receiver, containing three orthogonal coils, senses the magnetic field strength, as illustrated in Fig. 3. Typically, the transmitter is placed beside the patient and the receiver is mounted on the hand-held transducer.

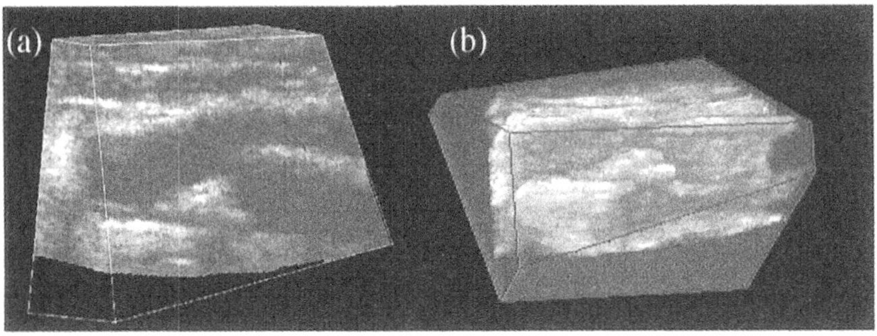

Fig. 4. 3D B-mode image of the carotid arteries obtained with the free-hand scanning approach using a magnetic position and orientation measurement. (a) The 3D image has been "sliced" to reveal a plaque at the entrance of the internal carotid artery. The plaque is heterogenous, with calcified regions casting a shadow. (b) The same image has been volume rendered to obtain a view of the vessel wall and the surface of the plaque.

Magnetic field sensors are small and unobtrusive, allowing the transducer to be tracked with less constraint than the mechanical approach. However, electromagnetic interference from sources such as CRT monitors, AC power cabling and electrical signals from the transducer itself can compromise the tracking accuracy. Geometric distortions in the final 3D image can occur if ferrous or highly conductive metals are located nearby. In addition, errors in the determination of the location of the moving transducer will occur if the magnetic field sampling rate is low and the transducer moved too quickly. By ensuring that the environment of the scanned volume is free of metals and electrical interference, and using an adequate sampling rate, high quality 3D images of the carotid arteries can be obtained, as illustrated by Fig. 4. These devices have successfully been used in many diagnostic applications, including echocardiography, obstetrics and vascular imaging.[18,25-29]

RECONSTRUCTION AND VIEWING

3D image reconstruction

Reconstruction of the 3D US image refers to the generation of a 3D representation of the examined structures from the acquired set of 2D images. A mathematical approach is used, in which the acquired series of 2D images are used to build a 3-dimensional voxel-based volume by placing each acquired 2D image in its correct location in the volume. The grey scale values of any voxels not sampled by the acquired 2D images are calculated by interpolation between the appropriate images. In this way, all 2D image information is preserved, allowing viewing of the original 2D planes, as well as other views. With modern computers, this process is very efficient allowing the user to view the 3D image within 1 or 2 seconds after the acquisition, or even during the acquisition process.

Viewing of the 3D US images

Cube view

After the 3D image has been reconstructed, it is ready for viewing. Our approach for viewing the 3D image is based on multi-planar rendering with a texture mapping. In this technique, the 3D image is presented as a polyhedron and the appropriate ultrasound image for that plane is "painted" on each face. The polyhedron can be rotated to obtain the desired orientation of the image and then any of the faces can be moved in or out (i.e., sliced), parallel to the original, or reoriented obliquely, while the appropriate ultrasound data is texture-mapped in real-time on the new face. In this way, the operator always has 3-dimensional image-based cues relating the plane being manipulated to the rest of the anatomy.[18,20,30,31] Examples of this approach are shown in Figs. 2 and 4.

Orthogonal planes

In this approach, computer-user interface tools provide three perpendicular planes, which are displayed on the screen simultaneously, with graphical cues indicating their relative orientations. The operator can then select any single or multiple planes, and move them within the 3D image volume, to provide a cross-sectional view at any desired location or orientation, including oblique views.[32,33]

Volume Rendering (VR)

The volume rendering technique presents a display of the entire 3D image after it has been projected onto a 2D plane, with the image projection typically being accomplished *via* ray-casting techniques.[34] The 3D image voxels that intersect each ray are weighted and summed to achieve the desired result in the rendered image. Although many VR algorithms have been developed, 3D ultrasound imaging currently makes use of primarily two approaches: maximum (minimum) intensity projection, and translucency rendering. Although this approach has been primarily used to display fetal anatomy, it can be also be used to display and vascular anatomy[17] as shown if Fig. 4(b).

Segmentation of carotid plaques from 3D US images

Accurate monitoring of carotid plaque progression or regression requires accurate and reproducible techniques to measure and analyze carotid plaque morphology. Thus, reliable segmentation of the vessel lumen, plaque surface and carotid wall borders is required. Segmentation by the manual planimetry approach requires that an operator must "cut" into a 3D US image, orient the revealed plane appropriately and then outline the plaque surface manually.[35,36] To obtain the morphology of the complete plaque, the operator must further “cut” into the volume and repeat the process until the complete plaque is traversed. Using the outlined surface, a mesh can be fitted to obtain a surface description and the volume of the plaque. Manual outlining of the plaque requires substantial experience is arduous and time-consuming, leading to inaccuracy and variability.

We have developed a semi-automated segmentation algorithm based on deformable models requiring minimal user interaction and processing time to obtain the surface morphology of plaques.[37-39] The deformable model is a discrete vector model, in which the boundary is defined as a series of vertices connected by vectors. This structure allows internal and external forces to be calculated at each vertex using local or image information, and moved iteratively until the forces acting on the vetices are zero.

The algorithm has of three major steps: (1) placement of the initial deformable balloon model (3D mesh) inside the lumen; (2) inflation of the model towards the plaque surface; and (3) automatic localization of the plaque surface. After the initial model is placed in the lumen near the plaque, it is rapidly inflated towards the plaque's surface. When equilibrium under the influence of inflation and general image forces is reached, the model approximately represents the shape the surface. Because the plaque's surface may not be smooth and uniform, we further deform the mesh by means of local image-based forces to localize the surface of the plaque. Equations describing the dynamics of the model and the forces acting on it are given below. The equation of motion for vertex i of the mesh is:

$$m_i x''_i (t) + v_i x'_i (t) + g(x_i (t)) = f(x_i (t)) \qquad (1)$$

where x_i (t) is the position of the vertex, x'_i (t) and x''_i (t) are its velocity and acceleration, respectively, m_i is its mass, v_i is the damping coefficient, $g(x_i (t))$ is the resultant surface tension at the vertex, and $f(x_i(t))$ is a "driving" force. Equilibrium is reached when both $x'_i(t)$ and $x''_i(t)$ become zero. To simplify and speed the calculations, the mass of each vertex is set to zero (no inertia) and the damping coefficient to one.

To compute the equilibrium position of each vertex, and hence the final shape of the model, the time in Eq. 1 is iteratively updated by Δt. After the driving force results in to an approximate segmentation of the plaque surface, the inflation is stopped and the driving force is replaced by an image-based force that causes the mesh to fit the fine details of the plaque. This is accomplished with a 3D potential function (P) constructed from the image data, which attracts the model to 3D gradients at the plaque's surface:

$$P(x_i(t)) = 1/[||\nabla G_\sigma * I|| + \varepsilon] \qquad (2)$$

where G_σ is a 3D Gaussian smoothing of width σ, and ε is a small constant to prevent division by zero. Minima of the potential function coincide with the plaque surface. The potential function produces a force field, which is used to deform the model thus: $f(x_i(t)) = -k\nabla P(x_i(t))$. The adjustable term k controls the strength of the force and must be optimized to allow the mesh to conform to the plaque's surface. In addition, the surface tension of the mesh must be also chosen appropriately. If the model is too stiff, then it will not deform into small ulcerations, but if it is too compliant, then it may extend into echo-lucent regions.

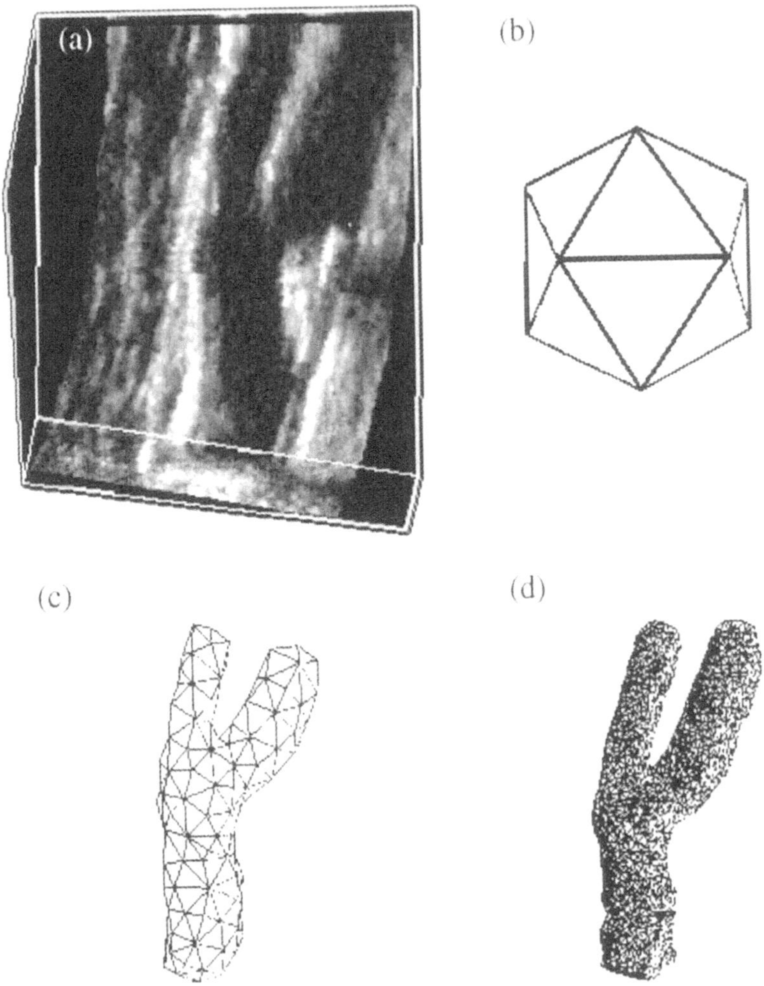

Fig. 5. Steps showing the segmentation process. (a) 3D US image of the carotid arteries showing a plaque in the common carotid artery. (b) Initial mesh placed within the lumen. (c) Appearance of the mesh after 50 iterations during inflation phase. (d) The final mesh after refinement phase.

Figure 5 shows segmentation results using a 3D image of a patient's carotid arteries with a plaque. Analysis of the segmentation showed that the mean separation between manual and algorithm segmentation was 0.35mm with a standard deviation of 0.25mm.[39]

3D US image registration with MRI

To allow sensitive monitoring of the progression/regression of plaque morphology requires registration of 3D B-mode images obtained at different times. In addition, it is also advantageous to correlate changes observed with MRI and US, requiring registration of B-mode 3D US to MRI images.

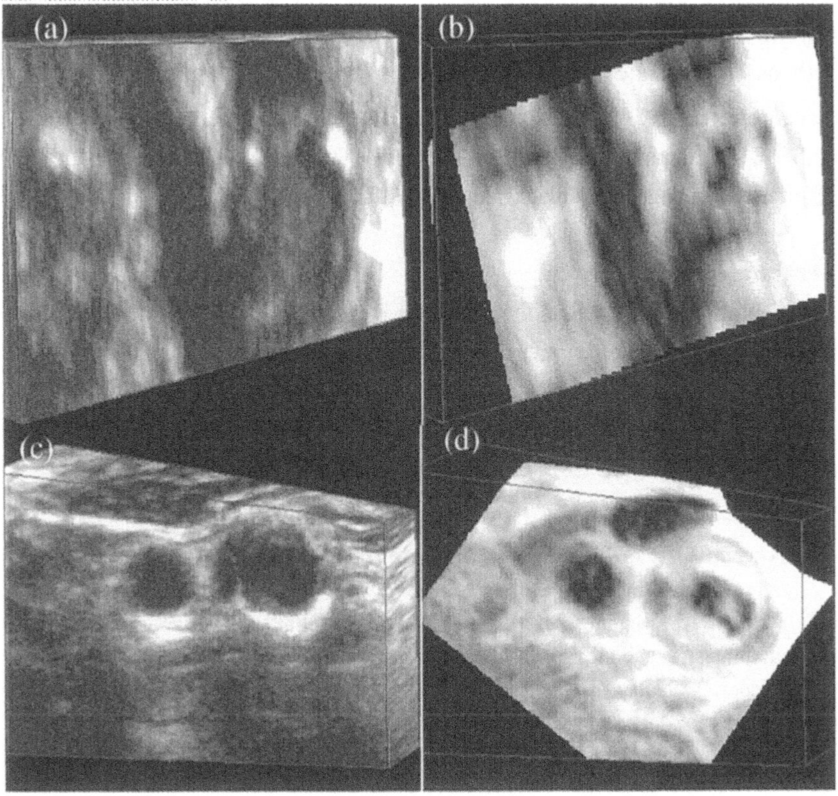

Fig. 6. Registration results of 3D B-mode US (left side) and 3D black-blood MRI images (right side) of a patient with a plaque in the internal carotid artery. The 3D images have been sliced to reveal the appearance of the plaque in the two modalities. The images on the top are sliced in the plain of the bifurcation, and the bottom transverse to the vessels.

Techniques for multi-modality 3D image registration have been well explored for use with MRI, CT, PET and SPECT images of the brain[40] and thorax.[41] However, multi-modality image registration of the carotid arteries and the use of ultrasound has been investigated to a lesser extent. Often, external markers cannot be used due to non-rigid organs or internal displacements, as is the case for the carotid arteries. Thus, a preferred automatic registration method of 3D US images of the carotid vessels should be based on image features.

We have developed a registration technique, which uses intrinsic image features to determine the spatial transformation between two 3D volumes. Using this approach, we have demonstrated the ability of the mutual information technique to register 3D power Doppler to MRA images. In our study, we quantitatively evaluated the sensitivity of this approach to the initial misalignments, and the effect of missing image data (shadowing in US, and signal loss in MRA). We demonstrated that 97% of images misaligned within a range of 40° and 40 pixels could be correctly registered using mutual information, with a mean error of 0.7 pixels along the axis of the vessel, and 1.6, 1.5 pixels in the other 2 directions to the vessel. In Fig. 6, we show the results of registration of 3D B-mode US images with black-blood 3D MRI images.

Acknowlegement
The authors acknowledge the financial support of the Canadian Institutes for Health Research and the Ontario R & D Challenge Fund.

References

1. Zwiebel WJ (1992) Duplex sonography of the cerebral arteries: efficacy, limitations, and indications. AJR Am J Roentgenol 158:29-36
2. Sterpetti AV, Schultz RD, Feldhaus RJ, et al (1988) Ultrasonographic features of carotid plaque and the risk of subsequent neurologic deficits. Surgery 104:652-660
3. Widder B, Paulat K, Hackspacher J, et al (1990) Morphological characterization of carotid artery stenoses by ultrasound duplex scanning. Ultras Med Biol 16:349-354
4. Jespersen SK, Wilhjelm JE, Sillesen H (1998) Multi-angle compound imaging. Ultrason Imaging 20:81-102
5. Nelson TR, Downey DB, Pretorius DH, Fenster A (1999) Three-Dimensional Ultrasound. Lippincott Williams & Wilkins, Philadelphia PA
6. Nelson TR, Pretorius DH (1998) Three-dimensional ultrasound imaging. Ultrasound Med Biol 24:1243-1270
7. Baba K, Jurkovic D (1997) Three-Dimensional ultrasound in Obstetrics and

Gynecology. The Parthenon Publishing Group, New York
8. De Castro S, Yao J, Pandian NG (1998) Three-dimensional echocardiography: clinical relevance and application. Am J Cardiol 81:96G-102G
9. Hatsukami TS, Thackray BD, Primozich JF, et al (1994) Echolucent regions in carotid plaque: preliminary analysis comparing three-dimensional histologic reconstructions to sonographic findings. Ultrasound Med Biol 20:743-749
10. Steinke W, Hennerici M (1989) Three-dimensional ultrasound imaging of carotid artery plaques. Journal of Cardiovascular Technology 8:15-22
11. Schminke U, Motsch L, Griewing B, Gaull M, Kessler C (2000) Three-dimensional power-mode ultrasound for quantification of the progression of carotid artery atherosclerosis. J Neurol 247:106-111
12. Yao J, van Sambeek MR, Dall'Agata A, et al (1998) Three-dimensional ultrasound study of carotid arteries before and after endarterectomy; analysis of stenotic lesions and surgical impact on the vessel. Stroke 29:2026-2031
13. Sameshima T, Futami S, Morita Y, et al (1999) Clinical usefulness of and problems with three-dimensional CT angiography for the evaluation of arteriosclerotic stenosis of the carotid artery: comparison with conventional angiography, MRA, and ultrasound sonography. Surg Neurol 51:301-308; discussion 308-309
14. Smith WL, Fenster A (2000) Optimum Scan Spacing for Three-Dimensional Ultrasound by Speckle Statistics. Ultras in Med and Biol 26:551-562
15. Smith WL, Fenster A (2000) Optimization of 3D ultrasound scan spacing using speckle statistics. SPIE Ultrasonic Imaging and Signal Processing 3982:56-67
16. Cardinal HN, Gill J, Fenster A (2000) Analysis of geometrical distortion and statistical variance in length, area, and volume in a linearly scanned 3D ultrasound image. IEEE Transactions on Medical Imaging 19:632-651
17. Downey DB, Fenster A (1995) Vascular imaging with a three-dimensional power Doppler system. AJR Am J Roentgenol 165:665-668
18. Fenster A, Tong S, Sherebrin S, Downey DB, Rankin RN (1995) Three-dimensional ultrasound imaging. SPIE Physics of Medical Imaging 2432: 176-184
19. Picot PA, Rickey DW, Mitchell R, Rankin RN, Fenster A (1991) Three-dimensional colour Doppler imaging of the carotid artery. SPIE Proceedings: Image Capture, Formatting and Display 1444:206-213
20. Picot PA, Rickey DW, Mitchell R, Rankin RN, Fenster A (1993) Three-dimensional colour Doppler imaging. Ultrasound Med Biol 19:95-104
21. Pretorius DH, Nelson TR, Jaffe JS (1992) 3-Dimensional sonographic analysis based on color flow Doppler and gray scale image data: a preliminary report. J Ultrasound Med 11:225-232
22. Guo Z, Fenster A (1996) Three-dimensional power Doppler imaging: a phantom study to quantify vessel stenosis. Ultrasound Med Biol 22:1059-1069
23. Guo Z, Moreau M, Rickey DW, Picot PA, Fenster A (1995) Quantitative investigation of in vitro flow using three-dimensional colour Doppler ultrasound. Ultrasound Med Biol 21: 807-816
24. Dabrowski W, Dumore-Buyze J, Cardinal HN, Fenster A (2000) A real vessel phantom for flow imaging: 3D Doppler ultrasound of steady flow. Ultras in Med and Biol

27:135-141
25. Hughes SW, D'Arcy TJ, Maxwell DJ, et al (1996) Volume estimation from multiplanar 2D ultrasound images using a remote electromagnetic position and orientation sensor. Ultras Med Biol 22:561-572
26. Leotta DF, Detmer PR, Martin RW (1997) Performance of a miniature magnetic position sensor for 3-D ultrasound imaging. Ultras Med Biol 23:597-609
27. Gilja OH, Hausken T, Olafsson S, et al (1998) In vitro evaluation of three-dimensional ultrasonography based on magnetic scanhead tracking. Ultras Med Biol 24:1161-1167
28. Detmer PR, Bashein G, Hodges T, et al (1994) 3D ultrasonic image feature localization based on magnetic scanhead tracking: in vitro calibration and validation. Ultrasound Med Biol 20:923-936
29. Hodges TC, Detmer PR, Burns DH, et al (1994) Ultrasonic 3-D reconstruction: in vitro and in vivo volume and area measurement. Ultrasound Med Biol 20:719-729
30. Rankin RN, Fenster A, Downey DB, Munk PL, Levin MF, Vellet AD (1993) Three-dimensional sonographic reconstruction: techniques and diagnostic applications. AJR Am J Roentgenol 161:695-702
31. Tong S, Downey DB, Cardinal HN, Fenster A (1996) A three-dimensional ultrasound prostate imaging system. Ultrasound Med Biol 22:735-746
32. Zosmer N, Jurkovic D, Jauniaux E, Gruboeck K, Lees C, Campbell S (1996) Selection and identification of standard cardiac views from three-dimensional volume scans of the fetal thorax. J Ultrasound Med 15:25-32
33. Kirbach D, Whittingham TA (1994) 3D ultrasound - the kretztechnik voluson approach. European Journal of Ultrasound 1:85-89
34. Levoy M (1990) Volume rendering, a hybrid ray tracer for rendering polygon and volume data. IEEE Computer Graphics and Applications 10:33-40
35. Delcker A, Diener HC (1994) Quantification of atherosclerotic plaques in carotid arteries by three-dimensional ultrasound. Br J Radiol 67:672-678
36. Allott CP, Barry CD, Pickford R, et al (1999) Volumetric assessment of carotid artery bifurcation using freehand-acquired, compound 3D ultrasound. Br J Radiol 72:289-92
37. Mao F, Gill JD, Fenster A (1999) Technique of evaluation of semi-automatic segmentation methods. SPIE- Image Processing 3661:1027-1036
38. Mao F, Gill J, Downey DB, Fenster A (2000) Segmentation of carotid artery in ultrasound images: method development and evaluation technique. Med Phys 27:1961-1970
39. Gill J, Ladak H, Steinman D, Fenster A (2000) Accuracy and variability assessment of semi-automatic technique for segmentation of the carotid arteries from 3D ultrasound images. Medical Physics 27:1333-1342
40. Maes F, Collignon A, Vandermeulen D, et al (1997) Multimodality image registration by maximization of mutual information. IEEE Trans Med Imaging 16:187-198
41. Dey D, Slomka PJ, Hahn LJ, Kloiber R (1999) Automatic three-dimensional multimodality registration using radionuclide transmission CT attenuation maps: a phantom study. J Nucl Med 40:448-455

Carotid elasticity measurements

Arnold P.G. Hoeks PhD, Jan M. Meinders PhD, *Ruben Dammers MD

Departments of Biophysics, Surgery*, Cardiovascular Research Institute Maastricht, Maastricht University, Maastricht, the Netherlands

Summary: The morphology of blood vessels, as expressed by (lumen) diameter and wall thickness, depends on the forces exerted on it. To what extent the balance and the interaction between wall shear stress and the transmural blood pressure and their absolute values are contributing to the adaptation of arteries is still a matter of debate, mainly because currently no adequate non-invasive measurement methods are available to answer these questions comprehensibly. However, recent ultrasound developments allow transcutaneous assessment of the morphological and functional characteristics of arteries. The latter depend on the elasticity as can be derived from the relationship between the time-dependent cyclic stress exerted on the vessel wall and the resulting pulsatile strain, i.e. the relative increase in circumference. This chapter will discuss the various techniques to quantify the elastic behavior of the carotid artery in combination with its morphology, the applicability and limitations.

Key Words: artery, elasticity, compliance, pulse pressure, wave velocity

Introduction

Aging of arteries manifests itself in many cross-related ways. The larger production compared to the removal of smooth muscle cells and the extracellular matrix results in a gradual increase in wall thickness, which might be aggravated by elevated transmural pressures. As a result the artery wall becomes stiffer, i.e. the relative increase in diameter for a given pulse pressure decreases. The wall thickness grows without compromising the lumen diameter, i.e., the wall grows outward [1] until at certain locations it exceeds the critical limits [2]. Because of the less elastic characteristics of the artery the afterload off the heart will increase, e.g. for a given stroke volume a higher systolic pressure is required. Since the peripheral resistance is not affected, the mean arterial pressure will not be affected and, henceforth, the diastolic pressure may decrease, resulting in an elevated pulse pressure.

Both the higher afterload and pulse pressure constitute important risk factors for cardiac failure.

The functional reasons for the gradual increase in wall thickness are still partly obscure. It has been shown in vivo that the wall thickness is larger in regions where the wall is exposed to a low wall shear stress, i.e. the drag exerted by the flowing blood on the endothelium [3-10], an observation that corroborates results obtained in cell cultures exposed to steady and pulsatile shear stresses. A possible explanation for the reduced proliferation of smooth muscle cells at higher than normal shear stresses might be the preparation for a gradual adaptation of the vessel lumen to eventually reduce the shear stress, but conclusive proof has still to be given. The reaction of the wall to increased transmural blood pressures is more transparent. Assuming that the cells in the wall can bear stresses only within narrow limits, higher (pulsatile) stresses should induce cell proliferation. But then the question remains whether high blood pressure precedes or is a reaction to increased wall thickness. In a longitudinal study Van Gorp et al. have shown that hypertensive rats [11] exhibit changes in elastic properties of the aorta wall before hypertension develops.

To reveal the interaction between genetic, hemodynamic, and metabolic factors and the mechanical characteristics of the wall, large scale cross-sectional and longitudinal studies are required. Because these studies should include the healthy population, either with or without risk factors for the development of atherosclerosis and cardiac failure as well as patients, there is a definite need for non-invasive methods to assess various parameters. The current ultrasound methods can provide direct information about wall thickness, lumen diameter change over the cardiac cycle and blood velocity patterns including the time dependent shear stress. Tonometers are not only suitable for the non-invasive assessment of blood pressure but also for the assessment of the pulse wave velocity. This chapter will focus on the interaction between the characteristics of blood pressure and that of the artery wall. Although most of it can also be extended to coronary arteries we will put main emphasis on the major peripheral arteries because of their accessibility for non-invasive methods.

Basic expressions for a segment of an elastic artery

The derivation for the pulse velocity is based on Newton's law, applied to a thin cross-sectional slice of an artery with an area A, an incremental thickness δx and a specific mass ρ, and the law of conservation of mass (Fig.1). The blood velocity is v_x with a velocity gradient δv_x while the pressure is p_x with a pressure gradient δp_x, both along the axis of the artery [12].

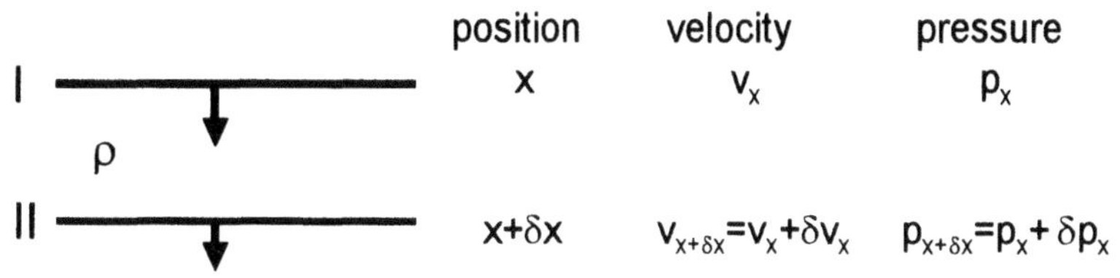

Fig. 1: Definition of basic parameters for the derivation of the pressure wave velocity.

In the incremental time interval δt mass $\rho A v_x \delta t$ and $\rho A(v_x+\delta v_x)\delta t$ enters and leaves the slice, respectively, resulting in an increase of the mass of the slice of $\delta m=\rho A\,\delta v_x \delta t$. Assuming that blood is incompressible and the contribution of a possible gradient in diameter is negligible and using the relationship $\rho A=m/\delta x$ (specific mass is mass divided by volume) results in:

$$\frac{\delta m}{m\delta t}=-\frac{\delta v_x}{\delta x} \tag{1}$$

The force exerted by the pressure on layer I is $p_x A$ while on layer II it is $(p_x+\delta p_x)A$. Assuming that the contribution of a possible gradient of the velocity is negligible the pressure gradient $-\delta p_x A$ causes an acceleration $\delta v_x/\delta t$ of the mass $\rho A\,\delta x$:

$$\frac{\delta p_x}{\delta x}=-\rho\frac{\delta v_x}{\delta t} \tag{2}$$

The final step is to link the change in volume to the change in pressure based on the material properties. For an elastic artery the distensibility D of a segment with initial volume V is defined as:

$$D=\frac{\delta V}{V\delta p}=\frac{\delta m}{m\delta p_x} \tag{3}$$

Substitution of (3) in (1) results in:

$$\frac{D\delta p_x}{\delta t}=-\frac{\delta v_x}{\delta x} \tag{4}$$

Differentiation of (4) and (2) with respect to t and x, respectively, and elimination of v results in the wave equation:

$$\frac{\delta^2 p_x}{\delta t^2} = \frac{1}{\rho D}\frac{\delta^2 p_x}{\delta x^2} \tag{5}$$

Substitution of a general harmonic solution $p_x = \hat{p}\cos(2\pi ft - 2\pi x/\lambda)$ with $c=\lambda f$, λ the wavelength, f frequency and c the propagation speed of the wave, results in the Bramwell-Hill equation:

$$c = \sqrt{\frac{1}{\rho D}} \tag{6}$$

As follows from (3) D, and therefore c, depends on the initial volume and, hence, on pressure: a larger volume is associated with a lower distensibility. The change in volume is mainly caused by a change in diameter. Because the artery is in vivo tethered to the surrounding tissues an elongation of the segment will hardly occur. Therefore, the distensibility D can be replaced by the distensibility coefficient DC, i.e., the distensibility per unit length. Moreover, if the change in diameter is small with respect to the initial diameter d the expression for the distensibility can be simplified further:

$$DC = \frac{\delta A}{A\delta p} \approx \frac{2\Delta d}{d\Delta p} \tag{7}$$

where the incremental change in diameter is replaced by the change in diameter Δd from diastole to systole and the pressure change by the pulse pressure Δp.

The elasticity coefficient (Young's modulus) E is defined as the change in stress S_w (force divided by area) divided by the strain (relative change in length). The wall stress, assuming that the stress gradient over the wall is negligible (thin walled tube), follows from the Lamé equation, which can easily be obtained by considering a plane through the axis of an arterial segment with length l. In equilibrium the forces in the wall counterbalance the downward force pld on this plane caused by the transmural blood pressure p:

$$2S_w hl = pld \qquad 2S_w h = pd \tag{8}$$

Using a first order Taylor series expansion the change in stress ΔS_w due to a change in pressure Δp can be approximated with:

$$2\Delta S_w = \frac{p}{h}\Delta d + \frac{d}{h}\Delta p - \frac{dp}{h^2}\Delta h = \frac{p}{h}\Delta d + \frac{d}{h}\Delta p - \frac{dp}{h}\frac{\Delta d}{d} = \frac{d}{h}\Delta p \tag{9}$$

where the relative change in wall thickness is approximated with relative change in diameter. For an increase in diameter of Δd the relative change in circumference is $\Delta d/d$:

$$E = \frac{\Delta S_w}{\Delta d / d} = \frac{\Delta p d}{2h \Delta d / d} = \frac{d}{hDC} \tag{10}$$

As stated above, the DC is pressure dependent. Consequently, a linear approximation is only valid for a limited pressure range, as indicated by the incremental E_{inc}.

The capacity to store temporarily blood for a given increase in pressure is given by the compliance C. Corresponding to the derivation for the distensibility coefficient the compliance coefficient CC is given by:

$$CC = \frac{\Delta V}{\Delta p} = \frac{\Delta A}{\Delta p} \approx \frac{\pi d \Delta d}{2 \Delta p} \tag{11}$$

Blood pressure measurements

From the previous section a few relevant quantities emerge, which form the basis for the calculation of the hemodynamic (compliance) and mechanical (distensibility, elasticity) parameters. Common parameters are the diameter and the change in diameter for a given change in blood pressure.

The (instantaneous) blood pressure is a common parameter and should preferably be measured at the same site where the other parameters are determined. For peripheral arteries this can be accomplished with applanation tonometery [13], where the blood vessel is flattened with a pencil-like probe, containing a high-fidelity micromanometer, against a bony structure balancing the circumferential stresses in the artery walls. If the probe is kept in a steady position the time-dependent change in blood pressure can be truthfully recorded. The recorded diastolic pressure, however, depends also on the pressure applied on the probe and has to be calibrated by a reference value. Because in diastole the pressure drop over the central arteries is negligible, any accessible location can be selected for the calibration procedure, e.g., the brachial artery. Under steady state conditions the diastolic blood pressure exhibits only minor fluctuations, allowing for an incidental reference measurement, for example with a conventional sphygmomanometer or an oscillometric blood pressure meter. Likewise, assuming that the mean arterial pressure (MAP) is also the same for the major arteries, the average of the tonometer waveform over a cardiac cycle can be equated to the MAP derived in another way. Unfortunately there is no physically sound algorithm to estimate the

MAP from oscillometric blood pressure readings. The reported values for the MAP, but also the recorded systolic and diastolic blood pressure, may vary considerably [14]. Since the diastolic pressure and MAP are used to calibrate the tonometer readings any error in the reference values will also emerge in the tonometer blood pressure values.

The approach as outlined above assumes that the tonometer is hold firmly in position, meaning that applied force should vary in accordance with the blood pressure to be measured. This requirement is difficult to comply with, resulting in distorted pressure registrations. The other pre-requisite is that a solid background supports the artery of interest. This is true for the radial, brachial and femoral arteries, but not for the carotid artery. That is why the blood pressure results for this artery are subject to dispute, especially since for ethical reasons verification measurements are scarce and virtually absent for the normal population.

Assessment of diameter and diameter change

A pre-requisite for vascular elasticity measurements is the assessment of vascular parameters, like luminal diameter and the diameter response to pressure change, with a minimum of physical interaction with the site of investigation. Ultrasound techniques comply with this requirement, offer a high temporal and adequate spatial resolution, and are widely available.

The spatial resolution of ultrasound systems is linked to the bandwidth of the received ultrasound signal: the larger the bandwidth, the better the resolution will be. Generally the bandwidth increases with the center frequency, at the expense of the penetration depth, restricting for peripheral artery applications the center frequency to the range of 5 to 10 MHz. Then the achievable resolution along the ultrasound beam (axial resolution) will be on the order of 0.2 mm. Perpendicular to the beam, in the plane of observation, the lateral resolution will be about 1 mm and will vary with the distance to the transducer. In the other direction (perpendicular to the plane of observation) the resolution will be worse because of the limited options to focus the beam.

Because of the shallow depth of investigation, a few cm, the pulse repetition frequency (PRF) of the ultrasound system can be high, and the time required to complete a single echo image will be low, resulting in echo frame rates above 50 Hz. Unfortunately the video recording frame rate is fixed at 25 Hz and echo frames will be dropped to synchronize the echo frame rate to the video frame rate. However, concurrent echo systems present and store the images in SVGA format, circumventing the timing problems associated with video recording. The ECG, presented simultaneously with the echo image, allows for the identification of the diastolic and peak systolic images. Subsequently the luminal diameter, averaged over an artery segment of 1 to 2 cm, in both phases of the cardiac cycle, and hence the change in diameter from diastole to systole, can be assessed [15]. The whole process can even be automated, resulting in a continuous diameter waveform, of course sampled at the video or echo frame rate [16].

Video image analysis has the disadvantage that the assessed values vary with the gain settings in combination with the attenuation of the ultrasound by intervening tissues. Moreover, the echo amplitude as displayed on the screen is compressed to accommodate the large amplitude scale and to emphasize minor amplitude differences. Consequently, echo signal transients behave differently at higher than at relatively low amplitudes. The echo amplitude is obtained from the underlying received radio frequent (rf) signal by rectification and lowpass filtering. The latter action causes a smoothing of edges, thereby degrading the axial resolution. Lastly, video processing can at best provide a coarse version of the change in diameter of time (distension waveform) because of the sampling in time (video frame rate) and spatial sampling (spacing video line in relation to the depth range). Because of the aforementioned reasons some research groups focussed on the processing of the rf signals rather than the video image, despite the basic problem to access it; traditionally the rf signal is not publicly available.

For artery diameter and distension assessment the echo system is operated in M-mode where the M-line is positioned perpendicularly to the blood vessel of interest using a two-dimensional B-mode image. To acquire an rf signal as function of depth and to reproduce the registration process over time (subsequent emissions) a trigger signal indicating the start of an emission is required. Synchronously with the emission trigger, a sample clock with a frequency of 3 to 4 times the ultrasound carrier frequency (beyond 20 MHz for peripheral applications) is generated. The sample clock is used to select the signal segment of interest and to digitize the rf signal as function of depth. These functions can be integrated on a PC plug-in card while the PC takes care for the storage of the signal segments acquired and for data processing, either off-line or on-line depending on the complexity of operations involved [17,18]. Figure 2 depicts an example of an rf signal segment covering the common carotid artery of a young healthy volunteer. Superimposed on the rf signal is the envelope. To obtain the instantaneous envelope the rf signal is converted to an analytic signal, using the Hilbert-transformation in either the time- or frequency-domain. This computational approach has the advantage that no smoothing or time shift occurs in the envelope with respect to the underlying rf signal. The region with low signal amplitude is the lumen while the high peaks originate from the adventitia. The media is a narrow region with a relatively low echogenicity with respect to the adjacent intima and adventitia.

As can be deduced from Fig. 2, there is no clear transition from the adventitia to the embedding tissue. That is why it is impossible to assess the thickness of the wall, composed of the three layers. However, changes in wall morphology due to cell migration and proliferation are more likely to occur in the media and intima, explaining the interest in intima-media thickness (IMT) as a surrogate for the wall thickness h in the equations 8-10. As explained above, echo signals are produced by tissue interfaces. Therefore, in figure 2 it is not clear where the echo from the adventitia-media interface of the anterior wall starts; only the trail into the media

276

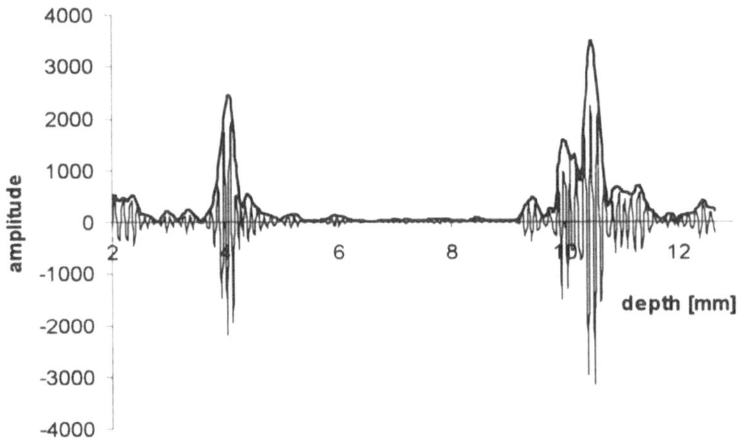

Fig. 2: RF-signal and its envelope (arbitrary scale) of the common carotid artery of a young healthy volunteer.

is seen. Moreover, the media-intima and/or intima-lumen interfaces have relatively low amplitudes. The combination of trailing echoes and a large difference in amplitude levels prohibits reliable assessment of the IMT at the anterior wall. At the posterior wall the leading echoes of the interfaces can clearly be distinguished resulting in a good estimate for the IMT [16,19,20].

The distance between the trailing edge of the anterior intima-lumen transition and the posterior lumen-intima transition is indicative for the lumen diameter but will be biased by the resolution of the echo-system. The latter can be deduced from the bandwidth of the rf signal of the anterior wall, converted to millimeters, allowing a depth-sensitive correction for the length of the echo-signal [18]. The same approach can be followed to establish the artery diameter, including the media. The latter, rather than the lumen diameter, is normally used in the evaluation of the arterial elasticity because the quality and amplitude of the low-level intima-lumen signals are quite susceptible for the angle of observation and noise.

The relatively sharp transitions in echo-amplitude in Fig.2 are indicative for the positions of the adventitia-media interfaces. This feature can be exploited to detect these interfaces automatically, and therefore the artery diameter, starting from within the lumen [18,21]. Repeating the detection procedure for each subsequent heart beat, synchronously with the R-top of the ECG, inherently corrects for motion artifacts. The reproducibility for diameter measurements is on the order of the resolution of the echo system employed [21].

Once the positions of the anterior and posterior artery walls have been identified synchronously with the R-top of the ECG, the change of the position over the subsequent cardiac cycle can be tracked by analyzing the rf signal from sample volumes placed automatically on the artery wall. This can be achieved by traditional Doppler processing, where a long emission burst (narrow bandwidth) is used to ensure that the center frequency of the received signal is a priori known despite the effects of frequency dependent attenuation by the intervening tissues [22,23]. But then the sample volume will be rather long in the axial direction compared to the wall thickness. From diastole to systole the wall thickness will decrease with about the same percentage as the lumen diameter increases to maintain its volume. The thinning will be more prominent at the inner than at the outer wall layers, resulting in a gradient in the velocity. The Doppler processing based on a long sample volume of about 1 mm will result in an average over the entire wall, discarding possible information about the inhomogeneity within the wall.

If the shape of the spectral distribution is known a priori, then also the displacement information can be extracted from the wide band rf signals (echo M-mode) of a sample window with a length on the order of the resolution of the echo system (for a 7 MHz system 200-300 μm). The change in position of a structure boundary follows from the argument of the complex cross-correlation function of corresponding rf signal segments with as unknown parameter the center frequency of the received rf signal. The latter can be estimated from the complex autocorrelation function [24], thereby accounting for possible frequency dependent attenuation effects. To improve the reliability of the displacement estimate, averaging is performed over a number of subsequent observations. The time window (emission interval times number of emissions) should be selected according the dynamic behavior of the structure examined. Since the most rapid phase from diastole to systole lasts on the order of 100 ms, a time window of 5 to 10 ms is appropriate. To avoid that the structure leaves temporarily the sample window its position is moved according to the observed cumulative displacement (tracking) of the wall boundary. Of course this adjustment is in increments dictated by the sample clock, while the detected displacements may be much smaller. The tracking procedure is also applied to the posterior wall and the difference between the cumulative displacements of the anterior (Fig. 3 top) and the posterior (Fig. 3 middle tracing) walls provides the change in diameter (Fig. 3 bottom) over time (distension waveform). Note that generally the excursion of the anterior wall should be much greater than that of the posterior wall because the spine blocks the motion of the latter. This difference in motion pattern of both walls results in a temporary translation of the artery unless also a firmly applied ultrasound transducer blocks the anterior wall. A translation of the artery along the ultrasound beam does not affect distension measurements. However, lateral translation due to the pulse wave, as frequently occurs in elderly people, may have unpredictable effects on the distension waveform. It is, therefore, advisable to select a line of observation along the translation direction of the artery.

Respiration will have a direct effect on the amplitude of the pulse wave and, therefore, on the distension. It is therefore recommended to have a registration duration of at least a ventilation cycle (5 seconds). Current high-resolution wall track

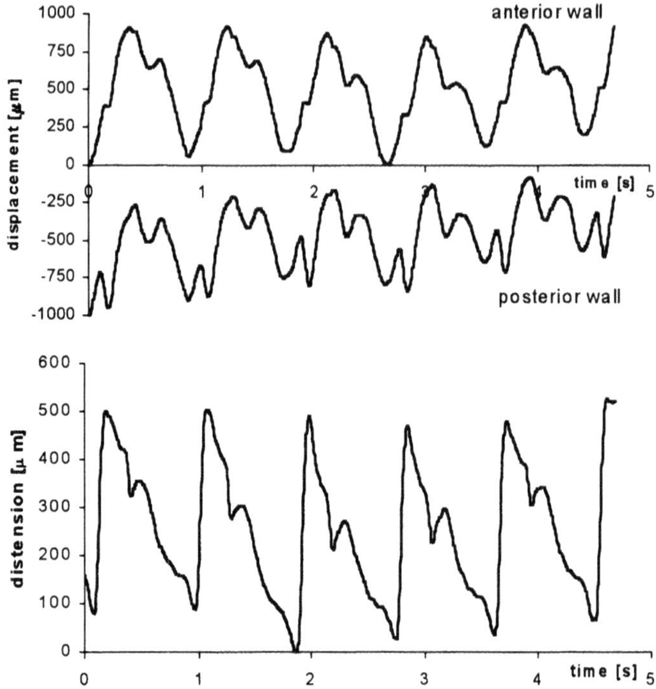

Fig. 3: Displacement of the anterior (top) and posterior (middle) wall of the common carotid artery. The bottom trace shows the resulting change in diameter over time.

systems (WTS) can accommodate much longer registrations, which are only limited by the patience of the subject and the skill of the investigator (steady probe position).

From the distension waveform some key parameters are extracted on a beat-to-beat basis, like the distension from end-diastole (minimum diameter) to peak systole, the delay between R-top and the onset of the distension waveform and the risetime from the 10% to 90% level of the upstroke. The delay can be used to estimate the pulse wave velocity, but it will be only a rough estimate because the time elapsed between the R-top of the ECG and the opening of the aorta valve (on the order of 40 ms) varies from subject to subject. Generally, to improve estimates and to eliminate confounding effects of ultrasound probe position, the average over a number of subsequent independent registrations is computed for each parameter of interest.

Pulse wave velocity

Expression 6 relates the distensibility coefficient to the pulse wave velocity, where it should be acknowledged that in this derivation the *DC* is a local measure while the pulse wave velocity is regional. The latter can be obtained by measuring the foot-to-foot delay between blood pressure curves assessed simultaneously at two sites with a known interspacing. For compliant arteries the blood velocity can be approximated by the time derivative of the pressure. Blood velocity has, therefore, a steeper upslope and can, moreover, be assessed non-invasively with Doppler ultrasound [25]. Since the estimate for the average transit time between two recording sites is based on the spectral distributions of the Doppler signals (update rate in the range of 5-10 ms) a large number (40-80) of heart beats has to be considered. The reproducibility is further affected by the problem to estimate the length of the arterial segment under investigation.

The tonometers provide an alternative to approach superficial arteries like brachial, carotid and femoral arteries. Because the main interest is on the onset of the pulse wave rather than a reliable pressure, the requirements for recording can be relaxed. As for the Wall Track System (WTS), a single tonometer can be used if the delay at two recording sites is timed with respect to the ECG (SphygmoCor, PWV Medical, Sydney). The difference results in the transit time over the trajectory between the recording sites, thereby canceling the individual variations in the electro-mechanical coupling of the heart. To avoid the effects of beat-to-beat variations in pulse wave velocity a pair of tonometers is used simultaneously (Complior, Colson, Les Lilas, France). It automatically excludes inadequate pressure waves within a recording segment of 15 seconds and has a good repeatability [26].

Recent studies on pulse wave velocity focus on the carotid-femoral trajectory and arrive at an estimated value in the range of 10-12 m/s, which is high compared to the direct measurements made by Ting et al. [27]. They have shown by invasive measurements that in hypertensives compared to normotensives the pulse wave velocity is higher in the iliac artery but, surprisingly, not in the aorta, where it is on the order of 4 m/s. The main reason for the deviation is that for the non-invasive approach trajectories are included with opposite pressure wave directions; as a result the transit time will be underestimated. Actually, if we assume that the aorta-carotid segment has the same pulse wave velocity as the thoracic aorta, the measured pulse wave velocity pertains to that of the segment between the abdominal aorta and the femoral bifurcation.

General Discussion

There are several reasons for the current interest in the mechanical characteristics of the common carotid artery. Although it is a segment that is not prone to artherosclerosis (the carotid bifurcation is the primary predilection site), its mechanical characteristics strongly resemble that of the coronary arteries. Because of its easy accessibility it is used to quantify the vascular changes as an independent risk factor for cardiovascular morbidity and mortality [15,28].

In general, the term elasticity creates confusion. In verbal terms a high elasticity applies to materials that easily elongates under stress, while the formal definition follows an opposite notion: a high elasticity E implies a stiff material (Eq. 10). To make matters worse, the literature does not adhere to a common concept about compliance. In this chapter we defined the compliance as the **absolute** rather than the **relative** change in volume due to a given change in pressure. Quite a number of publications follow the latter definition [29] for compliance, while we reserve this for the distensibility.

Distensibility is clearly a material property; it gradually increases from the heart to the periphery as the structure of the wall changes from elastic to muscular. Moreover, the distensibility and compliance are pressure-dependent, i.e., at higher mean arterial pressures the DC drops. To eliminate the pressure-dependence it is suggested to use the incremental DC around a fixed pressure, usually 100 mmHg (isobaric distensibility). It requires the simultaneous registration of the blood pressure with a tonometer and the (change in) diameter with a Wall Track System. After proper time alignment of the curves and correction for a phase mismatch between the waveforms [30], it can be concluded that normotensives and hypertensives exhibit the same isobaric DC [31].

For large population studies, the alternative is to execute a statistical correction for mean arterial blood pressure. Then both common carotid artery distensibility and aorta elasticity, the latter based on carotid-femoral pulse wave velocity, exhibit a strong association with carotid intima-media thickness, and the severity of plaques in the carotid artery and aorta [32].

The effect of ageing on the mechanical and hemodynamical properties of the common carotid artery, also after correction for mean arterial pressure, has been demonstrated by several studies [33-35]. Certain regions, like the carotid bifurcation, seem to age faster in terms of distension patterns than others [36].

Almost all of the studies, discussing the elastic behavior of the region around the carotid bifurcation, were performed in subjects without atherosclerotic plaques in this region. For M-line investigations, as previously described, the observed diameter, distension waveform and wall thickness would heavily depend on the position of the ultrasound beam with respect to the plaque. Evidently this restriction also applies for 2-dimensional B-mode investigations, although to a lesser extent. On the other hand, the Wall Track System is easy to operate, facilitating repeated measurements. Since the position of the M-line will vary from observation to observation, the average would be quite indicative for the regional characteristics. However, proper evaluation of focal lesions, as proposed for intravascular ultrasound systems, in terms of homogeneity in diameter, wall thickness and distension needs further clinical validation [21].

Elasticity measurement involves the assessment of the time-dependent diameter simultaneously with the local blood pressure. As has been shown the diameter waveform (and therefore the end-diastolic diameter and the distension waveform) is rather easy to obtain. However, the assessment of the local pressure waveform remains problematic, even with a tonometer. An alternative may be the regional pulse wave velocity, but at the moment no techniques are available to reliably detect the pulse wave velocity in vivo over a segment of about 1 cm [37].

References

1. Zarins CK, Weisenberg E, Kolettis G, et al (1988) Differential enlargement of artery segments in response to enlarging atherosclerotic plaques. J Vasc Surg 7:386-394
2. Bots ML, Hofman A, Grobbee DE (1997) Increased common carotid intima-media thickness. Adaptive response or a reflection of atherosclerosis? Findings from the Rotterdam study. Stroke 28:2442-2447
3. Tuttle JL, Nachreiner RD, Bhuller AS, et al (2001) Shear level influences resistance artery remodeling: wall dimensions, cell density, and eNOS expression. Am J Physiol Heart Circ Physiol 281:H1380-1389
4. Wentzel JJ, Krams R, Schuurbiers JC, et al (2001) Relationship between neointimal thickness and shear stress after Wallstent implantation in human coronary arteries. Circulation 103:1740-1745
5. Schmidt-Trucksass A, Schmid A, Brunner C, et al (2000) Arterial properties of the carotid and femoral artery in endurance- trained and paraplegic subjects. J Appl Physiol 89:1956-1963.
6. Kornet L, Hoeks APG, Lambregts J, et al (1999) In the femoral artery bifurcation differences in mean wall shear stress within subjects are associated with different intima-media thicknesses. Arterioscler Thromb Vasc Biol 19:2933-2939
7. Jiang Y, Kohara K, Hiwada K (1999) Low wall shear stress contributes to atherosclerosis of the carotid artery in hypertensive patients. Hypertens Res 22:203-207
8. Kornet L, Lambregts JAC, Hoeks APG, et al (1998) Differences in near-wall shear rate in the carotid artery within subjects are associated with different intima-media thicknesses. Arteriosel Thromb Vasc Biol 18:1877-1884
9. Gnasso A, Carallo C, Irace C, et al (1996) Association between intima-media thickness and wall shear stress in common carotid artery in healthy male subjects. Circulation 94:3257-3262
10. Gnasso A, Irace C, Carallo C, et al (1997) In vivo association between low wall shear stress and plaque in subjects with asymmetrical carotid atherosclerosis. Stroke 28:993-998
11. van Gorp AW, Schenau DS, Hoeks AP, et al (2000) In spontaneously hypertensive rats alterations in aortic wall properties precede development of hypertension. Am J Physiol Heart Circ Physiol 278:H1241-H1247
12. Fung YC (1984) Biodynamics: Circulation. Springer-Verlag, New York.
13. Kelly R, Hayward C, Avolio A, et al (1989) Noninvasive determination of age-related changes in the human arterial pulse. Circulation 80:1652-1659
14. Amoore JN, Geake WB, Scott DHT (1997) Oscillometric non-invasive blood pressure measurements: the influence of the make of instruments on readings? Med Biol Eng Comput 35:131-134

15.Riley WA, Barnes RW, Applegate WB, et al (1992) Reproducibility of non-invasive ultrasonic measurement of carotid atherosclerosis: the asymptomatic carotid artery plaque study. Stroke 23:1062-1068
16.Graf S, Gariepy J, Massonneau M, et al (1999) Experimental and clinical validation of arterial diameter waveform and intimal media thickness obtained from B-mode ultrasound image processing. Ultrasound Med Biol 25:1353-1363
17.Hoeks APG, Brands PJ, Smeets FAM, et al (1990) Assessment of the distensibility of superficial arteries. Ultrasound Med Biol 16:121-128
18.Brands PJ, Hoeks APG, Willigers J, et al (1999) An integrated system for the non-invasive assessment of vessel wall and hemodynamic properties of large arteries by means of ultrasound. Eur J Ultrasound 9:257-266
19.Hoeks APG, Willekes C, Boutouyrie P, et al (1997) Automated detection of local artery wall thickness based on M-line signal processing. Ultrasound Med Biol 23:1017-1023
20.Willekes C, Hoeks APG, Bots ML, et al (1999) Evaluation of off-line automated intima-media thickness detection of the common carotid artery based on M-line signal processing. Ultrasound Med Biol 25:57-64
21.Meinders JM, Brands PJ, Willigers JM, et al (2001) Assessment of the spatial homogeneity of artery dimension parameters with high frame rate 2-D B-mode. Ultrasound Med Biol 27:785-794
22.Stadler RW, Karl WC, Lees RS (1996) The application of echo-tracking methods to endothelium-dependent vasoreactivity and arterial compliance measurements. Ultrasound Med Biol 22:35-42
23.Stadler RW, Taylor JA, Lees RS (1997) Comparison of B-mode, M-mode and echo-tracking methods for measurement of the aterial distension waveform. Ultrasound Med Biol 23:879-887
24.Brands PJ, Hoeks APG, Ledoux LAF, et al (1997) A radio frequency domain complex cross-correlation model to estimate blood flow velocity and tissue motion by means of ultrasound. Ultrasound Med Biol 23:911-920
25.Lehmann ED, Gosling RG, Fatemi-Langroudi B, et al (1992) Non-invasive Doppler ultrasound technique for the in vivo assessment of aortic compliance. J Biomed Eng 14:250-256
26.Asmar R, Benetos A, Topouchian J, et al (1995) Assessment of arterial distensibility by automatic pulse wave velocity measurement. Validation and clinical application studies. Hypertension 26:485-490
27.Ting CT, Chang MS, Wang SP, et al (1990) Regional pulse wave velocities in hypertensive and normotensive humans. Cardiovascular Research 24:865-872
28.Riley WA, Barnes RW, Evans GW, et al (1992) Ultrasonic measurement of the elastic modulus of the common carotid artery. The atherosclerosis risk in communities (ARIC) study. Stroke 23:952-956
29.Lehmann ED, Hopkins KD, Gosling RG (1996) Multiple definitions of 'compliance'. Clin Sci (Colch) 90:433-4.
30.Hoeks AP, Willigers JM, Reneman RS (2000) Effects of assessment and processing techniques on the shape of arterial pressure-distension loops. J Vasc Res 37:494-500
31.Laurent S, Caviezel B, Beck L, et al (1994) Carotid artery distensibility and distending pressure in hypertensive humans. Hypertension 23:878-883
32.van Popele NM, Grobbee DE, Bots ML, et al (2001) Association between arterial stiffness and atherosclerosis: The Rotterdam study. Stroke 32:454-460.

33. Reneman RS, Van Merode T, Hick P, et al (1986) Age-related changes in carotid artery wall properties in man. Ultrasound Med Biol 12:465-471
34. Hansen F, Mangell P, Sonesson B, et al (1995) Diameter and compliance in the human common carotid artery - Variations with age and sex. Ultrasound Med Biol 21:1-9
35. Smulyan H, Asmar RG, Rudnicki A, et al (2001) Comparative effects of aging in men and women on the properties of the arterial tree. J Am Coll Cardiol 37:1374-1380
36. Van Merode T, Brands PJ, Hoeks APG, et al (1993) Faster ageing of the carotid artery bifurcation in borderline hypertensive subjects. J Hypertension 11:171-176
37. Brands PJ, Willigers JM, Ledoux LAF, et al (1998) A non-invasive method to estimate pulse wave velocity in arteries locally by means of ultrasound. Ultrasound Med Biol 24:1325-1335

Cross-Sectional Imaging of Elasticity around Atherosclerotic Plaque with Transcutaneous Ultrasound

Hiroshi Kanai[1], Hideyuki Hasegawa[1], and Yoshiro Koiwa[2]

[1]Department of Electronic Engineering, Graduate School of Engineering, Tohoku University, Sendai 980-8579, Japan. E-mail: hkanai@ecei.tohoku.ac.jp.
[2]First Department of Internal Medicine, Tohoku University Graduate School of Medicine, Seiryou-machi, Aoba-ku, Sendai 980-8574, Japan.

Summary: The minute change in thickness less than 100 μm of the arterial wall cannot be measured by conventional B-mode or M-mode images. This paper describes a novel noninvasive method (phased tracking method) for measuring such minute change in the arterial wall thickness during one cardiac cycle. A method for evaluation of the regional elastic modulus in the circumference direction, E_θ, from the resultant change in wall thickness is also described. The developed method was applied to subjects with hyperlipidemia, the spatial distribution of the regional elastic moduli, $\{E_\theta\}$, is evaluated for the carotid atherosclerotic plaques. The soft inclusion is found from every plaque and there are differences in elasticity for surrounding hard tissues. The novel method offers potential for new diagnostic techniques in vulnerability of plaques.

Key words: Phased tracking method, Atherosclerosis, Regional elastic modulus, Transcutaneous ultrasound, Plaque rupture, Vulnerability of plaque

Introduction

Increase in myocardial infarction and cerebral infarction, which are considered to be mainly caused by rupture of the atherosclerotic plaques, becomes a more and more serious problem even in Japan. It is, therefore, essential to supply the physical property of the atherosclerotic plaques in evaluation of its stability and vulnerability in clinical setting. For this serious problem, various approaches including the molecular biological approaches have been applied [1-6]. However, it has been still difficult to estimate the physical property of the atherosclerotic plaques for evaluation of its stability and vulnerability in clinical setting.

For measurement of the arterial elasticity, various methods have been proposed and developed previously in literature. Based on the Moens-Korteweg's equation, the elastic moduli are obtained by measuring the velocity of the pulse wave propagating along the artery [7]. From the measurement of change in diameter due to heartbeat, the elastic modulus is obtained [8-9]. However, the elastic moduli obtained by these methods show the properties averaged between two points where the pulse wave propagates or along the whole circumference direction and then the sufficient spatial resolution has not been obtained.

In our study, a novel high resolution measurement method, "phased tracking method" [10-11], is adopted to clarify the physical property (elasticity) at each 0.375 mm-thickness layer of the atherosclerotic plaque from intimal side to the outer side along the ultrasonic beam [12-14]. It is realized by measurement of the minute change in thickness of regional area in the arterial wall due to heartbeat. The size of each regional area corresponds to the focal region of the ultrasonic beam, several hundred micrometers. We also have proposed a novel method for calculation of the regional elastic modulus of the arterial wall from the measured change in thickness even if the artery has non-uniform wall thickness such like an atherosclerotic plaque [15]. The novel method is of use for estimation of the stability of the atherosclerotic plaques in clinical setting.

Principle

Though M-mode echocardiography offers an advantage in critically examining the motion pattern of the wall, its spatial resolution along the ultrasonic beam is limited to a few wavelengths, namely, at most about several hundred micrometers for ultrasound of 7 MHz because an M-mode image is displayed based on the amplitude of the reflected ultrasound. On the other hand, there have been numerous elaborate techniques proposed for noninvasive measurement of the velocity of the blood flow in the arteries based on the Doppler effect. Moreover, several methods, including the phase-locked-loop (PLL) techniques, have been proposed to measure rough changes in the diameter of the arterial walls by tracking arterial wall displacement in real time.

At the beginning of the ejection period, the arterial wall is affected by the pressure wave which comes from the left ventricle (LV) and expands the lumen. Consequently, the arterial wall becomes slightly thin during the ejection period. In the literature, measurement of the change in diameter, $\Delta d_{BA'}(t)$, of the lumen in Fig. 1, has been reported. From $\Delta d_{BA'}(t)$, the elasticity of the arterial wall can be diagnosed, but it is assumed that the arterial wall is homogeneously changed in the circumferential direction due to atherosclerosis.

On the other hand, measurement of the change in thickness of the wall makes it possible to evaluate the elasticity of the arterial wall in each local area, the locality in the circumferential direction being indispensable to diagnosis of the inner characteristics of atherosclerotic plaque. For this purpose, by tracking the

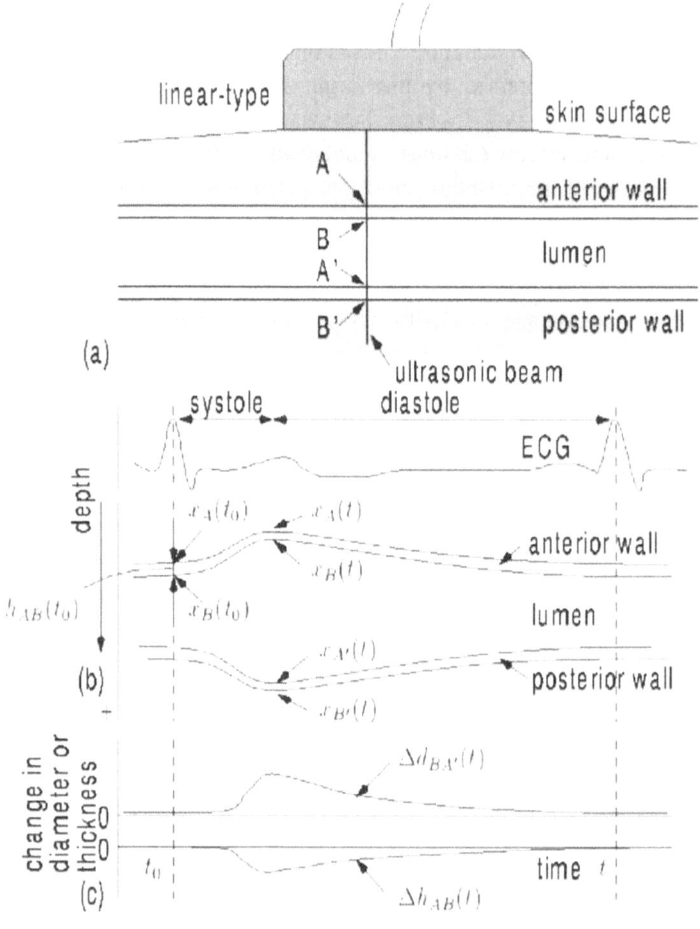

Fig. 1. An illustration explaining the procedure of measuring the change in thickness, $\Delta h_{AB}(t)$, of the arterial wall, and the change in diameter, $\Delta d_{BA'}(t)$, of the lumen by tracking the movement $x_A(t)$ and $x_B(t)$ of points A and B preset in the arterial wall along the ultrasonic beam. (a) cross-sectional image, (b) displacement $x_A(t)$ and $x_B(t)$, (c) change in thickness, $\Delta h_{AB}(t)$, of the arterial wall and change in diameter, $\Delta d_{BA'}(t)$.

instantaneous positions, $x_A(t)$ and $x_B(t)$, which are respectively preset at $x_A(t_0)$ and $x_B(t_0)$ at the end-diastole along the ultrasonic beam in the arterial wall as illustrated in Fig. 1(b), the small change in wall thickness, $\Delta h_{AB}(t)$, between these two points, A and B, is obtained as illustrated in Fig. 1(c) if the ultrasonic beam is almost perpendicular to the wall. From the ratio of the change in thickness, $\Delta h_{AB}(t)$, to the thickness $h_{AB}(t_0)$ between these points preset at the end-diastole, the deformation is obtained. If the deformation is sufficiently small and is in the linear regime, it shows the strain, and thus the regional elasticity of the wall can be approximately evaluated using the pulse pressure measured at the brachial artery. Therefore, the transcutaneous measurement of the change in thickness of the regional area of the arterial wall during each cardiac cycle provides the essential tools for diagnosis of atherosclerosis.

The accuracy required for the simultaneous measurement of the instantaneous positions $x_A(t)$ and $x_B(t)$ of points A and B is estimated as follows: For the carotid artery, the thickness of the wall is about 1 mm and the change in thickness, $\Delta h_{AB}(t)$, is less than 100 μm during one cardiac cycle in normal adults [12]. Therefore, the necessary spatial resolution in the measurement of instantaneous position as a waveform is at least 10 μm. If the velocity signals $v_A(t)$ and $v_B(t)$ of these two points, A and B, are detected based on the pulse Doppler method, the necessary accuracy of the velocity measurement is about 10 μm/200 μs =0.05 m/s when the pulse repetition frequency (PRF) of the transmission-pulse train is 5 kHz (=1/200 μs). If the equivalent sampling period of the velocity signal $v_i(t)$ is longer than 200 μs as in the FFT (fast Fourier transform)-based standard Doppler system, for example, more accurate velocity measurement is required.

Phased tracking method

For the accurate detection of the velocity signal, that is, the instantaneous movement, on or in the heart wall or arterial wall, we have developed the following "phased tracking method" [10-11]. In this method, by calculating the auto-correlation function with the constraint least-mean-square technique between the sequentially received echoes, the phase change caused by movement of the preset point (i) during the pulse repetition period ΔT (=1/PRF) is accurately determined and the average velocity $v_i(t)$ during the period is obtained. By adding the product of $v_i(t)$ and ΔT to the previous object position $x_i(t)$, the next position, $x_i(t+\Delta T)$, is estimated as $x_i(t)+v_i(t)\Delta T$. This method has been confirmed by experiments using a water tank and has been applied to the *in vivo* detection of small velocity signals, with sufficient reproducibility, at points in the interventricular septum (IVS) of the human heart [10]. The detected velocity signal shows rapid motion, including high frequency components with small amplitudes, which cannot be recognized by M-mode echocardiography.

Moreover, the method has been applied to multiple points preset along an ultrasonic beam in the LV wall so that the instantaneous object positions $\{x_i(t)\}$ and the velocity signals $\{v_i(t)\}$ are obtained for these multiple points $\{i\}$ in the LV wall [11]. The wall changes position and thickness with time. By making the location of the RV side of the IVS the reference point, the thickness change during myocardial contraction/relaxation can be detected. Then, their spatial distribution is obtained and is superimposed on the M-mode image using color coding.

Elasticity estimation

Letting us denote the thickness at the end-diastole by h_d, which is most thick during one heartbeat, the maximum change in thickness by Δh ($\Delta h<0$), and the pulse pressure during one heartbeat by Δp, the elastic modulus in the radius direction, E_r, is given by

$$E_r=\Delta p/(|\Delta h|/h_d) \tag{1}$$

By assuming that the Poisson's ratio to be 0.5 and the arterial wall is isotropic, the elastic modulus in the circumferential direction, E_θ, is estimated by

$$E_\theta=(1/2)(r_d/h_d+1)\ \Delta p/(|\Delta h|/h_d), \tag{2}$$

where r_d is the internal radius at end-diastole [15].

For the wall with non-uniform thickness

For the case when the artery has non-uniform wall thickness such like an atherosclerotic plaque, the above method should be modified as follows [15]:

We assume that the pressure distribution in the arterial wall changes linearly due to the distance from the inner surface of the arterial wall. When we consider the small region of the arterial wall divided into N layers, the pressure, $p_n(t)$, at the inner surface of the n-th layer is defined by

$$p_n(t)=\{(\mathrm{N}-n+1)/\mathrm{N}\}\{p_i(t)-p_a\}+p_a, \qquad (n=1, 2, ..., \mathrm{N}) \tag{3}$$

where $p_i(t)$ and p_a are the pressure in lumen and the atmospheric pressure. The elastic modulus of the n-th layer, $E_{\theta n}{}^{h}$, is obtained from the change in thickness by

$$E_{\theta n}{}^{h}=(1/2)[\rho_{n0}/(h_0/\mathrm{N})+(\mathrm{N}-n+1)]\Delta p(t)/\{-\mathrm{N}\Delta\varepsilon_{r\,n}(t)\}, \tag{4}$$

where the arterial wall is assumed to be incompressible (Poisson's ratio=0.5) and elastically isotropic, ρ_{n0} is the inner radius of the n-th layer at end-diastole, h_0 is the thickness of each layer preset at the end-diastole, $\Delta p(t)$ is the pressure increment in the n-th layer, and $\Delta\varepsilon_{r\,n}(t)$ is the radial incremental strain of the n-th layer [15].

Cross-sectional imaging of elasticity

By setting multiple points at the interval of h_d=0.375 mm from the intima to the adventitia along each ultrasonic beam as shown in Fig. 2, the change Δh in the regional thickness in each layer between the succeedingly preset points is measured and each elasticity $E_{\theta n}{}^{h}$ in the regional layer is estimated as described above. By controlling the position of the ultrasonic beam in the axial direction of the arterial wall, regional change in thickness and regional elasticity are estimated along each ultrasonic beam. The typical spatial resolution is 375 μm (=1.8×wavelength) along each ultrasonic beam and 300 μm in the axial direction of the arterial wall.

In this measurement, the minimum value of the measurable velocity is 0.5 mm/s and the time interval, ΔT, of the pulse repetition interval is about 200 μs. In this condition, the lowest value of the change in thickness is found to be about 0.1 μm in the measurement using a water tank [11]. Such a minute change in thickness cannot be measured by computer tomography (CT), magnetic resonance imaging (MRI), or standard ultrasonic diagnosis equipment. The measurement of the minute change of a several micrometer order in thickness by "phased tracking method" has recently been validated by expanding and contracting a rubber plate in a water tank [16].

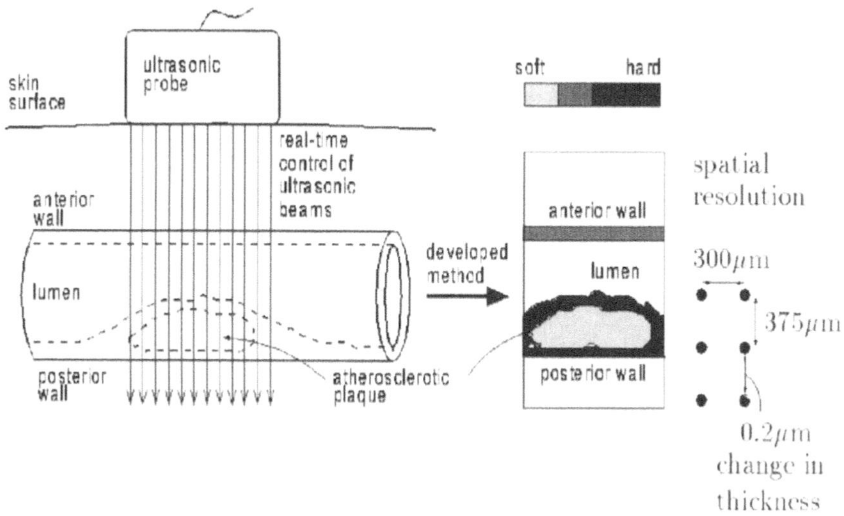

Fig. 2. An illustration for measurement and display of the elastic modulus for each point in the region of interest. The spatial resolution for each point in the developed system is shown in the bottom right.

Basic experiments for a tube with non-uniform thickness

The proposed formula of the elastic modulus, E_θ, of Eq. (4) is validated by the following basic experiments employing a silicone rubber tube (tube-A), which has non-uniform wall thickness and two layers, as a model of the artery. Figure 3 shows the schematic diagram of the experimental setup. In this system, the pulse pressure is generated by an artificial heart. The change in wall thickness caused by the pulse pressure is measured by the proposed method described above, and the pressure inside the silicone rubber tube is simultaneously measured by the pressure transducer (NEC, 9E02-P16). Figure 4(a) shows the cross sectional B-mode image of the silicone rubber tube obtained by the standard ultrasonic diagnostic equipment. The M-mode image shown in Fig. 4(b) is obtained by repeatedly transmitting the ultrasonic pulses to the silicone rubber tube. Figures 4(c) and 4(d) show the drive signal of the artificial heart and the pressure inside the silicone rubber tube. By manually setting the points, x_1 and x_2, at the timing of the rising of the drive signal (just before the artificial heart begins to pump), velocities, $v_1(t)$ and $v_2(t)$, of points x_1 and x_2 are measured as shown in Fig. 4(e) and 4(f), and the change in thickness, $\Delta h_1(t)$, of the inner layer is obtained as shown in Fig. 4(h) by integrating the difference between these two velocities.

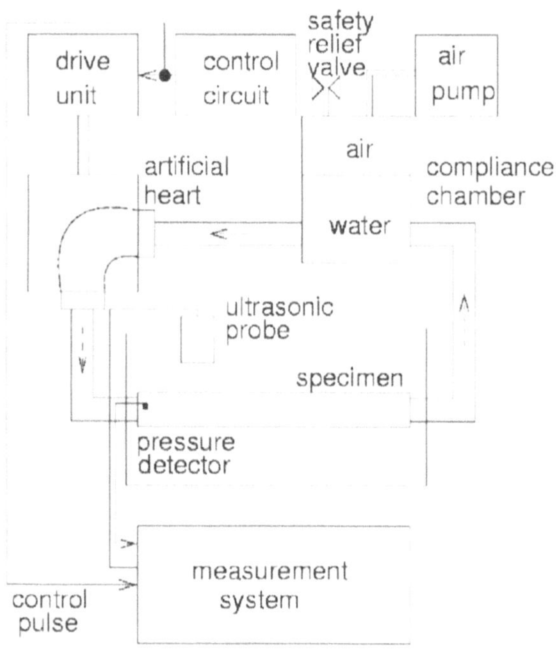

Fig. 3. The experimental setup of the basic experiment.

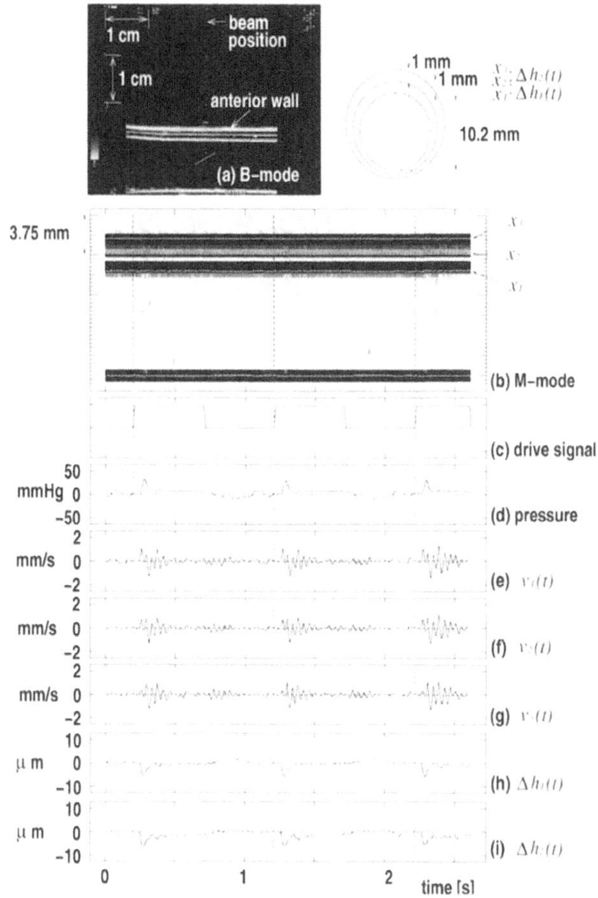

Fig. 4. Measurement of the change in wall thickness of tube-A. (a) B-mode image. (b) M-mode image. (c) The drive signal of the artificial heart. (d) The inner pressure. (e), (f), and (g) Velocities of points x_1, x_2, and x_3. (h) and (i) Changes in thickness of the inner and the external layer.

Simultaneously, the change in thickness, $\Delta h_2(t)$, of the external layer, which is shown in Fig. 4(i), is obtained by measuring velocities, $v_2(t)$ and $v_3(t)$, at the points x_2 and x_3. From the measurements of changes in thickness, $\Delta h_1(t)$ and $\Delta h_2(t)$, the elastic modulus of each layer is obtained at each insonification angle, θ, from 0 to 90 degrees as shown in Fig. 5. In Figs. 5(a) and 5(b), the elastic moduli, $E_{\theta 1}{}^{h}$ and $E_{\theta 2}{}^{h}$, of the inner and the external layer are plotted as diamonds. From the results in Figs 5(a) and 5(b), $E_{\theta 1}{}^{h}$ and $E_{\theta 2}{}^{h}$ of tube-A measured at each insonification angle agree well with the circumferential elastic modulus obtained from the static pressure-diameter test as plotted by dashed line. In the pressure-diameter test, the silicone rubber tube (tube-B) with uniform wall thickness made of the same material as tube-A is used as a specimen.

On the other hand, the pressure elastic modulus, E_p, is defined by

$$E_p=\{r_0/h_0\}\{\Delta p(t)/(\Delta d(t)/d_0)\}, \quad (5)$$

where d_0 and $\Delta d(t)$ are the inner diameter and the change in inner diameter, respectively. The pressure elastic modulus, E_p, which is obtained from the change in diameter in Fig. 4, is 1.8 MPa with respect to tube-A. It is different from the results of the pressure-diameter test. From these results, it is shown that the proposed method descried above has an advantage in measurement of the regional elastic modulus of the tube with non-uniform wall thickness.

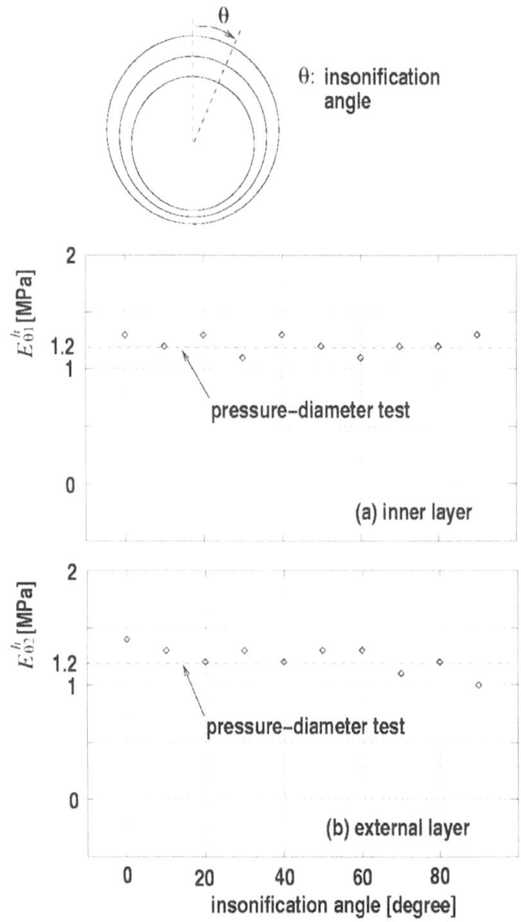

Fig. 5. Measured elastic moduli, $E_{\theta 1}^{h}$ and $E_{\theta 2}^{h}$, of the inner and external layer at each insonification angle, θ. (a) Results of the inner layer. (b) Results of the external layer.

In vivo experimental results

The proposed method is applied to *in vivo* measurement of a 54-year-old male patient with atherosclerotic plaque in the right common carotid artery. Figure 6(a) shows the cross-sectional (B-mode) image obtained by standard ultrasonic diagnostic equipment (Toshiba SSH-140A) with 7.5 MHz linear-type-probe. Figure 6(b) is obtained by scanning the ultrasonic beam around the atherosclerotic plaque shown by the region surrounded by the dashed line l_{10} in Fig. 6(a).

Figure 7(a) shows the M-mode image along the 10th scanned position l_{10} in Fig. 6(a). Figures 7(b) and 7(c) shows the electrocardiogram (ECG) and phonocardiogram (PCG), respectively. The arterial wall is divided into 6 layers by manually setting the points, where the velocities are obtained, at the timing of the R-wave of the ECG. From the velocities of these points in Fig. 7(d), the change in thickness, $\Delta h_n(t)$, of each layer is obtained as shown in Fig. 7(e).

By applying the same procedure to each scanned position, the elastic modulus $E_{\theta n}{}^{h}$ of each region is color-coded and superimposed on the reconstructed B-mode image as shown in Fig. 8. The employed color scheme goes from yellow (50 kPa) to red (2,000 kPa). The regional elasticity is measured at each point along the ultrasonic beams at 30 positions along the axial direction of the artery. The resolution in the depth direction and axial direction are 375 μm and 600 μm, respectively, in this experiment.

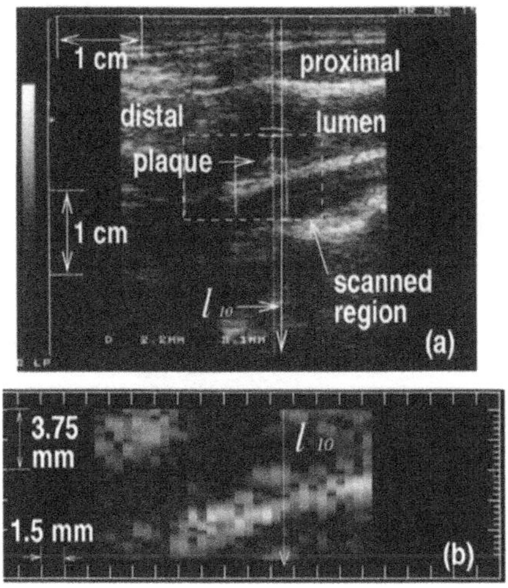

Fig. 6. (a) A cross-sectional B-mode image around the atherosclerotic plaque of the right common carotid artery in a 54-year-old male patient with hypertension obtained by the transcutaneous ultrasonic diagnosis equipment with a 7.5 MHz linear type probe. By the developed method, the elasticity of each point in the area surrounded by the dashed square is obtained and color-coded as shown in Fig. 6(b).

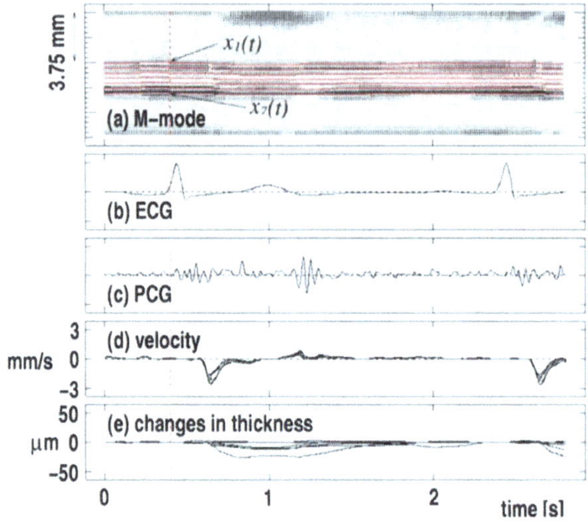

Fig. 7. (a) M-mode image along the 10th scanned beam. (b) The electrocardiogram. (c) The phonocardiogram. (d) Velocities of the points $\{x_i\}$ (i=1, 2, ..., 7). (e) The change in thickness of each layer.

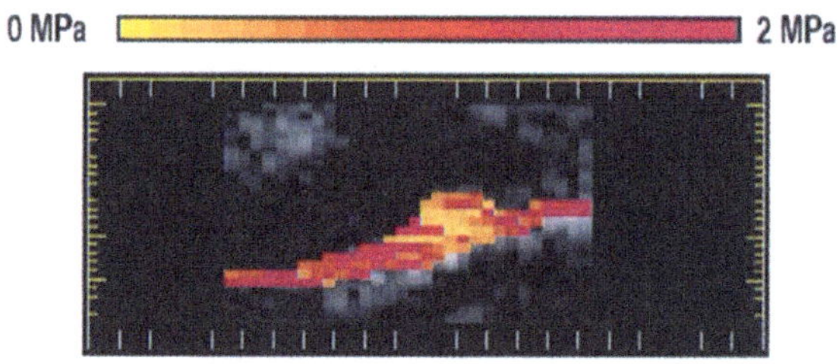

Fig. 8. Color-coded regional elastic moduli, $E_{\theta n}{}^{h}$, superimposed on the reconstructed B-mode image.

Discussion

In Fig. 8, the atherosclerotic plaque has been characterized as the compliant material interior surrounded by much elastic material. With extrapolation of the results in an in vitro study [17], we interpreted that the thin layer as the fibrous cap surrounds the soft tissue as the plaque. It should be investigated whether the internal soft tissue has the close relationship with the rupture of the atherosclerosis. However, the spatial heterogeneity in the elasticity will correspond to the points where the large stress is applied. In Fig. 8, the difference in the regional elasticity around the atherosclerotic plaque can be displayed by the proposed method and these results have never been obtained by any other methods.

Conclusions

The phased tracking method is unique in its extremely higher spatial and temporal resolution. This report shows the cross sectional image of the elasticity around the atherosclerotic plaque with this phased tracking method using transcutaneous ultrasound. This demonstrates the feasibility of the measurement of elasticity in the cross sectional image. The method is presently being evaluated in various subjects with atherosclerotic plaque.

In literature there have been many reports that the atherosclerotic plaques at the carotid artery have close relationship with the myocardial infarction and cerebral infarction [18-20]. By applying the angiography, MR angiography, or the CT, the atherosclerotic plaques can be measured. However, such a method to detect the vulnerability of the atherosclerotic plaques with sufficient accuracy has not yet be developed and reported.

The novel method described in this paper offers potential for new diagnostic techniques of the stability and the vulnerability of plaques based on the measurement of the physical property of the atherosclerotic plaques in clinical setting. The comparison of the estimated cross-sectional elastic image with the pathological finding is under investigation [21].

References

1. Falk E (1991) Coronary thrombosis: pathogenesis and clinical manifestations. Am. J. Cardiol. 68:28B-35B.
2. Loree HM, Kamm RD, Stringfellow RG, et al (1992) Effects of fibrous cap thickness on peak circumferential stress in model atherosclerotic vessels. Circ. Res. 71:850-858.
3. Moreno PR, Falk E, Palacios IF, et al (1994) Macrophage infiltration in acute coronary syndromes - implication for plaque rupture. Circulation 90:775-778.
4. Falk E, Shah PK, Fuster V (1995) Coronary plaque disruption. Circulaton 92:657-671.
5. Galis ZS, Sukhova GK, Lark MW, et al (1994) Increased expression of matrix metalloproteinases and matrix degrading activity in vulnerable regions of human atherosclerotic plaques. J. Clin. Invest. 94:2493-2503.
6. van der Wal AC, Becker AE, van der Loos CM, et al (1994) Site of intimal rupture or erosion of thrombosed coronary atherosclerotic plaques is characterized by an inflammatory process irrespective of the dominant plaque morphology. Circulation 89:36-44.

7. Hallock P (1934) Arterial elasticity in man in relation to age as evaluated by the pulse wave velocity method. Arch. Int. Med. 54:770-798.
8. Peterson RH, Jensen RE, Parnell R (1960) Mechanical properties of arteries *in vivo*. Circ. Res.8:622-639.
9. Hansen F, Mangell P, Sonesson B, et al (1995) Diameter and compliance in the human common carotid artery - variations with age and sex. Ultrasound in Med. & Biol. 21:1-9.
10. Kanai H, Sato M, Koiwa Y, et al (1996) Transcutaneous measurement and spectrum analysis of heart wall vibrations. IEEE Trans. on UFFC 43:791-810.
11. Kanai H, Hasegawa H, Chubachi N, et al (1997) Noninvasive evaluation of local myocardial thickening and its color-coded imaging. IEEE Trans. on UFFC 44:752-768.
12. Kanai H, Koiwa Y, Zhang J (1999) Real-time measurements of local myocardium motion and arterial wall thickening. IEEE Trans. on UFFC 46:1229-1241.
13. Hasegawa H, Kanai H, Chubachi N, et al (1997) Evaluation of the Elastic Modulus of the Arterial Wall by Accurate Noninvasive Measurement of Change in Its Thickness (in Japanese). J. Medical Ultrasonics 24:851-860.
14. Hasegawa H, Kanai H, Hoshimiya N, et al (1998) Measurement of Local Elasticity of Human Carotid Arterial Walls and Its Relationship with Risk Index of Atherosclerosis. IEEE Int. Ultrason. Symp. Proc. 1451-1454.
15. Hasegawa H, Kanai H, Hoshimiya N, et al (2001) Evaluation of Regional Elastic Modulus of Cylindrical Shell with Nonuniform Wall Thickness (in Japanese). J. Medical Ultrasonics 28:J3-J13.
16. Kanai H, Sugimura K, Koiwa Y, et al (1999) Accuracy evaluation in ultrasonic-based measurement of microscopic change in thickness. Electronics Letters 35, 949-950.
17. Lee RT, Grodzinsky AJ, Frank EH, et al (1991) Structure-dependent dynamic mechanical behavior of fibrous caps from human atherosclerotic plaques. Circulation 83:1764-1770.
18. Chambers BR, Norris JW (1986) Outcome in patients with asymptomatic neck bruits. N. Engl. J. Med. 315:860-865.
19. Salonen JT, Salonen R (1993) Ultrasound B-mode imaging in observational studies of atherosclerotic progression. Circulation 87(Suppl II):II-56--II-65.
20. Riley WA, Freedman DS, Higgs NA, et al (1986) Decreased arterial elasticity associated with cardiovascular disease risk factors in the young -Bogalusa heart study. Arteriosclerosis 6:378-386.
21. Kanai H, Hasegawa H, Ichiki M, et al (2003) Elasticity Imaging of Atheroma with Transcutaneous Ultrasoud –Preliminary Study–. Circulation 107:3018-3021.

Part 5

Acoustic Microscopy of Vascular Tissue

Quantitative Backscatter Acoustic Microscopy (30 to 50 MHz)

S. Lori Bridal

Laboratoire d'Imagerie Paramétrique, UMR CNRS 7623 – University of Paris VI, 15 rue de l'Ecole de Médecine, 75270 Paris Cedex 06, France

Summary: The goal of this work is to develop quantitative plaque mapping techniques for the evaluation of atherosclerotic plaque composition. Techniques for high resolution (30 to 50 MHz) radio frequency measurement of ultrasound backscatter and attenuation are described and applied in vitro. Local estimates of these parameters are presented in quantitative parameter maps (voxel dimensions on the order of 100's of microns), and their values are correlated to local atherosclerotic plaque content (normal media, collagen, lipidic infiltration, collagen-lipidic, lipids, calcification).

Comparison of quantitative ultrasound images to maps of magnetic resonance transverse relaxation time indicates that complementary information may be obtained by the two modalities. By lowering the frequency and increasing the size of the estimation volumes, quantitative techniques are adapted to image segments of carotid plaque. Best separation of carotid plaque types (lipid, calcification, hemorrhage and mixed) is obtained by using a multi-parametric analysis of attenuation and backscatter parameters.

This work demonstrates that quantitative and multi-parametric images hold potential for the characterization of atherosclerotic plaque content both at backscatter acoustic microscopy frequencies approaching those used for intravascular ultrasound and at lower frequencies (5 to 12 MHz) approaching those used for transcutaneous imaging of the carotid artery.

Key words: Atherosclerosis, Attenuation, Quantitative imaging, Spectral analysis

Introduction

The biomechanical properties, biochemical composition and structure of atherosclerotic plaque are important in evaluating potential lesion evolution and response to interventional therapy [1]. In particular, while highly fibrous lesions may remain stable, lesions with lipidic cores covered by thin fibrous cap are susceptible to rupture and thrombosis. An in vivo means to identify plaque composition and structure could thus provide valuable guidance for diagnosis and treatment of atherosclerosis.

Characteristics such as plaque echo-lucency and morphology have been related to plaque content and clinical outcome, but the level of interrater agreement for such evaluations is limited [2]. To overcome this subjectivity, methods have been proposed to relate texture, angle dependence of backscatter, spectral characteristics, ultrasound attenuation, and elasticity measurements to plaque content and structure. Many promising results have been obtained along these lines of research [3-8], and combining complementary information from several different quantitative parameters appears promising as a means to further refine the evaluation of plaque composition [9-11].

Results described in this chapter explore high-resolution mapping of local atherosclerotic plaque content using measurements of ultrasound backscatter and attenuation. Initial evidence that these quantitative parameters can distinguish between arteries burdened with different types of plaques was offered by measurements in vascular tissues between 1 and 15 MHz [12-15]. More local measurements (30 to 60 MHz) of the directivity, the backscattered power, the attenuation and the slope of attenuation have also been related to atherosclerotic plaque and artery wall content [8, 16, 17]. Our experience with in vitro human atherosclerotic plaque underlines the potential of quantitative spectral mapping for plaque composition evaluation and indicates directions for future developments.

Quantitative Spectral Analysis

Local attenuation and backscatter were estimated using backscattered radio frequency signals from in vitro human aortae. Specimens were frozen instead of fixed to better preserve their acoustic characteristics [17], and measurements were made in 0.9% saline solution. High resolution information was obtained by using ultrasound frequencies in the range of backscatter acoustic microscopy (30 to 50 MHz, wavelengths in soft tissue between approximately 50 and 30 μm). Each backscattered radio frequency signal, digitized at 400 MHz, was locally compensated for diffraction and then used to estimate attenuation and backscatter.

Compensation for Diffraction

Each radio frequency line backscattered from the specimen was gated using a 64-sample (120 μm) sliding Hamming window with 50 % overlap. The radio frequency signal in each gated region was Fourier transformed, squared in magnitude and averaged with the power spectra from the corresponding gated regions at the same distance z from the transducer of N neighboring radio frequency lines to provide a spatially averaged apparent backscattered power spectrum $\langle S_m(f,z)\rangle$. The use of high-frequency, tightly focused transducers can result in significant diffraction of the ultrasound beam [18]. To compensate for this, each average power spectrum was corrected for the system response and diffraction effects using the reference spectrum $S_p(f,z)$ obtained from a 64-point gated specular echo of a steel plate placed at the same distance z from the transducer:

$$\langle S_{cor}(f,z)\rangle = \frac{\langle S_m(f,z)\rangle}{S_p(f,z)}. \tag{1}$$

The distance between the transducer and the artery wall was adjusted so that the entire thickness of the plaque and artery wall of interest was within the depth of field of the transducer for which this simplified diffraction correction technique had been validated by comparison with a reference phantom diffraction correction method [19]. It should be noted that either of these diffraction correction methods may represent an upper limit to the correction required in tissue [20].

Attenuation

The attenuation at each frequency in the measurement bandwidth can be estimated using a multinarrow-band method [21, 22]. The estimation is based on the fact that the spatially averaged attenuated power spectrum compensated for diffraction and the system response can be expressed :

$$\langle S_{cor}(f,z)\rangle = B(f)e^{-2\alpha(f)2z_1} \tag{2}$$

in terms of the backscatter transfer function of the specimen, $B(f)$, and the amplitude attenuation coefficient, $\alpha(f)$ in Neper per unit length. The distance z is the entire one-way distance between the transducer and the region of backscatter, and $2z_1$ is the round-trip propagation distance in the attenuating specimen. The logarithm of Eq. 2 may be expressed as:

$$10\log_{10}\langle S_{cor}(f,z)\rangle = 10\log_{10}[B(f)] - \frac{20}{\ln(10)}\alpha(f)\cdot 2z_1. \tag{3}$$

Assuming that the measurement region is homogeneous, the attenuation may be estimated at each frequency from the slope of a linear regression fit to the log decay of the power at that frequency as a function of the propagation distance in the material. Velocities reported for normal human aortae of 1492-1534 m/s [15] and 1579 to 1628 ms [17] were used to estimate propagation distances at 20 °C and 37 °C, respectively.

The average attenuation in a selected bandwidth is calculated to find the integrated attenuation (IA) as:

$$IA = \frac{1}{f_{\max} - f_{\min}} \int_{f_{\min}}^{f_{\max}} \alpha(f)\, df \tag{4}$$

where the integral is performed numerically using the method of trapezoids.

Backscatter

For each spatially averaged attenuated power spectrum $\langle S_{cor}(f,z)\rangle$, the average power across a selected bandwidth was calculated to measure the apparent integrated backscatter (AIBS):

$$AIBS(z) = \frac{1}{f_{\max} - f_{\min}} \int_{f_{\min}}^{f_{\max}} \langle S_{cor}(f,z)\rangle\, df\,. \tag{5}$$

Although the AIBS has been compensated for the system response and diffraction, it remains apparent in the sense that the backscattered power has not been compensated for the attenuation of the signal across the intervening material between the transducer and the studied region.

A simple compensation for the attenuation across the two-way path traversed by the signal between the artery surface and the scattering volume can be applied to obtain an attenuation-compensated integrated backscatter IB_{AC}:

$$IB_{AC} = AIBS + \sum_{i=1}^{N} \frac{2}{f_{\max} - f_{\min}} \int_{f_{\min}}^{f_{\max}} \alpha_i(f) \cdot D_i\, df\,. \tag{6}$$

In our work, the thickness D_i, number N and types (e.g. collagen, media, lipid) of overlying tissue layers were estimated based on comparison between B-scan images and histological segments of the same specimen. The frequency-dependent attenuation used for each tissue layer was estimated based on the average frequency dependent attenuation calculated for the appropriate tissue type within the same specimen [23].

Volume Mapping to Quantitative Images

By analyzing the data from a windowed region of a small number of adjacent radio frequency signals forming a small volume V, a local estimation of an ultrasonic parameter can be obtained. Because of the stochastic nature of the echographic signals, a reduction in the size of the local measurement volume V leads to an increase in the variance of the estimated parameter [24].

Estimation volumes included an into-image-plane slice thickness of 0.5 mm. Each AIBS estimate originated from an arterial volume having lateral and axial (depth) image dimensions of 200 μm and 120 μm, respectively. Because the attenuation is estimated from spectral change with depth, the axial depth of the attenuation measurement volume was increased to 420 μm for each voxel. Volumes were displaced along the depth axis by 60 μm and did not overlap laterally. The division of a volume of backscattered radio frequency into spectral estimation volumes is illustrated schematically in Figure 1.

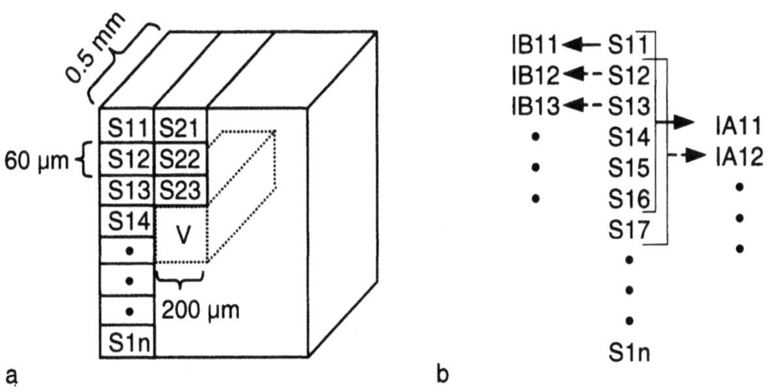

Fig. 1a, b. Estimation volumes and parameter mapping. **a** The entire cube represents the collected volume of radio frequency data. For each selected sub-volume V of data, an average power spectrum was calculated (labeled S: column number: depth number). **b** For each spectrum, one value of integrated backscatter (IB) was mapped to a pixel of the IB image. One integrated attenuation (IA) pixel was mapped from each series of 6 spectra.

Parameter Sensitivity to Plaque Composition

Example images of the ultrasound attenuation and apparent backscatter are compared to the corresponding histologic section in Figure 2. Lower values are represented as darker and higher values as brighter in the gray-scale of the parameter images. This segment presents a lipidic and collagen-lipidic core surrounded by a dense fibrous cap. The presence of microcalcifications is noted by the arrowheads on the histologic section in Figure 2c.

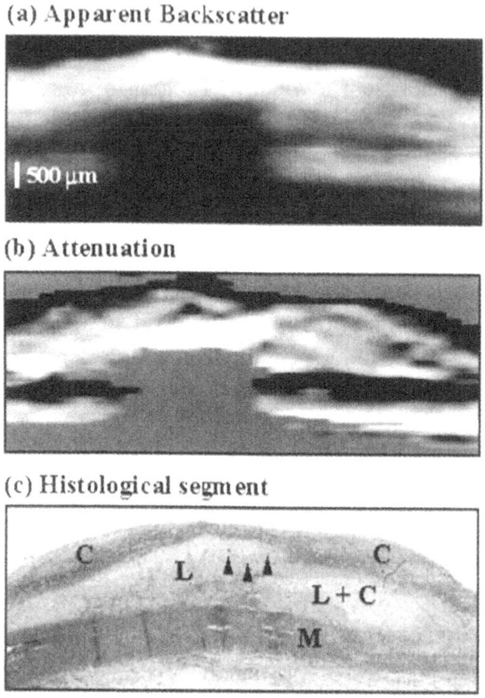

Fig. 2a, b, c. Segment of human aorta with atherosclerotic plaque. **a,** Apparent integrated backscatter (30 to 50 MHz). **b**, Integrated attenuation (30 to 50 MHz). **c**, Histological segment. The plaque consists of a collagen-lipidic (L + C) and lipidic (L) core containing microcalcifications denoted by the arrowheads and surrounded by a collagen cap (C). The media is denoted by M. In the quantitative images the level of gray corresponds to the value of the measured parameter with higher attenuation or backscatter being mapped to a brighter level of gray. (from [25], with permission of © Masson Editeur).

Dark bands at the surface of the plaque and the interface between the plaque and medial layers on the attenuation image result from artifactual values calculated at the tissue interfaces (estimation volumes are not homogeneous). In interpreting images of the local attenuation, it should be kept in mind that features having a narrow axial dimension may correspond to an interface between tissue types. Although such artifactual attenuation values do not characterize the structure or composition of the tissues at the interface, they do provide a useful delimiting mark between tissue types.

Bright levels of both attenuation and backscatter followed by shadowing are observed in the region containing microcalcifications. The collagen-lipidic core (L + C) presents elevated attenuation and decreased backscatter as compared to regions of the collagen cap. To more precisely explore the relationship between parameter values and plaque content, average values of attenuation and backscatter from local regions of quantitative images were systematically compared to plaque composition for two groups of specimens.

For each plaque segment studied with ultrasound, tissue composition was evaluated on histologic slices stained with orcein for elastic fibers and hematoxylin-eosin-saffron for nuclei, cytoplasm and collagen, respectively. Plaque regions were classified as: normal media, dense collagen, lipidic infiltration, collagen-lipidic, lipidic and containing calcifications. Results from two groups of specimens are summarized in Table 1. Group 1 included 124 regions of human aortae that were measured at ambient temperature. Group 2 included 54 regions of human aortae and iliac arteries measured at 36.8 ± 0.6 °C.

Table 1. Ultrasound attenuation and backscatter (30 to 50 MHz) as a function of local atherosclerotic plaque composition

Classification	Group 1	Group 2	
	IA (dB cm^{-1})	IA (dB cm^{-1})	IB_{AC} (dB)
Iliac media	–	84 ± 29	-21 ± 5
Aortic media	97 ± 20	88 ± 14	-16 ± 2
Collagen	107 ± 33	81 ± 23	-25 ± 3
Lipidic infiltration	–	99 ± 24	-23 ± 2
Collagen-lipidic	142 ± 51	–	–
Lipids	139 ± 53	132 ± 17	-28 ± 5
Calcified	245 ± 93	257 ± 40	-12 ± 4

IA: integrated attenuation [4, 23], IB_{AC} : attenuation-compensated integrated backscatter (IB_{AC} has also partially been compensated for the frequency-dependent beam volume [23]).

Quantitative Ultrasound and Magnetic Resonance Images

Ultrasound parameters are thought to vary as a function of plaque composition in response to differences in the material mechanical properties, organization and macromolecular composition. Magnetic resonance imaging offers an alternative means to probe the composition of atherosclerotic plaque that may be expected to be most affected by macromolecular composition as well as by the environment of the macromolecules within the plaque.

For specimens in Group 2, quantitative characterization of artery constituents by local measurement of ultrasound backscatter and attenuation parameters was compared to characterization by measurement of magnetic resonance transverse relaxation time T2 (8 echo-single-slice sequence at 3 Tesla, 110 × 220 μm resolution). Results of the comparison illustrated in Table 2 suggest that these different parameters may supply complementary information useful for the identification of tissue type. In particular, collagen free of lipids was not separated from collagen with lipidic infiltration using ultrasound parameters alone, but collagen with lipidic infiltration was not discriminated from highly lipidic regions by T2 alone. Only the comparison of parameters measured by the two techniques permitted the separation of these three specific groups.

Table 2. Comparison of magnetic resonance and ultrasound discrimination of atherosclerotic plaque constituents (from [23] © 1998 IEEE, with permission)

Measured Parameter	Collagen	Lipidic Infiltration	Lipids	Calcifications
US: IB_{AC}	–	–	–	↑[a], $p < 0.005$[c]
US: IA	–	–	↑, $p < 0.05$	↑, $p < 0.005$
MR: T2	↑, $p < 0.005$	↓, $p < 0.05_{\dagger}$	↓, $p < 0.05$	NR[b]

[a]Arrow direction indicates whether values are elevated or decreased relative to the media.
[b]Not Reported because calcifications do not provide magnetic resonance signal.
[c]The difference was not achieved between the plaque constituent and aortic media.

Extension to Carotid Plaque Characterization

By increasing the volume of tissue used for each parameter estimation, the techniques for local estimation of ultrasound attenuation and backscatter were adapted for lower frequencies approaching those used for quantitative imaging of plaque in peripheral arteries. Radio frequency data were acquired from 15 carotid endarterectomy specimens across two ultrasound bandwidths (5 to 12 MHz and 12 to 28 MHz). Volumes used for estimation of the average diffraction corrected backscattered power spectrum had a 2 mm slice thickness, 1 mm width and axial depths of 480 μm and 240 μm at 10 MHz and 20 MHz, respectively. For each estimation volume and bandwidth, the integrated bacscatter IB and the frequency dependence of backscatter n (slope of the least-squares linear fit of the log of backscater vs. the log of frequency) were calculated. These apparent backscatter parameters were not corrected for attenuation. The integrated attenuation IA and the slope of attenuation β were estimated from the power decay of spectra from six consecutive estimation volumes. Local parameter maps were constructed for each of these four parameters, and average parameter values in 59 regions of interest were correlated with plaque composition as assessed by histology (lipid, calcification, hemorrhage and mixed).

No single parameter was able to discriminate between all groups. Stepwise variable-selection retained the IB, n, IA and β from 5 to 12 MHz and the IB and n from 12 to 28 MHz, for separation of the plaques between the four groups. Values of three discriminant functions consisting of weighted linear combinations of these parameters are plotted in Figure 3. Some separation of the groups can be appreciated although no classification boundaries can be devised that entirely separate mixed plaques from intraplaque hemorrhage and lipidic plaques.

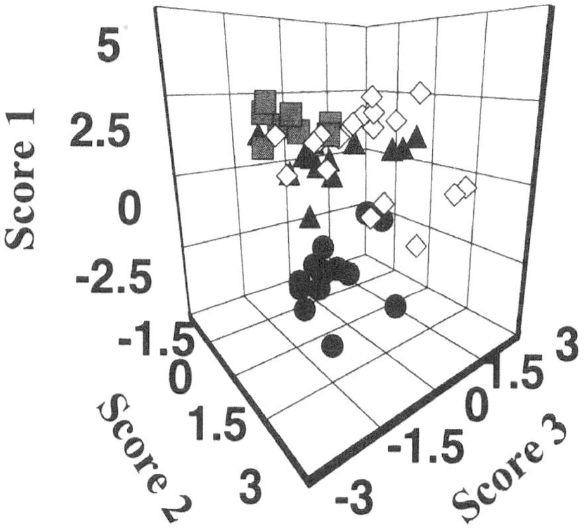

Fig. 3. Three-dimensional scatter plot showing the discriminant function scores for each of the carotid plaque slices divided by tissue groups of calcification (black circles), intraplaque hemorrhage (gray squares), lipid (black triangles) and mixed (white diamonds). From [11], with permission.

Conclusions

In this work, attenuation and backscatter were estimated locally within the arterial wall using backscattered radio frequency signals. Results demonstrate that high resolution quantitative ultrasound images can be obtained at frequencies comparable to those available with intravascular ultrasound (IVUS) probes and that the value of parameters in these images can be linked to plaque content [4, 23].

Thus, the potential exists to obtain more information linked to plaque structure than is currently offered by conventional IVUS. Although many challenges must be met to transfer such techniques to clinical application, by virtue of its direct acoustic path and high frequencies, IVUS offers a novel opportunity for quantitative tissue characterization. By relating in vivo images of plaque content to the plaque types which have previously been identified in histologic studies as the most likely to lead to clinical complications, plaque evolution and response to therapy could potentially be better predicted.

Magnetic resonance imaging represents an alternative modality that shows promise as a means for evaluating plaque content. We compared the sensitivity to plaque types obtained by high resolution measurement of quantitative ultrasound and magnetic resonance parameters [23]. Results indicate that the studied parameters offer complementary information which could be combined to provide a more precise plaque characterization than any one of the parameters alone.

Of the two modalities, the only that could currently provide in vivo resolutions on the order of those obtained in our in vitro backscatter acoustic microscope studies is intravascular ultrasound. In vivo application of magnetic resonance imaging for plaque characterization is most likely to be able to attain useful resolutions (pixel size on the order of 400 μm by 200 μm) for the carotid artery if 1.5 to 3 Tesla fields are used in conjunction with small surface coils. Quantitative transcutaneous ultrasonic imaging of the carotid artery cannot easily be anticipated to obtain pixel resolutions better than about a millimeter due to the lower frequencies necessary to penetrate intervening tissues. Despite these resolution limitations, noninvasive quantitative ultrasound imaging of the plaque appears to have potential for the identification of carotid plaque composition, and the separation of plaque types can be optimized by combining information from several quantitative ultrasound parameters.

The field of applications for these methods extends well beyond arterial characterization. Because most organs are accessible in echographic mode, these techniques may be adapted for many cases of clinical interest adding a new, more objective view to ultrasonic imaging.

References

1. Lusis AJ (2000) Atherosclerosis. Nature 407:233-241
2. Montauban van Swijndregt AD, Elbers HRJ, Moll FL, et al (1998) Ultrasonographic characterization of carotid plaques. Ultrasound Med Biol 24:489-493
3. Aly S, Bishop CC (2000) An objective characterization of atherosclerotic lesion an alternative method to identify unstable plaque. Stroke 31:1921-1924
4. Bridal SL, Fornes P, Bruneval P, et al (1997) Correlation of ultrasonic attenuation (30 to 50 MHz) and constituents of atherosclerotic plaque. Ultrasound Med Biol 23:691-703
5. de Korte CL, Pasterkamp G, van der Steen AFW, et al (2000) Characterization of plaque components with intravascular ultrasound elastography in human femoral and coronary arteries in vitro. Circulation 102:617-623
6. Jeremias A, Kolz ML, Ikonen TS, et al (1999) Feasibility of in vivo intravascular ultrasound tissue characterization in the detection of early vascular transplant rejection. Circulation 100:2127-2130

7. Spencer T, Ramo MP, Salter DM, et al (1997) Characterization of atherosclerotic plaque by spectral analysis of intravascular ultrasound: An in vitro methodology. Ultrasound Med Biol 23:191-203
8. Wilson LS, Neale ML, Talhami HE, et al (1994) Preliminary results from attenuation-slope mapping of plaque using intravascular ultrasound. Ultrasound Med Biol 20:529-542
9. Lee DJ, Sigel B, Swami VK, et al (1998) Determination of carotid plaque risk by ultrasonic tissue characterization. Ultrasound Med Biol 24:1291-1299
10. Noritomi T, Sigel B, Swami V, et al (1997) Carotid plaque typing by multiple-parameter ultrasonic tissue characterization. Ultrasound Med Biol 23:643-650
11. Bridal SL, Beyssen B, Fornès P, et al (2000) Multiparametric attenuation and backscatter images for characterization of carotid plaque. Ultrason Imaging 22:20-34
12. Barzilai B, Saffitz JE, Miller JG, et al (1987) Quantitative ultrasonic characterization of the nature of atherosclerotic plaques in human aorta. Circ Res 60:459-463
13. Picano E, Landini L, Distante A, et al (1983) Different degrees of atherosclerosis detected by backscattered ultrasound: An in vitro study on fixed human aortic walls. J Clin Ultrasound 11:375-379
14. Urbani MP, Picano E, Parenti G, et al (1993) In vivo radiofrequency-based ultrasonic tissue characterization of the atherosclerotic plaque. Stroke 24:1507-1512
15. Greenleaf JF, Duck FA, Samayoa WF, et al (1974) Ultrasonic data acquisition and processing system for atherosclerotic tissue characterization. IEEE Ultrasonics Symposium 738-743
16. de Kroon MGM, van der Wal LF, Gussenhoven WJ, et al (1991) Backscatter directivity and integrated backscatter power of arterial tissue. Int J Card Imaging 6:265-275
17. Lockwood GR, Ryan LK, Hunt JW, et al (1991) Measurement of the ultrasonic properties of vascular tissues and blood from 35-65 MHz. Ultrasound Med Biol 17:653-666
18. Laugier P, Berger G, Fink M, et al (1985) Specular reflector noise: Effect and correction for in vivo attenuation estimation. Ultrason Imaging 7:277-292
19. Bridal SL, Fornès P, Bruneval P, et al (1997) Parametric (integrated backscatter and attenuation) images constructed using backscattered radio frequency signals (25 - 56 MHz) from human aortae in vitro. Ultrasound Med Biol 23:215-229
20. Robinson DE, Wilson LS, Bianchi T (1984) Beam pattern (diffraction) correction for ultrasonic attenuation measurement. Ultrason Imaging 6:293-303
21. Fink M, Hottier F, Cardoso JF (1983) Ultrasonic signal processing for in vivo attenuation measurement: Short time Fourier analysis. Ultrason Imaging 5:117-135
22. Roberjot V, Bridal SL, Laugier P, et al (1996) Absolute backscatter coefficient over a wide range of frequencies in a tissue-mimicking phantom containing two populations of scatterers. IEEE Trans UFFC 43:970-978
23. Bridal SL, Toussaint J-F, Raynaud J-S, et al (1998) US backscatter and attenuation 30 to 50 MHz and MR T2 at 3 Tesla for differentiation of atherosclerotic artery constituents. IEEE Trans UFFC 45:1517-1525
24. Huisman HJ, Thijssen JM (1996) Precision and accuracy of acoustospectrographic parameters. Ultrasound Med Biol 22:855-871
25. Bridal SL, Saïed A, Chérin E, et al (1998) High-resolution quantitative echography with tissue parameters. Journal d'Echographie et de Médecine par Ultrasons 19:204-211

Evaluation of Atherosclerosis by Acoustic Microscopy

Yoshifumi Saijo

Department of Medical Engineering and Cardiology, Institute of Development, Aging and Cancer, Tohoku University

Summary: Acoustic microscopy can provide not only the morphology, but also the biomechanical properties of biological soft tissues. The biomechanics of atherosclerosis is important because the pathophysiology of atherosclerosis is closely related to mechanical properties and to mechanical stress. The initial event in acute coronary syndrome such as acute myocardial infarction or unstable angina is the rupture of the fibrous cap of atheromatous plaque. In addition to extrinsic physical stresses to the plaque, the intrinsic biomechanical property of the plaque is important for assessing the mechanism of the rupture. Two Scanning Acoustic Microscopes (SAM), operating in 100 to 200 MHz and in 800 MHz to 1.3 GHz frequency regions were equipped to measure the acoustic properties of atherosclerosis of human and mouse arteries. The values of attenuation and sound speed in the tissue components of atherosclerosis were measured by analyzing the frequency dependent characteristics of the amplitude and phase of the ultrasonic signals. Both values were highest in calcification and lowest in lipid pool. Although attenuation and sound speed were relatively high in intimal fibrosis, inhomogeneity of the acoustic parameters was found within the fibrous cap. Polarized microscopy of collagen stained with Picrosirius red showed that the attenuation of ultrasound was significantly higher in type I collagen with orange polarized color compared to type III collagen with green color. The SAM has shown that it is possible to detect plaque vulnerability and that it might improve our understanding of sudden rupture from a micro-mechanical point of view.

Key words: Atherosclerosis, acoustic microscopy, collagen, attenuation, sound speed, ultrasonics

INTRODUCTION

Optics has been the main observational mode in the microscopic world for many years. However, its domain is restricted to largely transparent media. Sokolov in the former Soviet Union, first proposed the concept of an "ultrasonic microscope" in order to visualize opaque media in 1949[1]. He pointed out that the wavelength of 3-GHz ultrasound in water is the same as the wavelength of green light in air. At that time, however, the technologies to produce 3-GHz ultrasound did not exist.

In 1959, Dunn and Fry at University of Illinois developed an "ultrasonic absorption microscope method" with 12-MHz ultrasound[2]. During the 1970s, high frequency techniques were developed for the microwave technology for radar and for satellite communications. In 1973, Quate and Lemons at Stanford University developed a "scanning acoustic microscope (SAM)"[3], now used in the biomedical field because of its high resolution and high quality imaging.

We have been applying SAM technology for medical and biological studies at Tohoku University since 1980[4,5] for the following purposes. First, SAM can be applied for intra-operative pathological examinations since it does not require special staining techniques. The University of California at Irvine group has determined that the obtainable resolution of 600-MHz SAM is sufficient to render a microscopic diagnosis[6,7]. We have shown that SAM can classify the types of cancer in stomach[8] and kidney[9] by quantitative measurement of ultrasonic attenuation and sound speed at 200-MHz.

Second, ultrasonic data obtained with the high frequency SAM can be used for assessing reflectability or texture in clinical echographic imaging. Density and sound speed c determine the characteristic acoustic impedance Z of the material as

$$Z = \rho c$$

On the assumption that the interface between two fluid-like media is infinite and plane, the relative reflected sound power, in dB, can be determined by the specific acoustic impedance of each medium as

$$dB = 10\log_{10}\frac{P_r}{P_i} = 10\log_{10}\frac{(Z_a - Z_b)^2}{(Z_a + Z_b)^2}$$

(P_r: sound power reflected at interface, P_i: incident sound power, Z_a: acoustic impedance of medium a, Z_b: acoustic impedance of medium b)

Although the assumption is overly broad for biological media, we have shown that the calculated reflection data and clinical observations are in substantial agreement for myocardium[10]. Chandraratna et al. also assessed the echo-bright area in myocardium by using 600-MHz SAM[11].

Third, in its simplest form, the relation between the sound speed and the elastic bulk modulus of liquid-like media is

$$c = \sqrt{\frac{K}{\rho}}$$

(c: sound speed, K: elastic bulk modulus, ρ: density)

As a biological tissue may be considered as a liquid-like material, this equation may be applied to assess its elastic properties. Recent biomechanical studies have suggested that the mechanical properties of tissues may not be sufficiently similar to liquids and should be treated as soft solid materials. However, the acoustical relations of solid materials can also be described by the following equation.

$$c = \sqrt{\frac{E(1-\sigma)(1+\sigma)}{\rho(1-2\sigma)}}$$

(c: sound speed, E: Young's modulus, σ: Poisson's ratio, ρ: density)

This relation shows that the Young's modulus of tissue and the sound speed are closely related.

The pathophysiology of atherosclerosis is closely related to the biomechanical properties as the arteries are always experiencing mechanical stress caused by blood pressure or blood flow. Both the dynamical and static mechanical properties of arteries are affected. As the term "arteriosclerosis" means hardening of the artery, atherosclerotic tissues increase in stiffness and hardness during the development of atherosclerosis. Thus, assessment of the mechanical properties of atherosclerosis is important, in addition to determining the chemical components of the tissue.

Coronary plaque disruption, with superimposed thrombi, is the main cause of the acute coronary syndrome such as unstable angina or acute myocardial infarction[12]. Pathological investigations have shown that plaque size, or degree of the lumen narrowing, is less important than the composition of the plaque; lipid-rich and soft plaques are more vulnerable than the collagen-rich hard plaques. In clinical settings, intravascular ultrasound (IVUS) imaging has been applied for detecting a vulnerable plaque, which is sometimes found in an angiographically normal coronary artery. The composition of atherosclerosis is defined as the collagen fibers, lipid and calcification in the intima, the smooth muscle and elastic fiber in the media, and the collagen fibers in the adventitia, respectively. However, the degree of staining for optical microscopy indicates that the collagen content or the chemical property is not homogeneous in fibrotic lesions in the intima.

In this chapter, two kinds of scanning acoustic microscope (SAM) systems were employed to characterize the tissue components in normal and in atherosclerotic arteries. The values of attenuation and sound speed in six types of tissue components in atherosclerosis were measured by 100-200 MHz SAM. The collagen type was assessed by 800-1300 MHz SAM and polarized microscopy for qualitative and quantitative analysis of the characterization of collagen fibers. In the chapter the potential for higher resolution imaging was also described by showing the examples on cultured cells.

MATERIALS AND METHODS

Acoustic microscopy in 100 to 200 MHz

Figure 1 is a block diagram of the SAM system developed at Tohoku University. The acoustic focusing element comprises a ZnO piezoelectric transducer with a sapphire lens. The ultrasonic frequency is variable over the range of 100 to 210

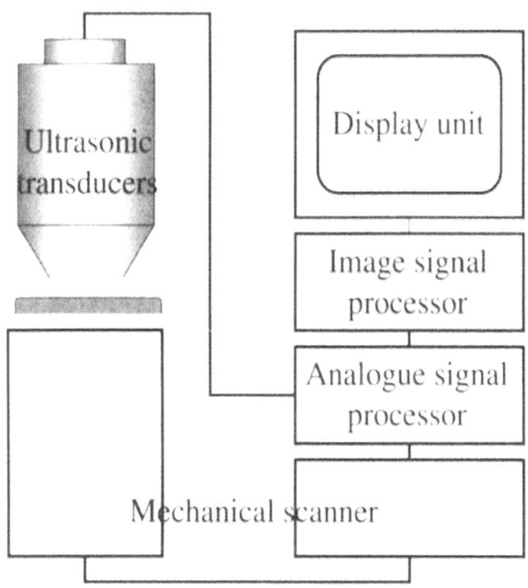

Figure 1. The block diagram of scanning acoustic microscope system developed at Tohoku University

MHz and the beam width at the focal volume ranges from 5 μm (at 210 MHz) to 10 μm (at 100 MHz). The focusing element is mechanically scanned at 60 Hz, in the lateral direction (x) above the specimen, which remains stationary on the specimen holder, while the holder is scanned in other lateral direction (y) in 8 sec, thus providing two-dimensional scanning. The mechanical scanner is so arranged that the ultrasonic beam be transmitted for every 4 μm interval over a 2 mm width. The number of sampling points is 480 in one scanning line and 480 x 480 points make one frame. Both amplitude and phase images can be obtained in a field of view 2mm x 2mm. Either reflection or transmission mode is selectable in the system. In the present study, the reflection mode was chosen. Distilled water, which was maintained at 37°C, was used for the coupling medium between the transducer and the specimen.

The region of interest (ROI) of the specimen was determined by observing the two-dimensional amplitude image. Both the amplitude and phase in the ROI were measured for 10 MHz intervals between 100 to 210 MHz to obtain the quantitative values of acoustic properties. Thus, the frequency varying change of the amplitude and phase were obtained.

The equation below shows the relationship between the frequency, amplitude and phase due to the interference between the reflection of the surface and bottom of the specimen. By fitting the obtained frequency varying curve with the calculated values, the thickness of the tissue in the ROI was obtained.

$$
\begin{aligned}
y_r &= y_s + y_b \\
&= e^{-(\alpha + j\frac{2\pi f}{c})d} \\
&= e^{-j\frac{2\pi f}{c}d} \cdot e^{-A_0 f^n d}
\end{aligned}
$$

(y_r: received signal, y_s: reflection from the surface of the specimen, y_b: reflection from the bottom of the specimen, f: frequency, c: sound speed of the specimen, d: thickness of the specimen)

Figure 2 shows graphs of the relationship between the frequency and the amplitude or phase in a 5 μm thickness specimen.

Once the thickness of the specimen was determined, the attenuation and sound speed were calculated by the following equations.

$$A = \frac{L}{f\,d}$$

(A: slope of attenuation, f: frequency, d: thickness)

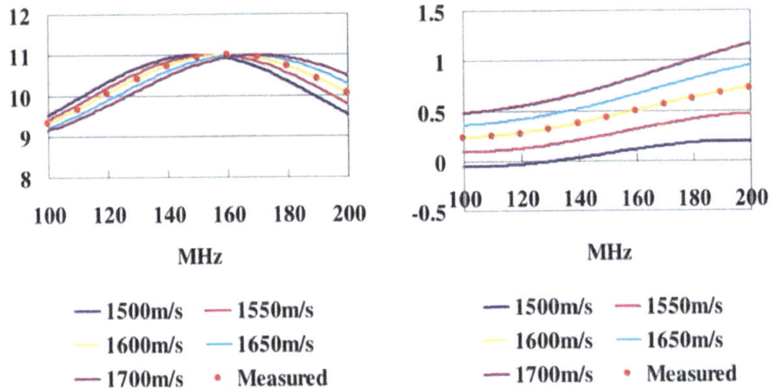

Figure 2. The graphs showing the relationship between frequency and the amplitude (left) or phase (right) in 6μm thickness specimen

$$c = \frac{1}{\frac{1}{c_w} - \frac{\theta}{2\pi f d}}$$

(f: frequency, d: thickness of the specimen, c: sound speed in the specimen, c_w: sound speed in the coupling medium, θ: phase shift)

Thus, quantitative values of slope of attenuation and of sound speed of each pixel can be obtained and this is the most important property of our scanning acoustic microscope system. The data of slope of attenuation and sound speed were converted into color signals and displayed on a color monitor as two-dimensional distribution patterns. Figure 3 shows the color bar scales of attenuation and sound speed.

Twenty-three human coronary arteries were involved in the study. A coronary artery was excised at autopsy. For SAM examination, the tissue was frozen and sectioned at 10-μm by a cryostat and mounted on a slide glass. For optical imaging, the tissue was formalin-fixed, paraffin embedded and sectioned at 4- μm by a microtome. The specimen was stained with Elastica-Masson's trichrome and

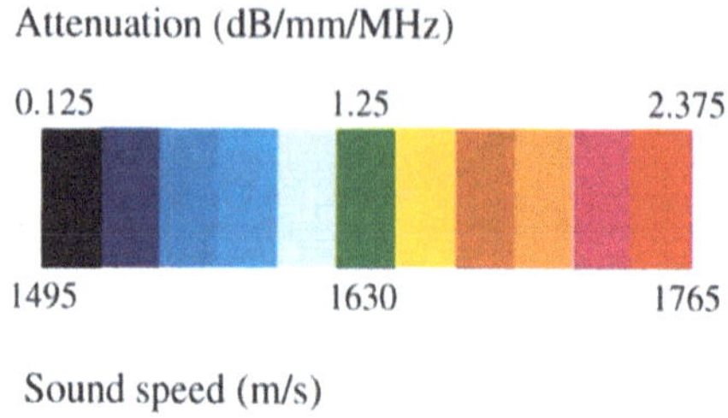

Figure 3. Color bar scales of attenuation and sound speed

used for optical microscopy. The ROI for acoustic microscopy was determined by optical microscopic observations. The tissue components in normal and atherosclerotic arteries were classified as the following tissue components; normal intima, fibrosis, calcification, lipid, normal media and normal adventitia. The ROI was so determined that only a single tissue component was included. Then, the mean values of attenuation and sound speed of each tissue component were averaged.

Acoustic microscopy in 800 to 1300 MHz

A scanning acoustic microscope (SAM2000; KSI, Herborn, Germany) was equipped for semi-quantitative ultrasonic measurements. In the present study, an acoustic lens with the optimal frequency range between 800 MHz and 1.3 GHz was used. The SAM frequency chosen in the experiments was 1.1 GHz because the image quality was best. Distilled water was used for the coupling medium between the transducer and the sample. The temperature of the transducer, coupling medium, sample and the glass substrate was maintained at 37 C° by a heating plate. The Rf output gain was fixed for all experiments. In the SAM system, an 80 dB difference of ultrasonic intensity corresponded to 256 gray scales (0-255). The reflection from the glass surface was set as level 255 (white). The average thickness of the specimen was determined by laser micrometer. Ultrasonic attenuation at each pixel was calculated from the number of the gray level and the thickness. All of the SAM images were stored in a personal computer for statistical analysis.

The optical and polarized images of atherosclerotic plaque were captured to make the same field taken by SAM. The images were analyzed using computer-assisted image analysis equipment (Leitz DMRBE light microscope; Leica

Wetzlar, Germany, JVC KY-F55 color video camera; Victor Co.Ltd., Tokyo, Japan, Scion CG-7 RGB color frame grabber, and ScionImage; Scion Corporation, Frederick, MD, USA, and Photoshop; Adobe Systems Inc., Seattle, WA, USA). Although the picrosirius red staining was not a specific staining technique, the area with collagen was specifically visualized by polarized microscopy. The areas were differentiated according to the polarized color; longer wavelength (red or yellowish color) or shorter wavelength (green).

For the study at 800-1300 MHz, 11 normal human coronary arteries and 23 atherosclerotic aortae in Apo-E deficient mice were involved. The specimen was perfusion-fixed at 100 mmHg with phosphate-buffered 4% formaldehyde (PH 7.2) and immersed in the fixative overnight. The specimen was sectioned at 4-μm and embedded in paraffin. The paraffin was removed by a coconut oil and graded alcohol process for study. Unstained sections were used for the SAM investigations. After the SAM investigation, the sample was stained with Sirius red and covered by a cover glass. The same field observed by SAM is also observed by normal and polarized optical microscopy for comparing the tissue components.

To test the potential of acoustic microscopy, living cells were also visualized. Human renal artery smooth muscle cells were cultured on 35 mm diameter dishes with the Dulbecco's modified Eagle's medium and 10% heat-incubated bovine serum. The incubator was maintained at 37 °C and filled with 95% air and 5% CO2. Phosphate buffered saline was used for the coupling medium between SAM transducer and the cells. SAM images of a single cell were recorded by 3 minute intervals in order to detect the cellular motility. The difference between two consecutive frames was considered as the cellular motility. Calculated difference was colorized and overlayed on a SAM image.

RESULTS

Figure 4 shows the optical (a), attenuation (b) and sound speed (c) images of atherosclerotic human coronary artery visualized with the 100-200 MHz range acoustic microscope. Normal coronary artery is comprised of three layers; intima consisting of endothelium and thin collagen fiber, media consisting of smooth muscle and elastic fiber, and adventitia consisting of collagen fiber. In this image, the intima is thickned and a lipid core (L) underlying a thick "fibrous cap" (F) is observed in the intima. The lipid shows very low values of attenuation and sound speed compared with fibrosis. Figure 5 shows the optical (a), attenuation (b) and sound speed (c) images of atherosclerotic human coronary artery with coronary thrombosis. Newly organized thrombosis (T) shows higher values of attenuation and sound speed than those of the fibrosis (F).

Figure 6 show the optical (a), 200MHz (b) and 1.1GHz (c) acoustic microscopy images of normal human coronary artery. The image quality is better in 1.1GHz image. For example, each elastic fiber in the media can be observed in the 1.1GHz

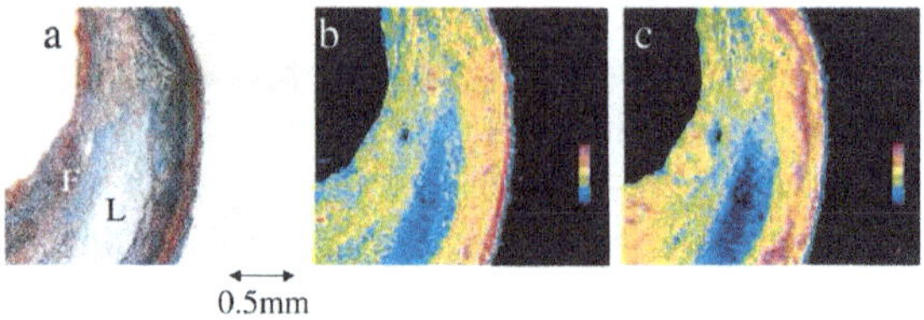

Figure 4. Acoustic microscopic images (a: optical, b: attenuation, c: sound speed) of an atherosclerotic coronary artery

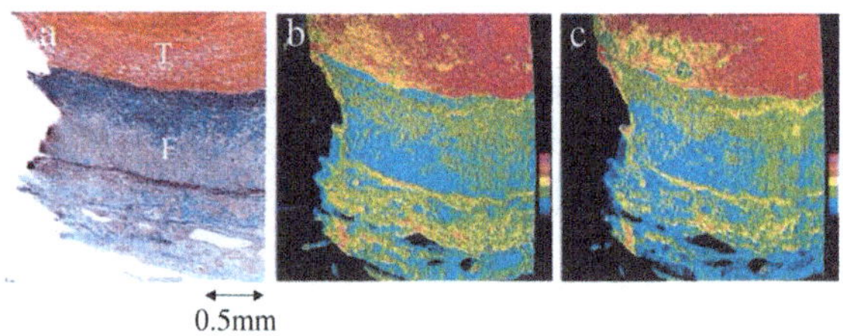

Figure 5. Acoustic microscopic images (a: optical, b: attenuation, c: sound speed) of an atherosclerotic coronary artery with thrombosis

image while the media is almost homogeneous in the 200MHz image. However, the three-layered appearance of coronary artery is more obvious in the 200MHz image. The intima and adventitia show higher ultrasonic attenuation than that of media.

Figure 7 shows the normal light (a), polarized light (b) and acoustic (c) microscopy images of a normal coronary artery. The ultrasonic frequency is 1.1

GHz. Endothelial cells aligning in a single layer and thin fibrosis comprise the intima. The smooth muscle cells and elastic fibers consisting of the media are shown by the acoustic image. Collagen fibers in the adventitia are highlighted by the polarized image. No pathological atherosclerotic lesions were found in this specimen.

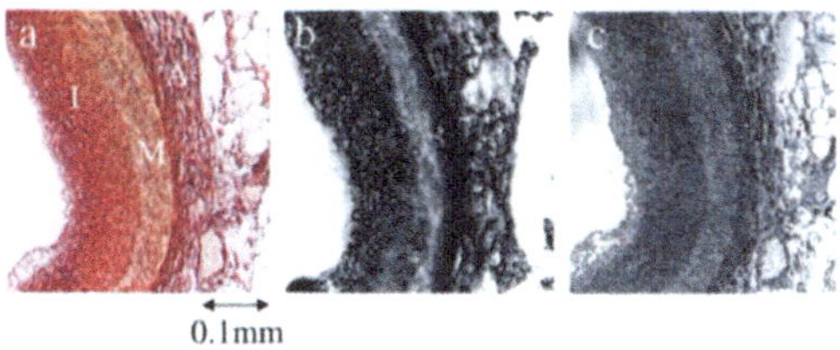

Figure 6. Optical (a), 200MHz (b) and 1.1GHz (c) acoustic microscopy images of normal human coronary artery.

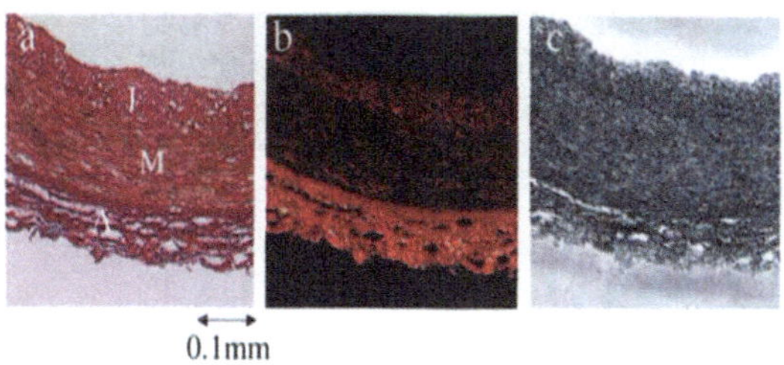

Figure 7. Normal light microscopic (a), polarized light microscopic (b) and acoustic microscopic (c) images of the normal coronary artery

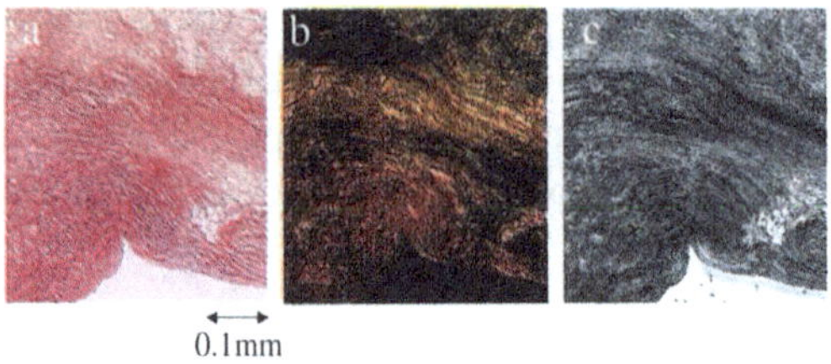

Figure 8. Normal light microscopic (a), polarized light microscopic (b) and acoustic microscopic (c) images of the atherosclerotic coronary artery

Figure 8 shows an atherosclerotic plaque found in a healthy young adult without a history of ischemic heart disease. The collagen network in the intima is shown. Although the lipid in the plaque was completely removed by alcohol dehydration, the remaining outer shape of the cholesterol crystal indicates that the cholesterol had existed in the spaces. The fibrosis in the plaque can be classified into two categories by optical and acoustical findings; weak polarization and low ultrasonic attenuation, and strong polarization and high ultrasonic attenuation.

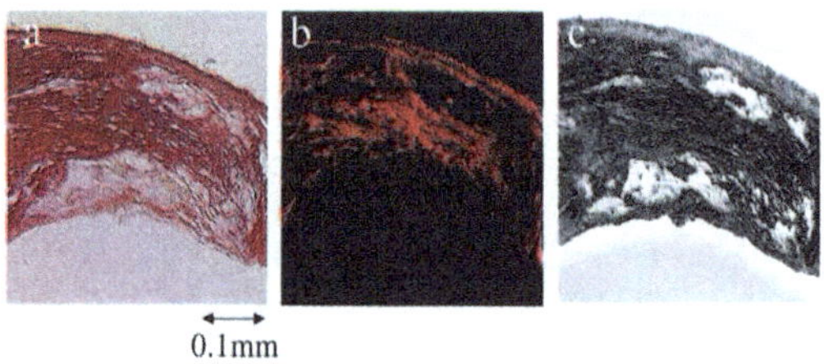

Figure 9. Normal light microscopic (a), polarized light microscopic (b) and acoustic microscopic (c) images of a thick fibrous cap in an Apo-E deficient mouse.

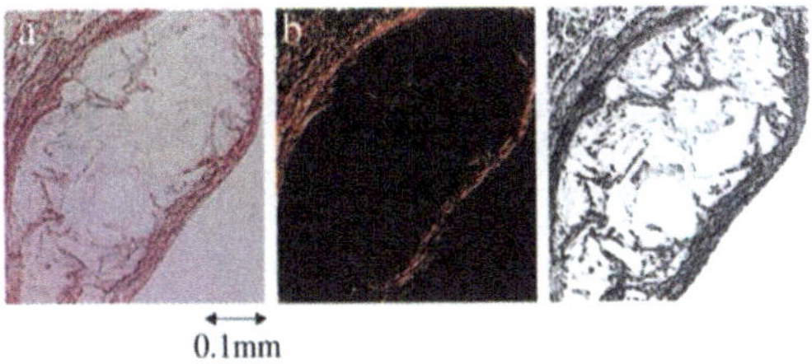

Figure 10. Normal light microscopic (a), polarized light microscopic (b) and acoustic microscopic (c) images of a thin fibrous cap in Apo-E deficient mouse.

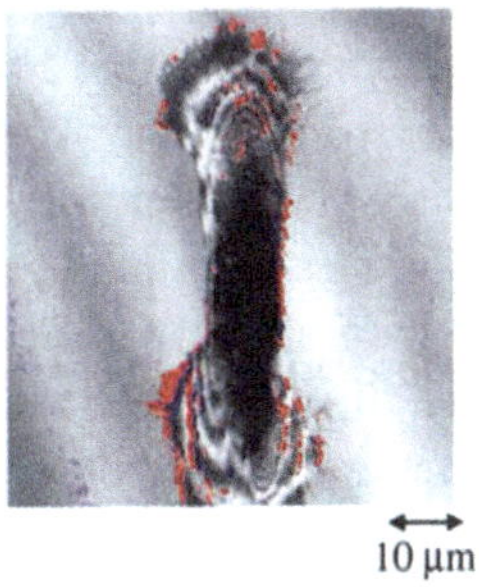

Figure 11. SAM image of a human renal artery smooth muscle cell. Blue indicates the difference between the previous and the present frames, while red indicates the difference between the present and the next frame.

Figure 9 shows a thick fibrous cap in an Apo-E deficient mouse. The fibrosis is strongly polarized in orange in the polarized image and shows high attenuation in SAM image. Figure 10 shows a thin fibrous cap in Apo-E deficient mouse. Thin fibrous cap overlying thick lipid pool is shown. In the lipid pool, some macrophages can be recognized by SAM image. The sparse collagen network in

Table 1. Acoustic properties of tissue components of atherosclerosis in Apo-E deficient mice

Tissue components	Attenuation at 1.1GHz (dB/mm@1.1GHz)
Collagen (Type I)	17.2±2.6
Collagen (Type III)	6.6±2.1
Smooth muscle	4.7±1.9
Elastic fiber	15.5±2.4
Adventitia	16.1±2.1

the lipid pool is polarized in apple green, which indicates type III collagen or thin collagen fibrils in polarized microscopy.

Table 1 shows the results of the study on Apo-E deficient mice. Type I collagen characterized by orange polarized color in polarized microscopy showed high attenuation in acoustic microscopy. In contrast, type III collagen characterized apple green, showed low ultrasonic attenuation by acoustic microscopy.

Figure 11 shows a SAM image of a human renal artery smooth muscle cell. The cellular motility was colorized. Striations observed in the cell are produced by the interference between the surface and the bottom (interface between the cell and the plastic plate) of the cell. The interference can be considered as the difference of thickness. Blue indicates the difference between the previous and the present frames, while red indicates the difference between the present and the next frame.

DISCUSSION

As the pathophysiology of atherosclerosis is currently understood, rupture of a fibrous cap overlying a lipid pool is considered to be the most common initial event to worsen the state. Biomechanical factors, such as hemodynamic shear stress[14], transient compression[15] and sudden increase in intraluminal pressure [16] have been postulated to play an important role in plaque rupture. As a result, the assessment of both chemical composition and the mechanical properties of the plaque have become an important aspect of evaluating atherosclerosis. In previous studies, pulsed wave velocity (PWV)[17] and stiffness parameter [18] were proposed to assess elastic properties of arteries. However, these methods provide information about the elasticity of the entire arterial wall. The mechanical properties of the tissue components in atherosclerosis should be measured because the pathological finding of plaque rupture is often found in the complex microstructure of atherosclerosis.

Since ultrasound has strong correlation to mechanical properties, acoustic microscopy provides not only morphological but also mechanical information on the biological tissues. In the present study, fibrosis showed greater values of attenuation and sound speed than those of normal media consisted of smooth muscle. This result suggests that the increase of collagen fiber in arteries would lead to hardening of the arteries [19]. PWV of the aorta is known to increase with aging and the phenomenon is explained by the increase and maturation of collagen fiber in the arteries.

In atheromatous plaques, lipid is covered with various thicknesses of collagen fiber. This structure is called "fibrous cap". A vulnerable plaque is characterized by a thin fibrous cap and infiltration of macrophages. In the previous studies assessing stress distribution on the fibrous caps by using finite element analysis, the circumferential stress was concentrated on the thin fibrous cap [20]. In those studies, a single set of values of mechanical properties was given to either a thick or a thin fibrous cap. As shown in the present study, the sound speed distribution was not uniform in the fibrous caps. We have previously applied the sound speed data in stress distribution analysis by the finite element method [21-22]. Our study suggested that the adaptive change of the elasticity of the thin fibrous cap prevented plaque rupture. Further evaluation is needed to prove the mechanical property change and the stress on these thin caps from both actual mechanical testing and computer simulation.

By comparing 200 MHz and 1.1 GHz images of SAM, the superior spatial resolution was confirmed for the latter. The following equation shows the relation between frequency and beam diameter.

$$BD = \frac{1.02 F c}{f D}$$

(***BD***: -6dB pulse-echo beamdiameter, ***F***: focal length, *c*: sound speed of the material, *f*: frequency, ***D***: diameter of the transducer)

This equation means that the beamwidth of 1.1GHz ultrasound is sharper than that of 200MHz if the geometry of the acoustic lens is the same. A single elastic fiber in media or collagen fiber orientation in intima was clearly visualized in 1.1 GHz SAM. However, the 200 MHz image was better in identifying the three-layered structure of the coronary artery.

Under polarized light, the collagen fiber stained with picrosirius red is highlighted by induced birefringence [23]. The collagen content is known to be proportional to the intensity of polarization. In other words, the intensity of polarization demonstrates "quantity of the collagen". In the present study, the thick fibrous caps showed high intensity polarization while the thin fibrous caps showed weak polarization. The results indicate that the collagen content is rich in thick fibrous caps and that thin caps contained l esser amount of collagen.

The wavelength of the birefringence light depends on the collagen type and thickness of the fibril; orange indicates thick collagen fibrils while green indicates thin fibrils[24]. Usually, the thick collagen fibrils compose type I collagen and the thin fibrils compose type III collagen. The color of polarization demonstrates "quality of the collagen". Collagen, predominantly fibrillar types I and III, is a major component of mature atherosclerotic plaques. Collagen is responsible for the mechanical strength of tissues and is believed to confer stability to plaques, protecting them against disruption and thrombosis. Immunohistochemical study showed that collagenolysis by collagen-degrading matrix metalloproteinases (MMPs) were increased and the content of type I collagen was decreased in the vulnerable human carotid plaques[25]. In addition, MMP positive macrophages were colocalized with collagenase-cleaved type I collagen. In the present study, macrophages were observed near the thin fibrous caps with green polarized color, which suggested that the green polarized collagen was affected by MMPs. Thus, ultrasound has a possibility to detect MMPs indirectly but without any special treatments to the tissues in vivo. The feature of ultrasound may lead to detect vulnerable plaques and to understand the pathophysiology of plaque rupture in clinical settings. A critical question is, however, whether chemical or physical factor is most important in preserving the integrity of the fibrous cap and preventing plaque rupture.

CONCLUSIONS

Two sets of acoustic microscope systems were equipped for visualizing the microstructure of coronary arteries. The difference of acoustic properties of each tissue component in atherosclerosis indicated the change of the mechanical properties according to the pathophysiological state in developing atherosclerosis. The ultrasonic tissue characterization of type I and III collagen was possible and this would be very important for detecting a vulnerable plaque leading to acute coronary syndrome. The results of the high frequency ultrasound should also provide the important information to understand the intravascular ultrasound imaging in clinical situations.

REFERENCES

1. Sokolov S (1949) The ultrasonic microscope. *Doklady. Akademi.i Nauk. SSSR. 64:* 333-335.
2. Dunn F, Fry WJ (1959) Ultrasonic absorption microscope. *J. Acoust. Soc. Am. 31:* 632-633.

3. Lemons RA, Quate CF (1973) A scanning acoustic microscope. *Proc IEEE Ultrason. Symp.*18-20.
4. Kushibiki J, Chubachi N. (1985) Material characterization by line-focused-beam acoustic microscope. *IEEE. Trans. Sonics. Ultrason. SU-32:* 132-135.
5. Okawai H, Tanaka M, Chubachi N, Kushibiki J (1985) Non-contact simultaneous measurement of thickness and acoustic properties of a biological tissue using focused wave in a scanning acoustic microscope. *Jpn. J. Appl. Phys. 26:*52-54.
6. Barr RJ, White GM, Jones JP, Shaw LB, Ross PA (1991) Scanning acoustic microscopy of neoplastic and inflammatory cutaneous tissue specimens. *J. Invest. Dermatol. 96:* 38-42.
7. Jones JP (1996) Applications of acoustical microscopy in dermatology. *Ultrasonic Tissue Characterization, Springer-Verlag:* 201-212.
8. Saijo Y, Tanaka M, Okawai H, Dunn F (1991) The ultrasonic properties of gastric cancer tissues obtained with a scanning acoustic microscope system. *Ultrasound in Med. and Biol. 17:* 709-714.
9. Sasaki H, Saijo Y, Tanaka M, Okawai H, Terasawa Y, Yambe T, Nitta S (1996) Influence of tissue preparation on the high-frequency acoustic properties of normal kidney tissue. *Ultrasound in Med. and Biol. 22:* 1261-1265.
10. Saijo Y, Tanaka M, Okawai H, Sasaki H, Nitta S, Dunn F (1997) Ultrasonic tissue characterization of infarcted myocardium by scanning acoustic microscopy. *Ultrasound in Med. and Biol. 23:* 77-85.
11. Chandraratna PAN, Whittaker P, Chandraratna PM, Gallet J, Kloner RA, Hla A (1997) Characterization of collagen by high-frequency ultrasound: Evidence for different acoustic properties based on collagen fiber morphologic characteristics. *Am. Heart J. 133:* 364-368.
12. Falk E (1992) Why do plaques rupture? *Circulation 86(6 Suppl):* III30-42.
13. Saijo Y, Sasaki H, Okawai H, Nitta S, Tanaka M (1998) Acoustic properties of atherosclerosis of human aorta obtained with high-frequency ultrasound. *Ultrasound Med Biol. 24:* 1061-1064.
14. Gerz SD, Roberts WC (1990) Hemodynamic shear force in rupture of coronary arterial atheroscleorotic plaques. *Am. J. Cardiol. 66:* 1368-1372.
15. Binns RL, Ku DN (1989) Effect of stenosis on wall motion. A possible mechanism of stroke and transient ischemic attack. *Arteriosclerosis 9:* 842-847.
16. Muller JE, Tofler GH, Stone PH (1989) Circadian variation and triggers of onset of acute cardiovascular disease. *Circulation 79:* 733-743.
17. Kanai H, Kawabe K, Takano M, Murata R, Chubachi N, Koiwa Y (1994) New method for evaluating local pulse wave velocity by measuring vibrations on arterial wall. *Electron Lett. 30:* 534-536.
18. Hayashi K, Sato M, Handa H, Moritake K (1974) Biomechanical study of constitutive laws of vascular walls. *Exp. Mechanics 14:* 440-444.
19. Greenberg SR (1986) The association of medial collagenous tissue with atheroma formation in the aging human aorta as revealed by a special technique. *Histol. Histopathol. 1:* 323-326.
20. Lee RT, Grodzinsky AJ, Frank EH, Kamm RD, Schoen FJ (1991) Structure-dependent dynamic mechanical behavior of fibrous caps from human atherosclerotic plaques. *Circulation 83:* 1764-1770.

21. Saijo Y, Ohashi T, Sasaki H, Kanno T, Miyakawa T, Nitta S, Tanaka M (2000) Prediction of plaque rupture by finite element analysis of IVUS imaging. *Ultrasound Med. Biol. 26, Suppl 2:* A88.
22. Saijo Y, Ohashi T, Sasaki H, Sato M, Joergensen CS, Nitta S (2001) Application of scanning acoustic microscopy for assessing stress distribution in atherosclerotic plaque. *Ann. Biomed. Engineering* 29: 1048-1053.
23. Junqueira LC, Montes GS, Sanchez EM (1982) The influence of tissue thickness on the study of collagen by the Picrosirius-polarization method. *Histochemistry 74:* 153-156.
24. Whittaker P, Kloner RA, Boughner DR, Pickering JG (1994) Quantitative assessment of myocardial collagen with picrosirius red staining and circularly polarized light. *Basic Res. Cardiol. 89:* 397-410.
25. Sukhova GK, Schonbeck U, Rabkin E, Schoen FJ, Poole AR, Billinghurst RC, Libby P (1999) Evidence for increased collagenolysis by interstitial collagenases-1 and –3 in vulnerable human atheromatous plaques. *Circulation 99:* 2503-2509.

Key Word Index

GPSR Compliance

The European Union's (EU) General Product Safety Regulation (GPSR) is a set of rules that requires consumer products to be safe and our obligations to ensure this.

If you have any concerns about our products, you can contact us on ProductSafety@springernature.com

In case Publisher is established outside the EU, the EU authorized representative is:

Springer Nature Customer Service Center GmbH
Europaplatz 3
69115 Heidelberg, Germany

Batch number: 10584501

Printed by Printforce, the Netherlands